AF754384

Painting with Fire

Painting with Fire

Sir Joshua Reynolds, Photography, and the Temporally Evolving Chemical Object

Matthew C. Hunter

The University of Chicago Press
Chicago and London

The University of Chicago Press, Chicago 60637
The University of Chicago Press, Ltd., London
© 2019 by The University of Chicago
All rights reserved. No part of this book may be used or reproduced in any manner whatsoever without written permission, except in the case of brief quotations in critical articles and reviews. For more information, contact the University of Chicago Press, 1427 E. 60th St., Chicago, IL 60637.
Published 2019
Printed in the United States of America

28 27 26 25 24 23 22 21 20 19   1 2 3 4 5

ISBN-13: 978-0-226-39025-3 (cloth)
ISBN-13: 978-0-226-39039-0 (e-book)
DOI: https://doi.org/10.7208/chicago/9780226390390.001.0001

Library of Congress Cataloging-in-Publication Data

Names: Hunter, Matthew C., author.
Title: Painting with fire : Sir Joshua Reynolds, photography, and the temporally evolving chemical object / Matthew Hunter.
Description: Chicago : University of Chicago Library, 2019. | Includes bibliographical references and index.
Identifiers: LCCN 2019018467 | ISBN 9780226390253 (cloth) | ISBN 9780226390390 (ebook)
Subjects: LCSH: Reynolds, Joshua, Sir, 1723–1792. | Photography—England—London—History.
Classification: LCC TR64.L6 H958 2019 | DDC 770—dc23
LC record available at https://lccn.loc.gov/2019018467

∞ This paper meets the requirements of ANSI/NISO Z39.48-1992 (Permanence of Paper).

CONTENTS

Introduction: Slow-Motion Mobiles *1*

1 "Pictures . . . in time petrify'd" *25*

2 Joshua Reynolds's "Nice Chymistry" in the 1770s *43*

3 "Rend'rd Imortal": The Work of Art in an Age of Chemical Reproduction *89*

4 Space, Time, and Chemistry: Making Enlightenment "Photography" in the 1860s *131*

Conclusion: Art History in/as an Age of Combustion *179*

Acknowledgments 185

Notes 189

Bibliography 241

Index 265

INTRODUCTION

Slow-Motion Mobiles

Early in the 1770s, Scottish-born banker James Coutts sat for a portrait with his friend Joshua Reynolds (plate 1). Loading his palette with ivory black, lead white, and red lake pigments, Reynolds blocked in Coutts's face. He scumbled wisps of hair up from a turgid, gray mud of paint framing the banker's head.[1] A fleshy geography of nostril, lip, and chin emerged in a second, thicker campaign of pigment, its inception almost visible as a purplish, vertical scar strafing Coutts's proper right cheek. Reynolds even appears to have marshaled his unprimed mahogany panel into representational service. The wood's umber hue reads up through the paint-film along the near side of Coutts's nose to act as the preliminary underlayer or "dead coloring" by which painters built complex tonal fields. Below, a "flame" of mahogany grain was held in reserve, kept visible below the sitter's cravat. Yet the picture is a wreck. Flakes tear through the forehead, eye, and cheeks; they pierce Coutts's visible ear and tatter his throat.

The Coutts portrait is not easy to reconcile with the ways in which we have come to talk about eighteenth-century British art and its leading exponent. First president of the Royal Academy of Arts from its founding in 1768, dean of Augustan aesthetics through the ubiquitous *Discourses on Art* (lectures he delivered semiannually at the Academy's prize giving ceremonies), Reynolds is often counted as a prop to Coutts and to other ruling-class friends he was paid, and paid amply, to flatter. Pilloried by poet/engraver William Blake in his own copy of the *Discourses*, that Reynolds is "Passive & Polite & a Virtuous Ass . . . Obedient to Noblemen's Opinions in Art & Science." He is a complacent tool of power, an agent for the retrenchment of hierarchical order amid the eighteenth century's rapid sociopolitical changes.[2] The painter's contemporaries could also entertain different ideas, however. Period viewers had much to say about the instability of Reynolds's paintings: their readiness to flake, to fade, to otherwise alter in time.

Such capacity for change was often linked to talk of Reynolds's secretive, volatile chemical experiments. "He is said to have been fond of trying experiments in colors," French observer Louis Simond explained in the early nineteenth century, "and thought he had found the secret of rendering them more lasting." Simond saw little evidence of those experiments' success: "His colouring . . . fades away, and disappears rapidly;—many of his pictures are now only black and white."[3] In short fiction published in 1807, painter James Northcote (who apprenticed in Reynolds's studio between 1771 and 1776) imagined a "disappointed genius" keen to make his mark on a London art world addicted to novelty. Exhausting every imaginable option (making pictures "in silks, in worsted, in wool, with bits of coloured rags, marble dust, sand, or a hot poker" are all considered), the painter develops a technical innovation: a "*Venetian ground* . . . by which means every painter should be enabled to paint exactly like Titian."[4] A noted chemical endeavor of the late 1790s, as Northcote knew well, the Venetian supports made by this fictive genius are taken up by a "very celebrated artist" commissioned to execute a portrait, which is then hung above a fireplace. That hothouse environment proves catastrophic; the ground becomes "soft to such a degree, that the eye floated down the face as low as the mouth."[5]

Reynolds is not named here, but apposite tales about his work abounded. Commentator J. T. Smith told of an Irish aristocrat who sat for Reynolds early in life, then departed for the Continent. There "he ran into excesses, became bilious, and returned to Ireland with a shattered constitution. He then found that the portrait and original had faded together, and corresponded, perhaps, as well as when first painted."[6] Extensively documented among modern picture conservators, the instability of Reynolds's paintings was thinkable by contemporaries in chemo-experimental terms.[7]

Reynolds's Coutts portrait would be made to deliver moral warning in just this way to a school of painting whose products appeared to be rapidly consuming themselves. So it was for Francis Charteris, Lord Elcho, who donated the Coutts panel to Scotland's fledgling national art gallery in 1858–1859. "I will make you a present of an unfinished sketch . . . which is full of life and power & which is interesting as being painted on mahogany," Elcho vowed, on the condition that no further cash was splashed on work by Reynolds. "The National Gallery should be a school for art as well an exhibition," Elcho declared, and "Sir Joshua is the worst model one can place before a student for with all his extraordinary beauties and power he has ruined the English School by his loose careless painting and inaccurate drawing."[8] Yet Elcho's talk of ruination belies a more unsettling possibility advanced by Reynolds himself. Rather than being the enemy of oil painting, chemical change in time might already—might *advantageously*—be built into the enterprise. Against the time-tested, professional ethos and protestations of his contemporaries, Reynolds positively courted that possibility. Not only did he acknowledge to patrons the susceptibility of his paintings to change, but he expounded to the Royal Academy's students principles for making artistic novelty through unexpected alloys of pictorial precedent. In his sixth discourse, from 1774, the painter appeals to processes by which the pure gold of new form could be forged after the manner of metals accidentally joined under fire. He called those processes "nice chymistry."

Fighting the ramifications of that fire with fire—appealing to chemistry as a means of allaying the deleterious effects of painting made unstable by conspicuous chemical experimentation—became an articulated goal among Reynolds's competing replicators. As the pages to come show, entrepreneurs such as Joseph Booth would tout the compendious benefits to be gained from a "mechanical and chymical process" for replicating grand-manner pictures.[9] Perfected over years of "chymical investigations," Booth's novel process was immune to the volatility besetting the bold, Reynoldsian experiments it preserved (see plate 17, fig. 3.4). Since his technique resisted "changing, cracking, peeling, or any other of those inconveniencies," Booth proclaimed in the late 1780s, "it will multiply pictures in such a manner as to perpetuate the genius, style, and effect of the most celebrated painters, to the most distant ages."[10] Booth was not the only practitioner seeking to capitalize on the chemical instability associated with the work of Reynolds and his allies. Forging the fine and industrial arts by fire, enamel painting was declared the "chymical practice" ideally suited for pictorial perpetuity. Because Reynolds's colors suffered "much from the use of Meterials he knew himself would soon rob his pictures of their beauty," so one votary would affirm, "There was no other Pencel in Enamel that reserved his tints before they faded then that of my own."[11] In the late eighteenth century's combustible era of political revolution, these copyists called not simply for Walter Benjamin's *technicshen Reproduzierbarkeit* to accelerate the image's spread; instead, they marshaled specifically chemical means to carry the rapidly decaying body of Reynoldsian painting into future time, to render it "immortal."

Still, the image of Reynolds as bulwark of stability persists, maybe nowhere more so than in proximity to the history of photography.[12] "Aesthetic discourse in England," Robin Kelsey has argued, "generally defined high art against the momentary, the indiscriminate, the mechanical, the particular and the accidental—all qualities associated with photography. The ideas of Sir Joshua Reynolds were a touchstone in this respect."[13] *Painting with Fire* does not deny that such values find support in Reynolds's *Discourses*; they do indeed. Rather, taking as axiomatic that neither were the *Discourses* univocal nor were Reynolds's "ideas" (or the idea of Reynolds) reducible to their pages, my proposition is that how we apprehend Reynolds has come to indicate much about what we understand the relations between painting and photography to be.[14] Consider Kelsey's argument: to leverage "openness to chance" as a distinguishing feature of photography, Reynolds is made to stand on the side of "higher typological truths instead of mere accident."[15] To argue that way is to occlude the Reynolds who would adjure architects in 1786 to "take advantage *sometimes* of that to which . . . the Painter ought always to have his eyes open, I mean the use of accidents; to follow where they lead, and to improve them, rather than always to trust to a regular plan."[16] Equally, highlighting Reynolds's remarks about the camera obscura as model for poor painting, interpreters from Kelsey to Svetlana Alpers take no note of the ingenious, collapsible object Reynolds actually kept in his studio (figs. 0.1–0.2).[17] Designed to look like an elephant-folio book when closed, the device opens to become a darkened space for optical projection. Given by Reynolds to a fellow artist, supposedly to replicate his works, such technical instruments commanded strong interest among the Academy's elite. Describing an improvement to the camera obscura in 1777, Horace Walpole relayed to a correspondent how Reynolds and Benjamin West (the Royal Academy's second president) "are mad with it, and it will be their fault if they do not exceed Rubens in light and shade, and all the Flemish masters in truth."[18]

If now lost to ambitious interpretation, artifacts of Reynolds's studio experiments had been assiduously collected and passed on by his would-be torchbearers. After his death in 1792, two canvases laden with Reynolds's experimental media and paint preparations were acquired by Sir Thomas Lawrence (see plate 10, fig. 2.12). Bought at the posthumous auction of Lawrence's effects by a leading family of picture restorers, the canvases passed through the hands of Sir Charles Lock Eastlake, who included a transcription of the processes, dates, and materials inscribed on them in his *Materials for a History of Oil Painting* (1847).[19] At the same time, royal portraitist Sir William Beechey was tracking Reynolds's financial ledgers through his surviving family in Plymouth (see figs. 2.10–2.11). For in the backs of those ledgers, the painter had made secretive annotations of recipes used to compound and bind his paint media. Publishing from those recipes in the mid-1840s, adepts such as Benjamin Robert Haydon saw truisms of stable facture immolated on the altar of sublime art: "Varnished three times with different varnishes, and egged twice, oiled twice, and waxed twice and sized once—perhaps in 24 hours." Of one such portrait-recipe from the mid-1770s, Hay-

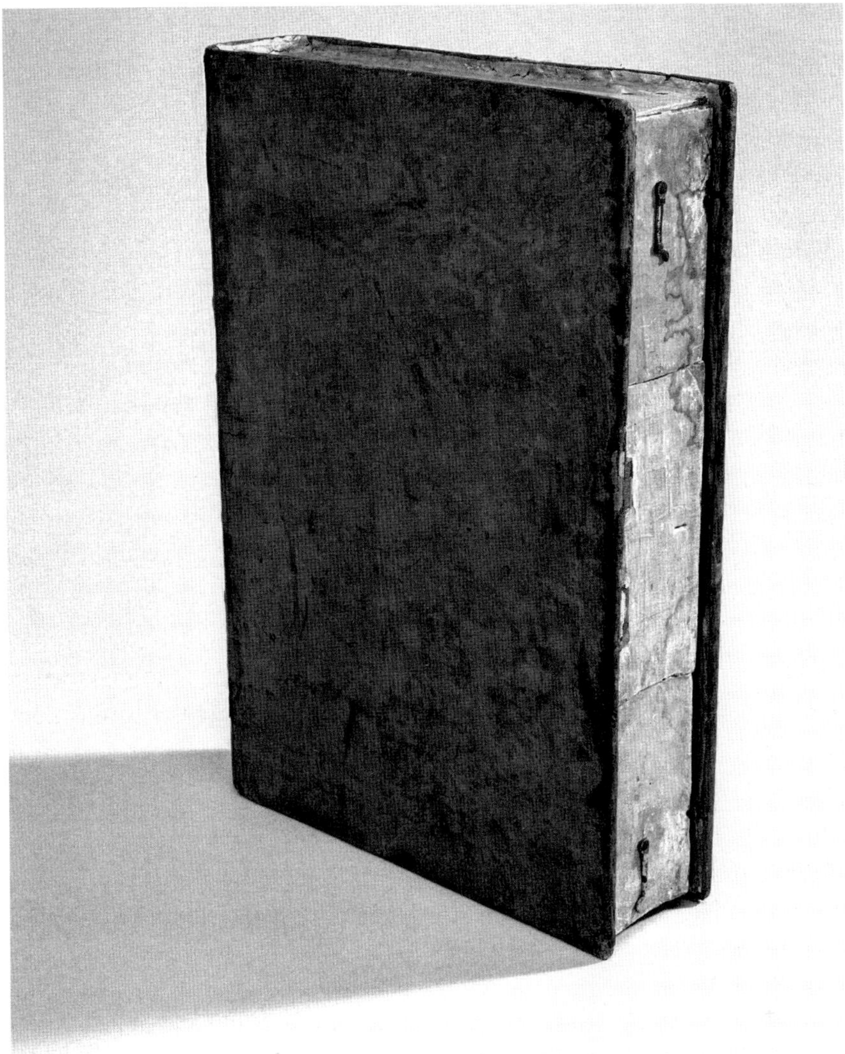

Figure 0.1 Folding camera obscura in the form of a book, previously owned by Sir Joshua Reynolds, as seen closed. Science Museum, London (acc. T 1875/28).

don enthused, "The surface Sir Joshua got was exquisite . . . and though the reward was worth the risk, in such extraordinary infatuation he must be a beacon."[20]

Far from dismissing it as irrelevant or hostile to their interests, mid-nineteenth-century Britain's practitioners in that chemo-mechanical enterprise we have come to call photography took such evidence of Reynolds's "nice chymistry" as guidance. Directing the would-be photographer's use of "light as it can be employed by lenses and chemicals," Henry Peach Robinson cast the painter's volatile techniques as equal parts inspiration and consolation.[21] "The fading of pictures did not originate with photography," Robinson observed in 1869. "Sir Joshua Reynolds' pictures were known to fade even in his lifetime; which means, that it is possible for paintings in oil

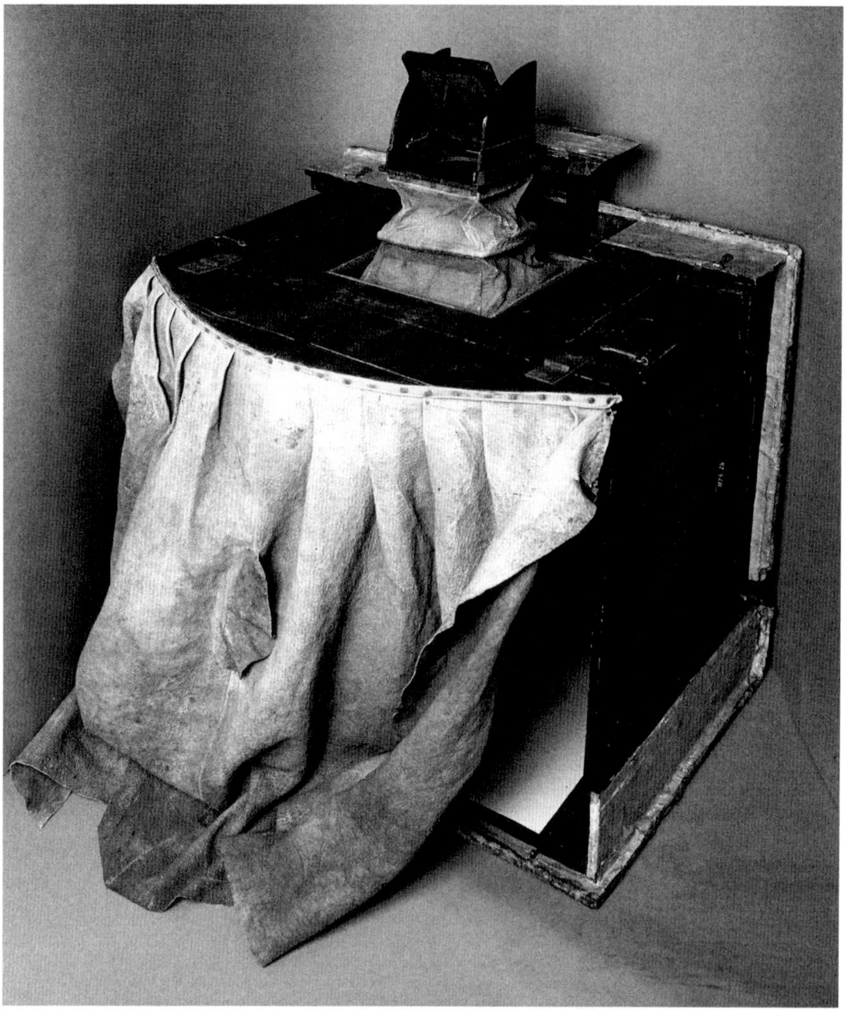

Figure 0.2 Folding camera obscura in the form of a book, previously owned by Sir Joshua Reynolds, as seen opened. Science Museum, London (acc. T 1875/28).

to deteriorate quite as quickly as photographs. It is not much consolation to the kettle to know that the pot is also black, but . . . there is no more necessity for photographs to fade than there is for paintings."[22] The line Robinson draws between fugitive pigments and fading photographs sounds strange because a claim dominating recent theory has insisted on fundamental—ontological—differences between painting and photography. Painting, according to this familiar characterization, is a manual, representational, intentional business. "No one doubts the relevance of the portrait painter's intentionality to the portrait": so claims one theorist in summation. "Everything on that canvas has been put there by him."[23] Photography is supposed to be an altogether different enterprise, one born from mechanical processes whose indexical, chance-hungry products somehow give us the world.[24]

What would happen, though, if we were to hold in suspension not only the usual stories told about Reynolds, but also the confidence with which we repeat these categorical assertions parsing photography from painting? What might become perceptible if we follow Robinson's hint and trace through *chemical substrates* a movement in photography, back to the replicas made after Reynolds by his chemical copyists, and on to Reynolds himself? Resisting talk of antipathy between Reynolds's "smooth cosmopolitan certainties, secured by the rules of international neo-classicism," and the embrace of science typically assigned to progressive intellectual circles in Britain's rising, industrializing provinces, how might those chemo-material supports allow us to map anew the territory of Enlightenment painting, photography, and larger stories about the Industrial Revolution?[25] And if we're determined to follow the chemicals rather than force of habit, where could such a story begin?

"Soe manie pleasant scenes"

In 1684 a collector named William Cole (ca. 1622–1701) launched a philosophical campaign. Learning from local women how to extract a mucilaginous film from the veins of *Murex lapillus* shellfish, Cole believed he could make a dye rivaling those used to tint ancient imperial garb. Sinking his fishy stain into satin and linen supports, Cole dispatched purple messages to his philosophical contacts.[26] A montage of those artifacts shows Cole's technique and his aspirations in full flight (fig. 0.3). At upper left, an alphabetic recital gives way to the date of fabrication, 1684, and concludes with the initials of its recipient: Robert Plot, first professor of chemistry at Oxford and keeper of the Ashmolean Museum, then newly opened.[27] A coarser support at center brings the Oxford-based Plot together with Cole in Bristol, their initials, R. P. and W. C., compressed between two stars. Below, a Latin quotation from the Roman satirist Persius carries an instructive gloss. Cole informs Plot that when the swatch's dye "was greene, being dryed by the fire [it] fuzed [?] to this colour."

Cole seeks to underscore the seven distinct hues he perceived as his purple dye matured:

> When the sunn is gotten higher, . . . [I] make a few lettres hastily and clap the clout in a booke whilest wett and white, then to take another clout, and lett the lettres dry soe as to become of a fine yellowish greene, then to carie out the next into the sunn and hold it till itt turnes of a faire deepe sea greene, and soe putt itt into the Booke, the next soe long in the sunn till itt be of a deepe watchett blew, the next untill itt be of a sullen purple, the last untill itt turne into a deep darke sanguine, and there twill rest untill washt in scalding water with soape, and then being presently dryed

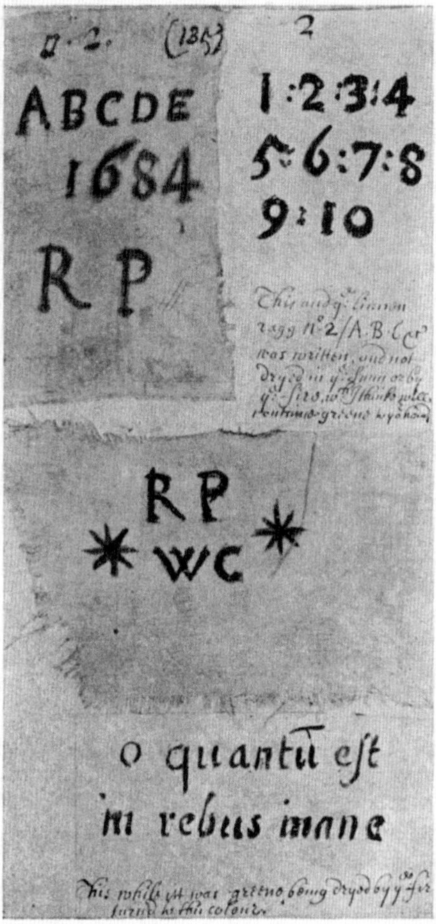

FABRIC AND PAPER DYED BY DR. PLOT ('R. P.') WITH DR. COLE'S ('W. C.') SHELLFISH IN 1684

Figure 0.3 A montage of William Cole's dyed messages, as reproduced in R. T. Gunther, *Early Science in Oxford*, vol. 4, *The Philosophical Society* (1925), between pp. 100–101.

> in the sunne will shew the bright and beautiful Tyrian purple, and this shall be the 7th [tint].[28]

Pliny and other classical authors had passed quickly over changes visible in anticipation of murex purple.[29] Writing to Plot, Cole aspired to join a conversation about dyeing shaped by the chemical research of leading figures in British experimental philosophy, including Sir William Petty and Robert Boyle. In *Experiments and Considerations Touching Colours* (1664), Boyle had marshaled chromatic transitions, ubiquitous in the laboratory, into service as indicators of chemical composition.[30] Commensurately, Cole expounds on the speed and durability of the color changes he observes. The best time to witness the dye's full color spectrum comes soon after the sunrise, since "when it is high and shines very hott the colours will passe on soe fast that you will not have time to discerne their distinction."[31] The intermediate tints can be halted if the swatches are removed from the elements: "I have severall months after, shewed the various colours distinct as aforesaid; yet by often opening the Book, and so exposing them to the *Air*, all the colours . . . will fade."[32] Although he provides no sustained analysis of the chemical principles by which "air" and the sun's heat inflect the dye's visible changes, Cole watches—indeed, he times—the parade of colors to observe what he calls "soe manie pleasant scenes."[33]

Why should a minor naturalist's smelly fumblings with fish dye in late seventeenth-century Bristol bear any relation to the "nice chymistry" practiced secretively in the octagonal painting room on London's tony Leicester Fields by the Royal Academy's first president? A recent interdisciplinary argument provides a compelling rationale. In a Protestant-dominated land skeptical of images and bereft of the fine-arts institutions consolidating among its Continental rivals, the Royal Society of London (Britain's premier scientific institution, established at the Restoration of the Stuart monarchy in 1660) embraced visual representation. As Ann Bermingham has put it, the prestigious Royal Society contributed "to the legitimation of the arts insofar as it saw drawing as a process of visual record keeping, which coincided with a practical concern with technological problems surrounding scientific observation."[34] Just as pictorial conventions learned in portrait painters' studios shaped how seventeenth-century experimentalists saw, so the institution's fellows translated Continental theories of art and architecture.[35] They collected the Old Masters and commissioned new works from London's artists, while acting as conduits for mezzotint and other innovative techniques. Researching pigment-making, dyeing, and other arts under the vaunted, Baconian banner of the "history of trades," the Society's empirical epistemology undergirded and informed Enlightenment aesthetic thought.[36] "It was the Royal Society," Craig Hanson has claimed, "that provided the first 'public' institutional basis for the arts in England."[37]

If that is so, then residue of such an experimental legacy should plausibly be perceptible within Britain's emergent school of painting. *Painting with Fire* began from the proposition that there would be no better candidate for testing that experimental hypothesis than Reynolds himself. Leading painter of the British School, Reynolds was both a fellow of the Royal *Society* and founding president of the Royal *Academy*. Elucidating the experimental milieus in which Reynolds moved in 1750s London, tracing his connections to technical cultures of metal-forging and precision instrumentation in his native environs of Plymouth—as the pages to follow will do—goes some way to highlighting the force of an older experimental tradition acting within eighteenth-century British pictorial art.

But drawing a line from Cole's dye experiments to Reynolds's chemistry also opens matters reaching far beyond Restoration science and the seemingly staid domain of Anglo-American painting. For, to some interpreters, Cole's work has suggested an alternative conception of photography: one that arches away from the names of Joseph Nicéphore Niépce, Louis Joseph Mandé Daguerre, and William Henry Fox Talbot, as well as from 1839 and other pivotal dates in the first third of the nineteenth century. Where one such interpreter sees Cole's dye research as continuous with the "discovery of the effect of light on chemicals and, consequentially photographic technology," another plots it into "an alternative narrative of photography's origin, one that did not begin and end with the camera, that did not privilege the black-and-white image, and did not proceed in a linear, progressive or teleological manner."[38] To this way of reasoning, perhaps Reynolds's coupled interests in the camera obscura and pictorial chemistry should also make him count as an early expression of what Geoffrey Batchen has called the "desire to photograph" as it emerged in the years around 1800.[39]

I make no arguments such as this whatsoever. Instead, the fundamental assertion of this book is that we need to remove photography as organizing *telos* or culminating point from our histories of chemical image-making. *Painting with Fire* contends that only by deflating the Whiggish, appropriative ambitions once voiced in the name of photography can we apprehend the projects of Thomas Wedgwood and other supposed "proto-photographers." Moreover, placed into a longer history of making and thinking with dynamic chemicals, the endeavors of Niépce, Talbot, and other figures who count significantly for the history of photography proper will come to be seen in fundamentally new ways. Just as their "local" imbrication with Reynoldsian painting and its replicative legacies can in this way be drawn out, so too may the truly global stakes of their projects be made clear at planetary level. Yet how could a history be written that would integrate Talbot's calotypes with Cole's dye supports, Northcote's "Venetian" grounds, and Reynolds's chemical vehicles? By what means can we meaningfully apprehend the host of chemical beds, baths, films, and fluids known by practitioners and commentators in the long eighteenth century to change visibly in time, alongside artifacts as patently unphotographic as steam engines and airplanes (all of which will appear in these pages)? That history would not be about photography directly. Instead, it would take as its target what I call "temporally evolving chemical objects."

Temporally Evolving Chemical Objects in the British Enlightenment

Experimental philosopher Robert Hooke never defined temporally evolving chemical objects, nor did he use the phrase. Instead, he supplied recipes for making them. Watching the skies in the late 1670s, Hooke explained to his reader how to study comets at home:

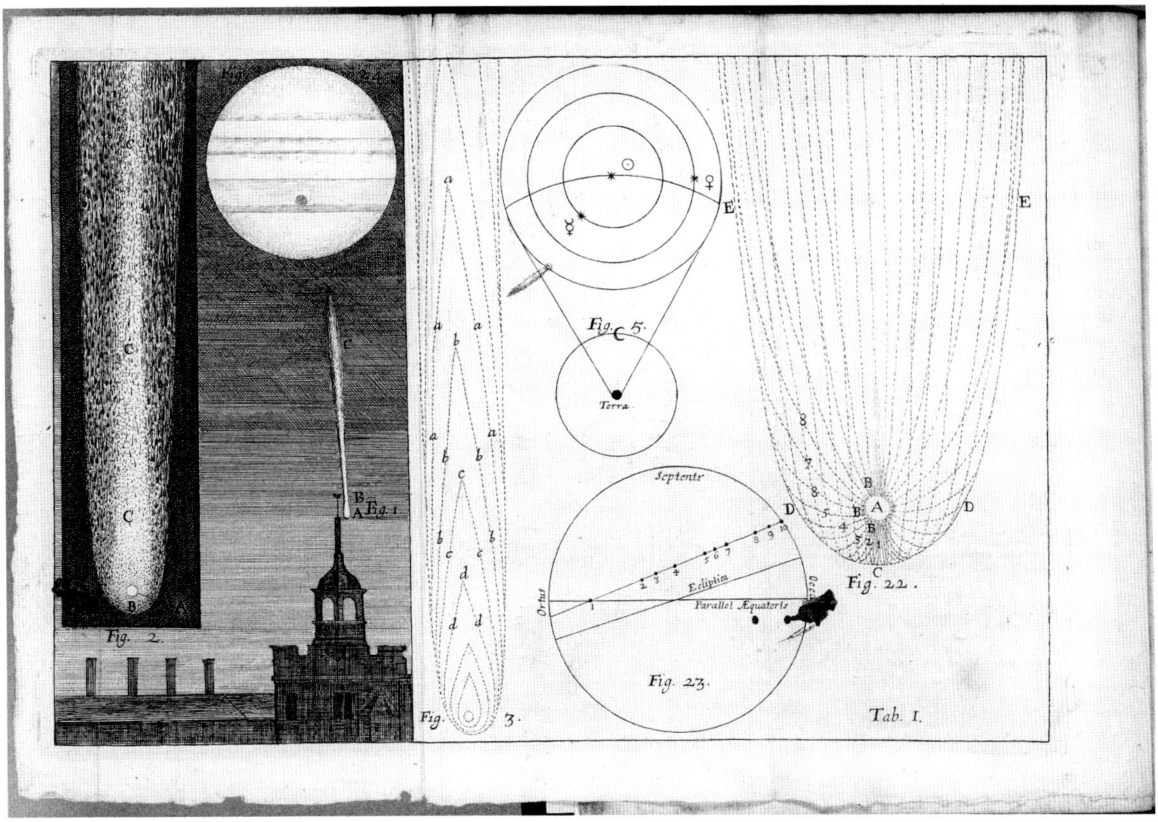

Figure 0.4 Etched and engraved figures of comets from Robert Hooke, "Cometa," in *Lectures and Collections* (1678).

> Take a very clear long Cylindrical Glass, which may hold about a quart of water; fill it three quarters full with water, and put into it a quarter of a pound of Oyl of Vitriol [sulphuric acid], and in the midst of this suspend by a small silver wire, a small wax-ball, rould in filings of iron or steel, and you may plainly observe a perfect representation of the Head, Halo, and Beard of the Comet; for the *menstruum* falling on, or dissolving the iron, there is a continual eruption of small bubbles, and dissolv'd particles from all the sides of this body; and after the eruption they all ascend upwards from the center of the earth; for being of a much lighter consistence than the anbient liquour, they are by the greater gravity of that, continually protruded upwards.[40]

The undergirding chemistry of Hooke's preparation can be understood easily enough. Diluted sulphuric acid forms hydrogen gas as it reacts with ferrous particles embedded in the wax ball; that reacting ball can be guided through the fluid bath when stuck to a nonreactive silver wire. But were Hooke's proportions accurate? What did their visual effects look like (fig. 0.4)? Would his chemical recipe have yielded a result resembling the de-

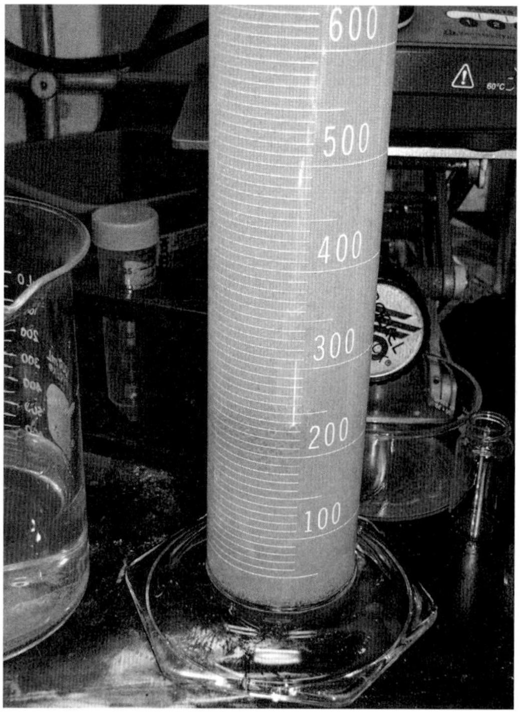

Figure 0.5 Cloudy comet: a preliminary attempt to duplicate Robert Hooke's "Cometa" demonstration. Produced by Dr. Janek Szychowski and Matthew C. Hunter at California Institute of Technology's Tirrell Lab, fall 2009.

pictions of the comet's head published with his 1678 text—prints made in part by cutting a copper plate's "*Vernish*, razing and *Scalping* as it were, the Superficies of the Plate a little, which afterwards the *A.F.* [i.e., aqua fortis, or diluted nitric acid] corrodes and finishes"?[41] That is to say, did Hooke's chemical recipe produce visual results comparable to those yielded by etching and other chemical processes within the art historian's ken?

These are not questions that art history is well equipped to pose, let alone to answer. *Painting with Fire* argues that we can begin to apprehend the legacies of Reynoldsian chemistry, their crossings with photography, and their broader implications only by making steps toward a different kind of art history. As Hooke's comet shows, the making matters. To understand this "perfect representation," I pursued several trials of Hooke's recipe with Dr. Janek Szychowski at California Institute of Technology's Tirrell Lab.[42] Converting Hooke's schematic proportions into consistent units of measure, our modest reconstructions quickly encountered practical problems.[43] What, for example, was the texture of seventeenth-century iron filings? After melting a cake of beeswax with a heat-gun to form the stipulated ball, we found that the chiplike flakes of iron sourced from an industrial supplier failed to produce the "continual eruption of small bubbles" Hooke describes. Rather, fat globules of hydrogen clung to our filings' flat sides. Results were worse when the ball was coated with iron powdered to the texture of flour; that caused a reaction so dispersed as to mask the ball altogether in a milky cloud (fig. 0.5). Only when the particles were of a size like coarsely ground

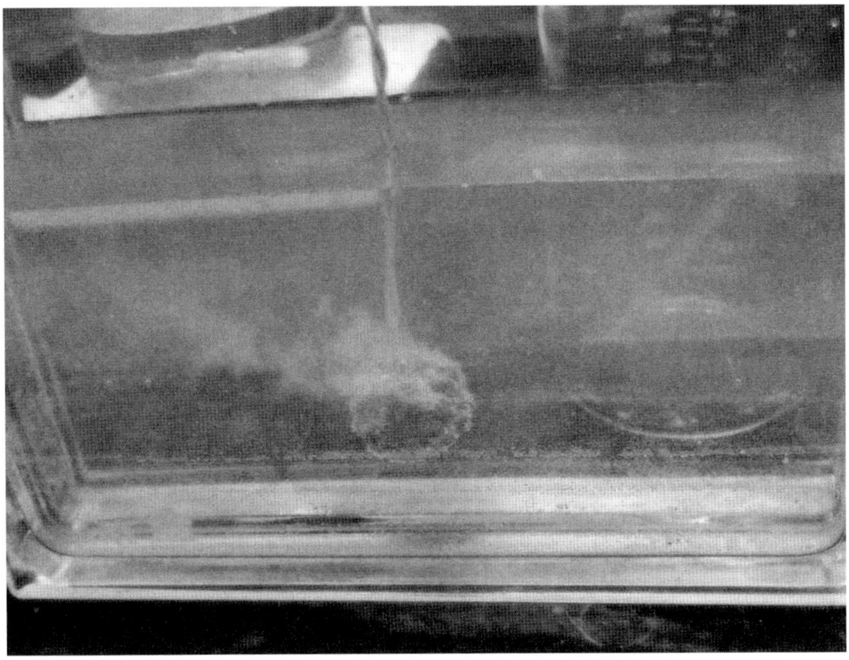

Figure 0.6 Hooke's Comet, Mark II. Produced by Dr. Janek Szychowski and Matthew C. Hunter at California Institute of Technology's Tirrell Lab, fall 2009.

coffee did any salient results begin to appear (fig. 0.6). Even then, they did not "erupt" quickly: at least thirty minutes were required for any significant activity to materialize; effects comparable to those Hooke reported consistently appeared only after forty-five minutes. The reaction then began to slow as the wax ball broke up. The whole enterprise took on representational life and decayed out of it in some four to six hours.

A deflationary reading might protest that this unstated temporal lag could well have been an artifact of the reconstruction, a tic of translation jeopardizing any strong conclusions drawn from the trial. But the existence of the chemical comet in time was central to its point. Hooke knew—and could easily have used—far speedier reagents. In 1676 medical doctor Nehemiah Grew (who Hooke had proposed as a Royal Society fellow in 1671) had explained to the institution how "the filings of Iron or Steel, with Oyl of Vitriol make a fair Ebullition. . . . But Spirit of Nitre [nitric acid] make them boil with much celerity."[44] Forgoing nitric acid's rapid reactions, Hooke's 1678 recipe envisions not some passive observation of chemicals erupting, but a mode of manual and cognitive engagement by which the reader-cum-experimentalist makes materials changing visibly in time into a tool for celestial understanding.[45] A compound of materials and thinking, this changing chemical visualization moves too with broader thoughts about the nature of temporality. Hooke's views on comets' internal instability had been shaped by controversial lectures on the "deep time" of the earth that he had first delivered to the Royal Society in 1668. Earthquakes, volcanoes, fossils—all those, he argued, manifested ways in which "Sublimations, Distillations, Petrifactions," and other volatile, chemical processes *had to have* transformed the earth radically since a receding moment of divine creation.[46]

A stronger argument puts the point this way: material reenactment and other elementary means of harnessing evidence unconventional to art history can change as much the disciplinary tools as the targets of analysis.[47] For these trials demonstrate in powerful ways just how generously a maker has to overlay conceptual allowances onto the chemical reaction to yield Hooke's "perfect representation."[48] The reenactment also underscores the kinds of tools needed for tracking chemical images more broadly conceived. The comet's "figure" only becomes visible in the presence of a particular kind of chemical "ground."[49] The representation is utterly reliant for its fugitive visibility on a field of physical forces and chemical materials, which are themselves not observable. Supports to the long eighteenth century's additive and subtractive techniques (oil painting's priming layers, paper manufacture and sizing, the preparation of copper plates for etching) do have their art-historical languages, of course.[50] But the passivity implied by the phrase "blank canvas" holds little purchase in the domain of temporally evolving chemical objects. Instead, as summoned by Hooke's comet, the analytic instruments we need must be capable of highlighting maneuvers made as much in visible space as through experimental time. We need tools to clarify how catalytic effects were variously anticipated, accelerated, or restrained by chemical makers' manipulations of figuring body and supporting nidus (to borrow a term we will see employed by Tom Wedgwood), with all components active and mutually constituting.[51]

Forging such tools even provisionally can make the art-historical landscape morph and open anew. The proposition of this book is that Hooke's comet, Reynolds's paint films, and Northcote's grounds can be meaningfully apprehended as a history—as a self-conscious relay of temporally evolving chemical objects—that moved into and out of grand-manner painting, among other domains, in the British Enlightenment. It is worth briefly clarifying what this book takes "temporally evolving chemical objects in the British Enlightenment" to be. Taking it from the top: "temporally." No shortage of ambitious restructuring of temporality has been assigned to the long eighteenth century, but Hooke was hardly alone in using chemicals as what Michael Baxandall has called "slow-motion mobiles" to think time.[52] By the mid-eighteenth century, chemical enhancements to encaustic, enamel, and competing techniques that would "fix" the privileged artistic medium of painting in oils (itself mythically born from alchemical pursuits, of course) against decay had prompted pan-European reflection on art's material susceptibility to time.[53] An "Epigram on the Paintings of Sir Joshua Reynolds" in 1782 put the problem this way:

> Sir Joshua, master of the mimic art,
> Paints in the face the passions of the heart;
> Sketch'd by his hand, the rudest portraits please.
> The strength of Rembrandt with Correggio's ease!
> Bu[t] soon the transitory tints decay,
> A morning sun that melts at noon away;
> Yet candour owns, say critics what they will.
> That he goes off with flying colours still.[54]

Where advocates for the chemical processes of patina would sound time's visible changes as sites for the gain of aesthetic and economic value rather than its loss, others would take ongoing chemical transformations as tools for politico-conceptual action.[55] Chemical supports unfixed provided Tom Wedgwood, as chapter 3 argues, with means to militate against a capitalist industrialization of factory time, one embodied keenly at the Staffordshire pottery manufactory called Etruria established by his father, Josiah Wedgwood. Even if the artwork in general and photography in particular are apprehended as veritable compression units for temporality, *Painting with Fire* highlights the reciprocities between making with chemical materials evolving visibly in time and the ways they would prompt agents in Britain's long eighteenth century to stage confrontations *with* time.[56]

Any talk of materials "evolving" in time is likely to prompt association with the theory of biological development by random mutation linked to the name of Charles Darwin.[57] That is not the term's only meaning, and it is not the one used here. Evolution bears a specific, chemical sense: a combination of materials that yields an emission of light, heat, or gas. "The voltaic circle being thus completed," so William Henry Fox Talbot claims in one of his early engine patents, "immediately . . . oxygen and hydrogen gases are evolved, which rise through the water, and fill the space above the surface of the liquid. . . . This evolution of gases continues during the greater part of the revolution of the axis."[58] Evolutions can also be changes that unroll or unfurl in time, per the Latin root *evolvere*. Darwin's grandfather, Erasmus Darwin, used the term with that sense to describe the opening, disentangling, and untwisting of life in the womb—a matrix crucial, as we will see, for the period's chemical speculations. In his *Zoonomia; or, The Laws of Organic Life* (1796), the elder Darwin relayed a "preformationist" account of sexual reproduction whereby all potential beings "have existed in miniature in the animal originally created; and that these infinitely minute forms are only evolved or distended, as the embryon increases in the womb."[59] Forged between his own robust chemical education and Darwin's supposed anticipations of his grandson's mature theory, Tom Wedgwood voices the capacious sense of evolving employed in this book when he claims, "I know the nature of a thing or of an event when <you call up in me> the vis. id. [visual idea] of it, & of what lie before & behind it in the perpetually evolving series of visual perception."[60] Of course, evolution in a weaponized, declensionist sense became an explicit target for nineteenth-century thinkers at the core of disciplinary art history. Internalizing evolutionary thinking in his notion of *Kunstwollen*, Alois Riegl railed against what he called a "materialist interpretation of the origin of art [which] is nothing other than Darwinism imposed upon an intellectual discipline."[61] As its last chapters make most explicit, this book uses chemical materials that evolve—that change—in broader, nonteleological senses to reclaim the possibilities of those materialist methods of interpretation foreclosed to build the foundations of disciplinary art history.

How that discipline has figured connections between eighteenth-century art and matters "chemical" might well be emblematized by Jacques-Louis David's double portrait of Antoine-Laurent Lavoisier and his wife, Marie-Anne-Pierrette Paulze (plate 2). Opponent of phlogiston, master of precision measurement, reinterpreter of combustion, hero of a "chemical revolution," Lavoisier turns from his philosophical apparatus and writings to gaze at his wife, an accomplished artist who produced visualizations for her husband's books.[62] Her height elongated by the fluted pilasters and veined marble revetment behind her, Paulze-Lavoisier looks out from the immaculately finished picture plane as much toward an implied beholder as to that beacon of artistic modernism with whom she had trained: David himself.[63] My story looks less at depictions of chemistry by artists than it excavates dynamic chemicals in art's material and conceptual fabrication; so too do the art and chemistry I analyze give themselves less easily to stories about modernism. In part, this is a matter of scale. Few domains changed as profoundly as did chemistry during the two-hundred-year relay examined in this book. Surveying only the half century between 1675 and 1725, historian Lawrence Principe has characterized its shifts as "so significant and so sudden that they bring the word revolution almost naturally to mind . . . [and] are at least as fundamental as those experienced in an equivalent time-period straddling the celebrated Chemical Revolution—say 1760 to 1810."[64] The eighteenth century would witness what Principe and William R. Newman diagnose as a splitting of chemistry proper from those practices of metallic transmutation and pursuit of the philosopher's stone that have since come to define alchemy in modern, pejorative conception. Instructively, the archaic spelling of "chymistry" Newman and Principe have influentially promoted for signaling a pre-Enlightenment, unified domain of medicine-making, metallurgy, natural philosophy, and chrysopoeia would linger on among the protagonists examined here.[65] While he may have headed a new school of painting in Britain comparable to Continental models, Reynolds was clearly behind the times (at least by those standards) when in 1774 he described "chymistry" as the transmutation of base metals into gold that had become *alchemy's* defining act.

That cloven character of chemistry—its disputed identity, its lags between theory and practice, its perceived threats—also serves to indicate the heft of its hold in Enlightenment Britain. Broadly defined, chemistry was the science of materials, their compositions and transformations.[66] Hardly novel to material encounters with the visual arts, those chemical transformations had acquired marked importance across early modern Europe.[67] Against the grain of entrenched Galenic medicine, European alchemists had promoted artificial means for curing disease and prolonging human life at least since the thirteenth century.[68] Expanded by buccaneering alchemist Paracelsus von Hohenheim (1493–1541) and his numerous followers, that iatrochemical project gained a significant foothold in Britain with the abrogation of medical orthodoxy under the Protectorate of Oliver Cromwell, when it was firmly allied to Francis Bacon's vision of progressive improvement of the human lot through collective enterprise.[69] Burnished in status by Thomas Sydenham and other elite practitioners at the early Royal Society, chemistry as panacea to illnesses of mind and body would be widely sounded in the later eighteenth century's era of industrial expansion—reaching dizzying heights at Thomas Beddoes's Pneumatic Institution in Bristol circa 1800, where Tom Wedgwood, Samuel Taylor Coleridge, and others would receive nitrous oxide treatments administered by Humphry Davy on equipment designed by James Watt.[70] In a fomenting, iatrochemical ambit circa 1774, a reader of London's *Morning Post* some two months before Reynolds's sixth discourse could find the mind "compared to an apothecary or chymist; whose materials indeed are furnished by Nature; but for the purposes of his art, he mixes compounds, dissolves, evaporates, and sublimes them, till they put on a quite different appearance."[71]

Chemistry's ability to make one material into another placed it at the center of British industrialization. By the mid-1740s, the quarter pound of sulphuric acid required for Hooke's comet was dwarfed by the output of vitriol factories established at Twickenham and Birmingham. Producing acids at industrial scale may have helped hatters, tanners, and papermakers, but it was fundamental to the metal-refining and cloth-bleaching operations at the core of Britain's Industrial Revolution.[72] Whether driven by a consumer revolution's demand for cheaper, lighter fabrics imitating Indian calicoes or enabled by a still-earlier "industrious revolution," British textile production only accelerated with the application of Watt's steam engine in the nineteenth century's early decades.[73] In turn, that production spurred the growth of an alkali industry serving the sprawling mills of Lancashire and the broader northwest as they processed cotton harvested by slave labor in the American south. Alloyed in a volatile compound of desire, ingenuity, greed, and exploitation, chemistry was, as one historian memorably put it, "the most fundamental science of the industrial arts."[74]

Painting with Fire is a narrative of making and thinking with visibly changing materials known by their users as chemical, not a history of chemistry. A science of materials in transformation often catalyzed by fire, chemistry is thus pursued not in some systemic purity; it is studied here frequently in relation to other, interpenetrating domains of natural philosophy, always as incarnated in visualizing practices made at the joins of the fine and industrial arts privileged through much of the period.[75] In doing so, I seek to amass as much evidence as I can about what practitioners did with what they called "chemicals," how they thought about them, and what they were understood to be doing, but without forcing the issue. Key protagonists in this book such as Joshua Reynolds and Benjamin West talk about "chymistry"; they make experiments in pictorial facture apprehended by their contemporaries in "chymical" (or "chemical") terms, while their enterprises could suggest (and, indeed, have been interpreted in light of) familiarity with broader "alchemical" interests. Yet the frustratingly fragmentary evidence of their libraries and exact working procedures prohibits any easy reconciliation between what Herman Boerhaave wrote, what Reynolds painted, and how Samuel Johnson thought about both.[76] These evidentiary lacunae are worse still with shadowy figures such as Thomas Rennell (a painter "fond of chemistry," as Northcote would declare him in his biographical writings on Reynolds) or with the "mechanical and chymical process" kept as a trade secret by replicator Joseph Booth.[77] My procedure has been to retain spellings used by the period sources, while keeping in mind the longer chymical heritage disclosed by critical historians of alchemy; to appeal to reconstructions of eighteenth-century practices in modern chemical terms and operations whenever available, while retaining historical emphasis. As should be clear, the guiding aim is not to reveal what the chemistry of each artifact is "really" (i.e., in our terms), but to elucidate a chemical penumbra capable of accommodating well-documented agents recognized by contemporaries and subsequent interpreters as chemists (i.e., Boyle, Priestley, James Watt), as well as a larger, looser array of practitioners (e.g., Reynolds, West, Rennell).[78]

The artifacts of that chemical penumbra are dubbed "objects" for three reasons. First, most art-historical research tends to approach post-Renaissance paintings, prints, and so on as "pictures": as two-dimensional representations or figures whose material encrustation into a third dimension may be acknowledged, but is treated to extended analysis only in exceptional cases.[79] *Painting with Fire* shows how substantially the chemical thickness of the pictorial arts was thought in the period, an aim advanced by the defamiliarizing description of its targets as objects. Second, that usage helps to disclose historical ways in which chemical experiments with materials that change in time both entered into and exited from Britain's grand-manner painting tradition. The final chapter examines debates in the 1860s that turn on technical studies of early chemical replicas of Reynoldsian paintings. I argue for those discussions' consequence for both photochemical preparations and pursuits of chemical explosives used in combustion engines. Those connections would be severed (or at least understood in very different ways than those I pursue) if the artifacts implicated were apprehended as "pictures" or "images."[80] Third, vexed as talk of objects may be to scholarship inflected by strains of existential phenomenology, retaining the offending term offers an opportunity for interdisciplinary intervention.[81] In *Objectivity* (2007), their widely influential shadow-history of subjectivity, Lorraine Daston and Peter Galison center their analysis on scientific atlases, sources wherein "pictures are the alpha and the omega of the genre."[82] Operating by those principles, Hooke's chemical comet would be all but invisible to interpretation; the significant gaps between prints' epistemic virtues and dynamic chemicals' challenges to the thinking subject would be effaced entirely. Since exploring and exploding subjective time via chemical supports is a recurrent concern among the book's protagonists, I retain the term object as subtle provocation for readers within and beyond art history.

Finally, I treat the period examined in this book less as a long eighteenth century than as a "British Enlightenment." Critiquing any view that would figure "the Age of Reason as torch-bearer in the great relay race of human progress," Roy Porter once stressed how the Enlightenment in England had to be approached as "the ideology of particular articulate elites with defined interests."[83] In its own relay, *Painting with Fire* dispenses with progress to track materials and their conceptual affordances over time among a small, intricately intertwined elite, many of whom were Scottish (or Scottish-educated), Irish, or Anglo-American, not only English. Sociable, variously progressive, informed about developments in the sciences, and broadly answerable to the cosmopolitan interests associated with the Enlightenment, the agents implicated in this relay were closely linked to one another—often consciously so—despite the centuries separating them. Departing Reynolds's studio for his belated Grand Tour in 1776, Northcote wrote a letter to his brother from the library of the Rev. Leonard Troughear Holmes on the Isle of Wight. "A man of great fortune, heir to Lord Holmes, nearly related to admiral Holmes, and descended from a late Governor of the Isle of White," as Northcote put it, the Rev. Holmes was a relative of Sir Robert Holmes, the man who seduced Grace Hooke, Robert Hooke's niece and lover, on the Isle of Wight in the fall of 1677.[84] In Plymouth, Northcote had been friendly with James Yonge (1748–1799), a descendant of Plymouth naval surgeon James Yonge (1646–1721), who had contributed to Hooke's publications in the 1670s.[85] While Reynolds's own familial relations to Hooke's experimental-philosophical networks are noted in chapter 2, multiple generations of Wedgwoods, Watts, and Darwins figure in later chapters. And none other than the grandson of Matthew Boulton emerges from the shadows in the 1860s to intervene into the interpretive fate of his grandfather's chemical image-making. In short, mine is not a comprehensive survey of intersections between art and science in Britain's long eighteenth century. Instead, I use the fraught label Enlightenment to get at the threshold of what would become a global problematic as reckoned with by a small, self-conscious network.

The Relay

The following chapters seek to dramatize that small group's movement in and out of synchronicity through a simple geometry. Each chapter begins with a vignette positioned at least three decades before the main action of the chapter. Several of these vignettes endure substantially longer than the two to three paragraphs' allowance of scholarly convention. The chapters, in turn, often tarry over a single decade, or even a shorter span. Possibly prompting the reader to reflect on the point of time at the vignette and the line of the chapter's narrative, this twin-speed geometry aspires to a motion that is "digressive, and it is progressive too, —and at the same time," in the immortal words given to Tristram Shandy by Reynolds's friend Laurence Sterne.[86] In practical terms, the story moves this way. Chapter 1, "Pictures . . . in Time Petrify'd," highlights a new nexus between chemical materials known to change conspicuously in time and visual art forged at the Royal Society of London circa 1680. I situate experiments on the cold fire of artificial phosphorus and other "self-shining" substances within broader research on the chemistry of life, light, and combustion. A quintessential temporally evolving chemical object, artificial phosphorus becomes pivotal to experimentalist conceptions of time. Ignited into storable flame, artificial phosphorus will thence pass through each of the following chapters—as subject of depiction, as object of chemical preservation, as reputed source of photographic origin—in relay, a movement itself then transitioning into a conception of art.[87]

The second chapter, "Joshua Reynolds's 'Nice Chymistry' in the 1770s," moves to the mid-eighteenth century and the consolidation of Britain's national school of painting in oils. I highlight Reynolds's protracted and various access to practical chemistries: among instrument makers in his native milieu of greater Plymouth, at the Society of Arts near London's Charing Cross, and in his own studio. The chapter centers on Reynolds's public embrace of chemistry at the Royal Academy in 1774 and its rapid, contested response. Framed against Joseph Wright of Derby's 1771 rendering of phosphorescent invention (see plate 8), the chapter explores a moment in the 1770s when time and painting's capacity to represent it had achieved particular urgency. I then read Reynolds's "infant portraits" of the early 1770s through contemporaneous debates about the Paracelsian homunculus—a life born from purely chemical means—as forging a prescient temporality of fine art from unstable alloys of chemistry and ambitious painting.

The third chapter, "'Rend'rd Imortal': The Work of Art in an Age of Chemical Reproduction," moves to the revolutionary decade of the 1790s, when several leading entrepreneurs and industrialists explored new, explicitly chemical means by which to "fix" or otherwise intervene into a body of British grand-manner painting made chemically unstable through Reynolds's influential model. Highlighting the central role of the second Royal Academy president, Benjamin West (1738–1820), the chapter places the decade's most acclaimed chemical invention, lithography, in relation to the "Venetian system" of painting, and "pollaplasiasmos" alongside enamel and encaustic painting as competing, chemical instruments for preserving the Reynoldsian project. From aspiring echelons of London's art world, the chapter expands to consider James Watt's copy press, Josiah Wedgwood's forays into aquatint, and other modes of replication via chemical change in time among the Lunar Society of Birmingham. That mingling of fine and industrial arts, channeled together at opposing speeds to remap temporal metaphysics, finds supreme expression in the work of Tom Wedgwood (1771–1805), Josiah Wedgwood's youngest son and the "first inventor" of photography per the famous pronouncement of William Henry Fox Talbot. I argue that Tom Wedgwood's so-called photographic project is better apprehended as a contradictory drive to visualize a fundamentally antivisual conception of time via chemical means. That star-crossed project also affords return to the figure of the Reynoldsian chemical homunculus, as reconfigured in Mary Shelley's *Frankenstein; or, The Modern Prometheus* (1818), imaginative fiction penned by the daughter of Wedgwood's radical friends.

The fourth chapter, "Space, Time and Chemistry: Making and Disfiguring Enlightenment 'Photography' in the 1860s," examines a moment in the subsequent century when those indestructible, chemical replicas of Reynoldsian painting were rediscovered *as* photographs. The chapter follows the recovery of what were then called "sun pictures" and their role as evidence for a history of photography achieved in 1780s Birmingham, a tale promoted by Francis Pettit Smith (an engineer-cum-curator at London's then newly formed Patent Museum) and Edward Price, his chief supplier at Matthew Boulton's derelict Soho Manufactory. The chapter follows these sun pictures up to their largely successful presentation before a leading photographic body in November 1863. Thereafter, Smith's revisionist history would meet vociferous opposition from Matthew Piers Watt Boulton, the Soho industrialist's grandson and namesake. Seeing M. P. W. Boulton's opposition to the sun pictures not as an attempt to shake off photography's disfiguring, industrial shadow, but as a means to wrest chemical secrets for driving combustion engines from a rival technologist, the chapter offers larger lessons for rethinking chemical replicas, Reynoldsian painting, *and* photography. A brief conclusion reads the British Enlightenment's quest for fire as much for its methodological lessons toward a different kind of art history as a groping in the dark with what we can see now as a global threat—as a kind of "native anthropology" of the Anthropocene.

As will be clear, *Painting with Fire* builds upon the brilliant insights of scholars working in several fields: from the critical historiography of alchemy and studies of the material lives of artworks to the kaleidoscopic panorama of the "chemistry of substitution" in German industrial history and critical theory offered by Esther Leslie's *Synthetic Worlds* (2005).[88] No work has been more important, though, than "Photography and Liquid Intelligence," a brief, poetic essay by contemporary photographer Jeff Wall around which this project long centered. In four paragraphs, Wall imagines an alternate lineage for photography's "dry" rationality of optics and machinery through what he calls "liquid intelligence." Literally present in the fixative baths and washes of chemical photography (while displaced to the generation of electric power by the turbines that support digital media), liquids constitute an "archaism . . . a memory-trace of very ancient production processes—of washing, bleaching, dissolving and so on, which are connected to the origin of *technè*."[89] Once I had imagined that my story might make good on that gestural tracing; I hoped I might excavate an intellectual genealogy of making and thinking with liquid chemicals that changed in time. Maybe that would build upon my previous work to offer, say, a counterhistory to narratives about artistic skill and deskilling over the long eighteenth century.[90] Yet, as the project progressed, liquidity as such began to feel epiphenomenal to the story. And mine, I came to realize, was not a story about photography directly.

Prefacing a different approach to Enlightenment Britain, novelist John Fowles once described a germinal image. "A small group of travellers, faceless, without apparent motive, went in my mind towards an event," he writes. "They simply rode along a skyline, like a sequence of looped film in a movie projector; or like a single line of verse, the last remnant of a lost myth."[91] My story is less chivalric, closer to the ground: more elemental. In darkness, a torch alights; it is passed in relay from one site of combustion (itself never original) to a waiting material substrate and eager torchbearers. Gestures toward a critical history of temporally evolving chemical objects, not Promethean myth-making, these four points offer a different story of Enlightenment art and its crossings with the Industrial Revolution.[92] So too do they argue for another way of understanding painting, photography, and the relations we draw between them. But this story is also after something older, something like an archaic or primordial way of imagining a history of art. As I found myself writing a book stretching over some two centuries and sticking my fingers into any number of scholarly beehives, I came to realize that those were not intelligent procedures nor was mine a story about intelligence. Fowles describes his story as propelled by "obsession with a theme."[93] Overextended in scope, excessive, given to risky behavior with uncouth tools: those were equally the objects and the affects of my inquiry, not intelligence. Playing with fire became, with a nod to Yves Klein, *Painting with Fire*.

1

"Pictures . . . in time petrify'd"

Esther swoons. She collapses sidelong into the scarlet arms of a female attendant (plate 3). As she falls, the backs of her fingers brush the cinched splendor of iridescent green velvet that carpets the steps to her husband's throne. Made moot by her insensibility, the attractions of those lustrous surfaces are turned outward to the beholder. They are slowed—magnified—as the drag of her cape's flushed-gold lining drapes over that emerald step, replaying her fingers' movement at amplified volume. Rising from his seat above, Persian king Ahasuerus leans forward. Right arm wriggling with agitation, his looming body casts shadow onto the crush of courtiers who crowd forward to see protocol and, now, posture broken. According to the biblical story, Esther had concealed her Jewish identity from her husband. Learning of a plot to massacre the Jews hatched by royal minister Haman (who contorts at upper left), she violates taboo by entering Ahasuerus's throne room unbidden, only to faint in the act. The picture bristles with luminous cues—pearly dots crossing at Esther's waist, glints of light off lolling eyes at upper right, the receding floor tiles' pasty whites—to draw the beholder to its own centripetal insentience. The beholder is guided by lights as Esther blacks out.

Painted on canvas in the later 1540s, Tintoretto's vision of Esther's faint anticipates its own fate unknowingly. In the spring of 1628, the picture was packed into the London ship *Margaret*, which set sail from Venice, homeward-bound, on April 15. The ship contained a prodigious prize: a collection of paintings by Raphael, Titian, and other Renaissance masters. Valued at some fifteen thousand pounds sterling (equivalent to the cost of a disastrous naval siege operation then being run against the French at Île de Ré) and bought at perhaps twice that sum, Tintoretto's *Esther* and other artworks had been acquired from the bankrupt Gonzaga family of Mantua. In Venice, they were crated for travel under the supervision of courtier Nicholas Lanier, acting on behalf of the purchaser, British king Charles I.[1]

The *Margaret* docked at Antwerp in mid-June. There, her cargo was found to be in satisfactory condition by Lanier, who had traveled overland through Switzerland and France to intercept the ship. But an unpleasant surprise awaited when the crates were unloaded in London one month later. As recounted by royal physician and connoisseur Sir Theodore de Mayerne (1573–ca. 1655), the *Margaret* had been packed at Venice with cargo that included not only fine art, but a load of currants and several barrels of mercury sublimate (a white salt used to treat venereal disease, among other applications). In the fetid depths of the ship's hold, those incongruous components had come together to disfiguring effect. Some blamed a storm that rocked fishing boats in the Gulf of Venice soon after the *Margaret*'s departure. Mayerne focused on a vapor excited by the heated currants that blackened the precious paintings, turning them as dark as ink.[2]

Practical means for pictorial restitution were to hand, however. Recording that the king's oil pictures had been successfully restored by washes of milk, Mayerne went on to recommend a graduated scale of interventions, from modest care with egg yolks and white Venetian soap to sterner salves. Aqua fortis (nitric acid), a preparation made with salt and alum, various sulphuric acids: all had their uses.[3] Although it is unclear if or how Tintoretto's *Esther* was then treated, the doubled presence of Haman now visible in the picture (he appears both as the figure in purple and as a ghostly apparition closer in to the leaning king) betrays the extensive reconstruction enacted upon it.[4] Should pictures be defaced by vapors still stronger than the pong of putrefying currants that besmirched Charles's paintings, Mayerne could recommend a solution made from salt and sulphur (two thirds of the *tria prima* promoted by Paracelsus von Hohenheim). "Like goes with like," he reasoned.[5]

Auctioned off by Parliament after Charles I's execution in 1649, Tintoretto's *Esther* was bought back by his son Charles II following the restoration of the Stuart monarchy in 1660. So too was Mayerne's research rediscovered by Restoration-era philosophers at the Royal Society of London, later seventeenth-century England's leading learned institution. With the onset of London's devastating plague in the spring of 1665, the Royal Society read extracts from Mayerne's writings on the growth of worms in human teeth and gums, along with a recipe for salting beef.[6] Their conversation ranging widely in July 1668 between vegetal saps' circulation and the role of iron in the production of copper salts, the Society was referred to "certain papers about chemistry . . . from Sir Theodore Mayerne," documents then turned over to a chemical subcommittee. Dyeing and coloring in the late 1660s, fermentations of ale by 1679, mixtures of metals in 1680, the staining of agates in 1681: all entered experimentalists' conversations from Mayerne's papers.[7]

On the face of it, the Royal Society's regard for Mayerne—patron of Peter Paul Rubens, confrere of Anthony Van Dyck, artistic advisor to Stuart princes—as a repository of chemical lore would confirm the darkest suspicions of an older historiography. "A genuine taste for art did not exist among the English virtuosi of the seventeenth century," as literary critic Walter Houghton once claimed. "On the contrary, they looked at painting in the same way they looked not only at coins, but even at nature and mechanical inventions."[8] This chapter proceeds from the proposition that Mayerne's plotting of pictorial volatility amid expansive research bearing on living bodies and metallic transformations is tactically instructive for a history of temporally evolving chemical objects. For where the restitution of darkened pictures by chemical means had attracted Mayerne's interest in the early Stuart era, fine art made to bear an unearthly, chemical glow became a subject of the Royal Society's investigations in the later 1670s. Those spectacular experiments too looked back in time, to trials by Bolognese craftsman Vincenzo Cascariolo in the years around 1603. Mythically pursuing the philosopher's stone, Cascariolo had then developed a method for calcining or roasting local stone to a fine powder. Exposed to sunlight, Cascariolo's "Bononian stone," or Bolognese phosphorus, could store and return illumination in the darkness. By the mid-1670s, competing chemical preparations using other materials (frequently human urine and feces) were not only being parlayed across Europe but were being physically applied to art and other sensitive surfaces.[9] From Bologna, physician Marcello Malpighi reported how beholders could witness the art of a local collector appliqued with "the *Bononian* Stone calcined, [to see] Statues and Pictures variously shining in the dark."[10] In London, experimentalist Robert Hooke possessed his own recipe for making "Figures and Representations with this Light, as if often done . . . shine like the Stone."[11]

Contrary to hopes raised by art-historical interest in shine and gleam, phosphorescent glow made but a modest visible mark upon Restoration London's fledgling artistic cultures.[12] We possess no contemporaneous depictions of British phosphorus experiments; what artifacts do survive give little to see on account of the volatile preparations' time-sensitive effects. Yet, as expressed supremely by Joseph Wright of Derby's *The Alchymist, in Search of the Philosopher's Stone, Discovers Phosphorous, and prays for the Successful Conclusion of his operation, as was the custom of the Ancient Chymi-*

cal Astrologers (first exhibited in 1771), "cold fire" would become an important catalyst in the subsequent form and fabric of British picture-making (see plate 8). Known to and imaginatively replayed by Enlightenment-era makers, Restoration-era phosphorus research opens a history of temporally evolving chemical objects, as it were, before the era of British art.[13]

To enter that story, this chapter uses three operatives and a counterpoint. Meet the agents: Thomas Willis (1621–1675), Royalist physician and anatomist; Robert Boyle (1627–1691), chymist and natural philosopher, as well as one of the wealthiest men in Europe; and Robert Hooke (1635–1703), former assistant to both Willis and Boyle, who served in the late 1670s as the Royal Society's powerful secretary. All three were practitioners of what historian of science Robert G. Frank Jr. has called the "physiological tradition" of natural philosophy that emerged from Civil War–era Oxford in the long shadow cast by William Harvey (1578–1657).[14] Habitués of Oxford's evolving philosophical clubs, Willis, Boyle, and Hooke all became fellows of the Royal Society and prominent, international exponents of the "new science" of experiment. Refracting trans-European interest in artificial phosphorus through these English protagonists, this chapter foregrounds how chemical materials changing visibly in time (and modeling the nature of time thereby) moved with research bearing on questions of light, life, and fire. It maps chemical connections between ephemeral pictures, bodies, and combustion, bonds that Enlightenment-era makers would subsequently root to seventeenth-century phosphorus research.[15]

The chapter also implants a methodological operation key to the book writ large, and takes a cue from Tintoretto's *Esther* in doing so.[16] Overcome, Esther loses sentience and collapses to the floor, no longer able to see. So too would many chemical processes made and used by Enlightenment practitioners leave little for the art historian to apprehend visually. All too often, there is no figure to square with the chemical grounds under experimental investigation. Yet Esther's fall from upright sight into horizontal blindness can also guide us into the elemental; it can take us from the vertical picture plane to the chemical beds, baths, and other real but invisible agents subtending visibility. Equally, by following the chemicals binding life to combustion and the time of clocks to transitory visual effects, we can reroute artificial-phosphorus research in the Royal Society circa 1680 away from available endpoints in the history of photography. The chapter concludes on that historiographical juncture, prompted by the countervailing, period voice of physician-experimentalist Nehemiah Grew (bap. 1641–1712), as it passes into its own recursive tale.

Shining

In early April 1676, Somerset virtuoso John Beale had a fat pig slaughtered for a family meal. The pig was gutted. The intestines were boiled with its feet and the chitterlings kept in a briny solution in a cool, dark larder. On the fourth night of their submersion, a strange light was seen in the darkness. "All those parts of the guts, and the claws of the feet, which floated on the top of the pickle, began to shine," Beale reported, "and the parts immersed under water gave no light; the light increased daily in all the parts that floated." Some ten days after the pig had been killed, the light "seem'd as bright as the brightest Moon-shine." A hand dipped into the transformed liquid would retain its weird glow.[17] By the start of April's third week, the light had vanished entirely. Beale was nothing if not modest about the interpretation he could offer for such fleeting phenomena. Only had he been moved to publication, he insisted, by the precedent of Robert Boyle, author of writings on a neck of mutton that shone with a greenish-blue light sufficient for reading at night from the pages of Henry Oldenburg's *Philosophical Transactions of the Royal Society of London*—the very journal to which Beale submitted his findings.[18] In Beale's estimate, such matters awaited treatment by "Expert Chymists . . . [who] deal with such fickle agents, as Fire and Flame."[19]

Beale's wish quickly came true. By the mid-1670s, several streams of research gathered around experiments with heatless substances capable of yielding light. Adept Johann Daniel Krafft demonstrated one such marvelous material at Boyle's London laboratory in September 1677.[20] Extracting a luminous, smoking preparation from a glass vial, Krafft scattered tiny fragments "without any order about the Carpet, where it was very delightful to see how vividly they shined; . . . they seemed like fixt Stars of the sixth or least magnitude, but twinkled also like them." Not content at representing nature's heavens, Krafft then turned his phosphorescent means upon the human arts. "Calling for a sheet of Paper and taking some of the stuff upon the tip of his finger," the chemist "writ in large Characters two or three words, whereof one being DOMINI . . . shone so briskly and lookt so oddly, that the sight was extreamly pleasing, having in it a mixture of strangeness, beauty and frightfulness."[21] In a commensurate trial then being made across Europe, a preparation that would "not offend a Ladies hand" was applied to a visage (likely Boyle's), an act that "made not only his own Face to shine, but the lustre of his Face discovered three or four other faces not far distant."[22]

Why did the flickering sparks, flashes, and fumes of artificial phosphorus command such interest in philosophical circles circa 1680? Historians of science have proposed several plausible explanations. Made from bodily wastes, phosphorus promised an imminent realization of chymistry's claims to ameliorate human health. "The cold fire, which contravened the paired qualities of hot and dry, cold and wet," as Pamela H. Smith has argued, "raised hopes that the much expected stone of the philosophers in the animal realm had been discovered."[23] Local context also mattered. A wondrous emblem of the Royal Society's nascent public culture of chemistry in Jan Golinski's reading, self-shining substances' coolness acquired significant, rhetorical value to British practitioners amid sectarian reprisals after the Test Acts of 1673 and 1678.[24] "At the epicenter of anti-Catholic hysteria," so claims Simon Werrett, the Royal Society fellows "exploited spectacular effects to assert their cosmologies but made great effort to show how their experiments undercut hot-blooded drama and put nature in the colder light of matters of fact."[25]

Such considerations are important. But it is also worth reconstructing the matrix of elemental questions through which phosphorus research grew in the 1670s. For what *was* fire? How did it relate to organic vitality? What was consumed and yielded when combustible materials ignited? By following such questions down into the putrid stuff of unstable, cold fire, I mean to survey the ground subsequent operators would cite as binding the ephemeral spectacle of chemicals changing conspicuously in time to questions of life, combustion, and temporal ontology.

*

A novel replication by art of effects made visible in natural materials, synthesized phosphorus resounded with echoes of older research.[26] "The Learned *Willis* (were he alive)," so declared an assistant to Boyle in 1681, "would rejoice to see such a Product out of our own Bodies, who was very confident of something igneous or flammeous or very analogous to fire, that did kindle and impregnate our blood."[27] As recently as his *De Anima Brutorum* (1672), Thomas Willis had drawn from comparative anatomy to explain cognitive and physiological function through the heating and circulation of fluids distilled off from the blood as it "silently burns with a gentle and friendly heat, like a Fire shut up in *Baneo Maria*."[28] Willis built such assertions from a decades-old program of research based on fermentation of five chemical principles (spirit, sulphur, salt, water, and earth) that he used to explain fire, along with many other transformations in nature and human art: "Hay or Dung laid up wet, the Particles of Sulphur very much abound . . . heaped together, they unfold themselves more largely, and begin to breaks the Dens of the Subject, and so produce a burning."[29] If living bodies were heated by a process akin to the ignition of flame, then a phosphorescent glow made from fermented, bodily waste gave reciprocal support to Willis's physiology of inner fires.

In his classic study of the Oxford-associated philosophers, historian of science Robert G. Frank Jr. identified Willis's concept of fermentation as paradigmatic for the group's apprehension of respiration, combustion, and a host of related problems.[30] More recent scholars have recalibrated. The quasi-Paracelsian, active "principles" informing Willis's concept of fermentation became a target, so this argument goes, for Robert Boyle's influential, corpuscularian chemistry, where shape, motion, and position of particulate matter explained bodies' phenomenal properties in "mechanical" terms.[31] Nonetheless, fermentation remained a process to which Boyle appealed when confronted with the ephemeral glow of phosphorus in the later 1670s. Resembling little more than dirty water when sealed, phosphorus uncorked in the dark would bloom with "a *Light* or *Flame* in the Cavity" as it erupted into an encompassing vial.[32] "Air, either by some subtle Salt that it contain'd, or upon some such account, excited in the fumes, it mingled with a kind of *Fermentation*, or (if you please) a *Commotion*," so Boyle proposed in 1680, "by which means the matter acquired so brisk an agitation, as to propagate the motion to the eye, and *there* make an impression, the sense whereof we call *Light*."[33] But how did air cause this spectacular, luminous ferment?

To test the role of atmospheric air in producing phosphorescent light, Boyle made numerous experiments using the air pump that Hooke had designed and built for him in collaboration with London instrument maker Ralph Greatorex (fig. 1.1). That device and Boyle's research on phosphorescence grew from his chemical work at Oxford in the mid-1650s. Then pursuing the elusive target of "aerial niter," Boyle had developed an experimental program around saltpeter (potassium nitrate), a key ingredient in the manufacture of gunpowder. Melting four ounces of crystallized saltpeter in a crucible, the chemist and his assistants cast burning coals into the mixture until the mass would no longer combust. Taken off the heat, a portion of this "fix'd Nitre" was dissolved in water, and drops of nitric acid added until bubbling ceased; then, it was placed on a windowsill exposed to the air. Within a day, the sample teemed with the crystalline growth of saltpeter, as it were, renewing itself.

Boyle's deflagration of potassium nitrate in ambient air offered significant opportunities for his theoretical ambitions.[34] Where "vulgar Philosophy" talked of "Forms" fashioned in matter by inscrutable means, Boyle saw his experiment as giving evidence of the steps by which a body "may again result or be produc'd after it's dissipation and seeming destruction, by the re-union of the same component particles, associated to their former disposition."[35] The experiment demonstrated how a material body apparently destroyed could rise again in terms explicable by corpuscularian chemistry.[36]

From that bold theoretical thrust and its ready theological implications, Boyle withheld definitive address to air's role in saltpeter's "Redintegra-

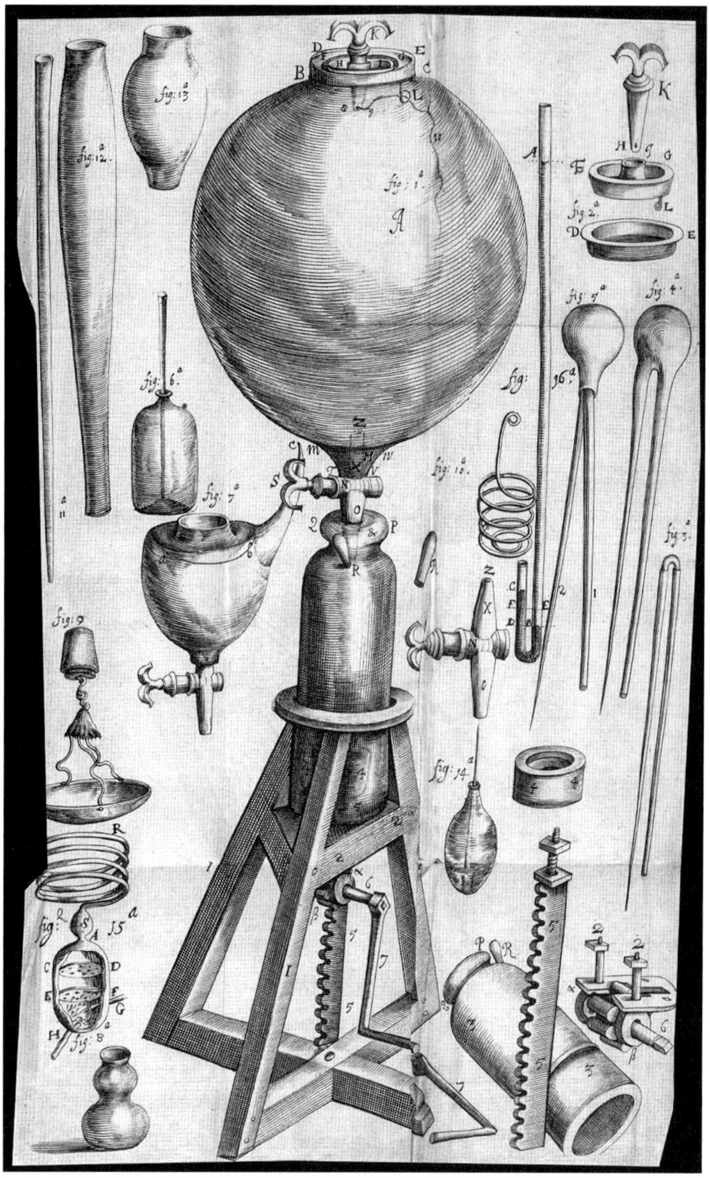

Figure 1.1 Engraving of Robert Hooke's design for the air-pump, from Robert Boyle, *New Experiments Physico-Mechanicall . . .* (1660).

tion."[37] Hooke took it up. As Willis's assistant from 1655 who then worked for Boyle until 1662, Hooke expanded aerial inquiries into a broader account of combustion and the making of light.[38] At the Royal Society in early 1665, Hooke placed nitrous salts heated in a red-hot crucible inside the air pump's receiving vessel. After the glass chamber had been evacuated by laborious, manual pumping, sulphur was lowered into contact with the heated niter through an aperture at the vessel's crown (BCDE in figure 1.1). There, the sulphur was "seen to flame as freely, as if it had been in the open air."[39] Paradoxical as it might appear, flame igniting inside a chamber evacuated of ambient air actually lent support to a theory of combustion Hooke had then recently articulated in his *Micrographia* (1665). Air, as it were, melts bodies as flame.[40] That is, the normally diaphanous mixture of chemical components called air acts as a solvent to all "Sulphureous" or combustible bodies. Niter is the crucial agent to those dissolutions. Gesturing back to Boyle's work, Hooke identified niter as "a substance inherent, and mixt with the Air, that is like, if not the very same, with that which is fixt in *Salt-peter*."[41] Combustion results when nitrous particles in the air dissolve sulphureous components of bodies agitated by heat into volatile motion. Rather than being an element of the classical tradition, fire is a "shining transient body" that can appear in a vacuum where niter, sulphur, and heat are present, even if ambient air is not. Said the other way around, fire will ignite in an airless void so long as the components needed for combustion are present in another state.[42]

Living bodies are heated by fermentation. Shining requires air. Fire counts as a chemical operation. That web of research binding life, light, and combustion informed conversations about self-shining substances in the years around 1680.[43] It framed an especially brilliant exposition on the nature of time that Robert Hooke delivered in his "Lectures of Light" at London's Gresham College between 1680 and 1682. Where Willis's fermented fire of hay and dung had modeled bodily heat, so Hooke would ask in the fourth of those lectures: how could such a process explain the cold light of phosphorescent preparations? Long had he claimed light to be a vibration propagating outward in orbital pattern through the dense, fluid medium of the ether; hammering, rubbing, and similar heat-yielding operations might too produce light in bodies through vigorous, mechanical motions. Phosphorescent substances troubled that account. Materials such as "the *Bononian stone*, and the Preparation lately found out of common Chalk by Dr. *Baldwin* [i.e., Christian Adolph Balduin (1632–1682)]" were best understood as *relays* of light. They "receive such a Power from the Influence of Light, that being carried into a Dark Room . . . they then appear to shine like a Cole of Fire, and continue to do so for a pretty while, but will by degrees lose their Light."[44] Beale's glowing pig parts fell within a still different category, one including "such Bodies as shine without Heat, by an inward Fermentation . . . [such as] the *Phosphorous* made out of the *Caput Mortuum*, or the Rob of Urine found out by Dr. *Kunkell* [i.e., Johann Kunckel (ca. 1630–1703)], and many others."[45] Whether stimulated externally by means of motion or yielded internally by chemical fermentation, these various phosphorescent preparations showed that bodies could be made to shine with light's "prodigious swiftness, or rather Instantaneousness." Heat need not be present.[46]

If vital heat, ignition of flame, and the production of light were all functions explicable through fermentation, then so too was time. That is what Hooke proposes in the final installment of his "Lectures of Light," with artificial phosphorus playing a key role.[47] Hooke begins with the Scholastic mantra whereby nothing can be in the intellect that has not previously been in the senses. Since all sensorial impressions are momentary, time cannot be detected by the senses directly.[48] Where, then, does time come from? Hooke's solution to this storied philosophical puzzle is to make time a product of the physiology of human thought. "We shall find a Necessity," he proposes, "of supposing some other Organ to apprehend the Impression that is made by Time. And I conceive this to be no other than that [organ] we generally call Memory."[49] In the dazzling account Hooke gives, memory is construed as a material instrument comparable to the eyes, ears, and other sensory organs; therein, perceptual data are gathered, encoded by the "Directive and Archiectonical Power" of the soul, and fabricated into time.

"Certainly the greatest Mechanick this day in the World," as one contemporary described him, Hooke has played a key role in apprehending how the early Royal Society helped to elevate the status of instruments and devices from the mechanic's dirty, hands-on workshop into the lofty domain of natural philosophy.[50] But research in the critical history of alchemy also suggests significant ways in which a longer chymical tradition interpenetrated and even drove the vaunted project of mechanical philosophy central to familiar narratives of the Scientific Revolution. Against the venerable impulse to laud Boyle's corpuscularian mechanization of chemistry, this research exposits the debts of mechanical philosophy to the chymical tradition.[51]

Hooke's lecture on the physiology of memory and the ontology of time is an apt expression of that chemo-mechanical reciprocity. On the one hand, Hooke envisions the soul using the apparatus of memory to fashion physical objects called ideas, which represent fleeting signals transmitted by the senses through nervous channels.[52] The soul makes these "material and bulky" ideas in every instant of its cognitive activity (that is, in every second of time) and links them together in a sequential chain.[53] What we call time is not caused by perceptible stimuli, but by the soul's inner acts of measurement:

> Time, as understood by Man, is nothing else but the Length of the Chain of these Ideas, between any two that are at any time apprehended together; And according to the Number of the Links in this Chain, so is the Impression made to the Soul that apprehends it, of a longer or shorter time interposed; and the Notion of Time is the Apprehension of the Distance of Ideas from the Center or present Moment. And so Time comes to be apprehended as a Quantity, and so falls under the consideration of Geometry and Mensuration.[54]

Imagined thus by a professor of geometry at London's Gresham College, mensuration of time and the precision mechanics capable of parsing it were consuming passions for Hooke (fig. 1.2). Plausibly, he had been in posses-

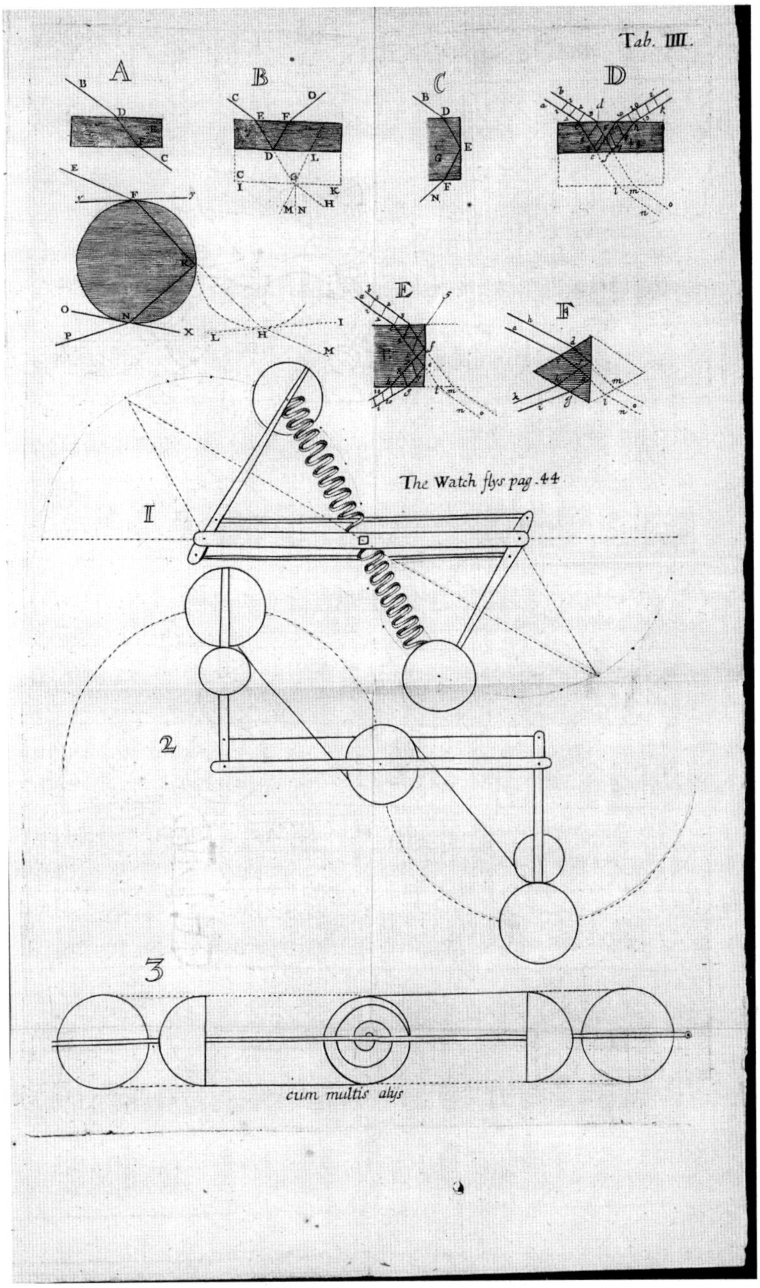

Figure 1.2 Robert Hooke's design for a spring-balanced watch mechanism, from his *Lampas; or, Descriptions of Some Mechanical Improvements of Lamps & Waterpoises: Together with Some Other Physical and Mechanical Discoveries* (1677).

sion of a design for a spring-balanced longitude clock by the mid-1660s, a design he variously hoped to patent and otherwise monetize.[55] When lecturing on the physiology of memory in 1682, Hooke was also some seven years beyond an acrimonious priority dispute with philosopher Christiaan Huygens, who too laid claim to balance springs in marine timekeeping.[56] That clock maker's vision of temporality registered clearly enough to auditors at the Royal Society in summer 1682, who charged that Hooke's account of memory and time "seemed to tend to prove the soul mechanical."[57]

Yet Hooke's 1682 lecture is also marked by chemical technics that inflected his temporal conceptions more broadly. Since the late 1660s, he had been arguing a strong case for fossils as unambiguous artifacts of previously living creatures. Against the grain of a Neoplatonic *vis plastica* shaping matter below the earth's surface or Aristotelian tradition whereby "seeds" containing beings' specific forms germinated in the subterranean depths, Hooke cited "Liquefactions, Vitrifications, Calcinations, Sublimations, Distillations, Petrifications, Transformations, Suffocations," and other processes as means by which organic tissues (or their impressions) could be turned to stone.[58] Terrestrial history imagined through those chemical operations became as convulsive as it was deep.[59] The earth and all that live upon it must be subject to profound transformations wrought, as Hooke put it in explicitly chemical terms, by "the universal Menstruum of Time."[60]

Those chemical operations would resurface in 1682, as Hooke sought to explain how material linkage of mnemonic ideas could yield time in practice. His proposition was that the organ of memory is equipped with various "Elements out of which Ideas are made; among which Variety there are principally five sorrs fitted and adapted to receive the Impressions from the five Senses."[61] These "Elements" are not the earth, air, fire, and water of the classical tradition; instead, they are inflected by practical knowledge of chemical instruments and materials. In a book review published the previous year in his short-lived journal *Philosophical Collections*, Hooke had reproduced a diagram of Marc Antonio Cellio's chemical furnace, while translating instructions on how to calcine Bononian stone (fig. 1.3).[62] Hooke brings those self-shining materials to the fore as he seeks to explain visual memory.[63] Receiving and retaining visual stimuli is akin to

> the Matter of the Phosphorous made of the *Bononian* Stone, or that found out by *Baldwinus*, made of Chalk and Niter; which Matters are so made and adapted by the Chymical Preparations of them by the force of Fire and Mixtures made in their Processes, that they, so soon as exposed to the Impressions of Light, receive and retain those Impressions, though for no long time, yet enough to shew us a Specimen of a certain Qualification . . . , which may yet possibly be done much more powerfully and effectually the Chymistry of Nature in the wonderful Elaboratory of the Animal Body.[64]

Materials can shine chemically through fermentation; that was what Boyle and Hooke had both argued, taking up an operation Willis had elevated among the Oxford philosophers. The generation of light and heat often coincide; frequently, they require ambient air. Hooke's combustion experiments of the 1660s had demonstrated, however, that light could be produced in the artificial absence of atmospheric air so long as the heat needed to generate material motion and the niter required for combustion were

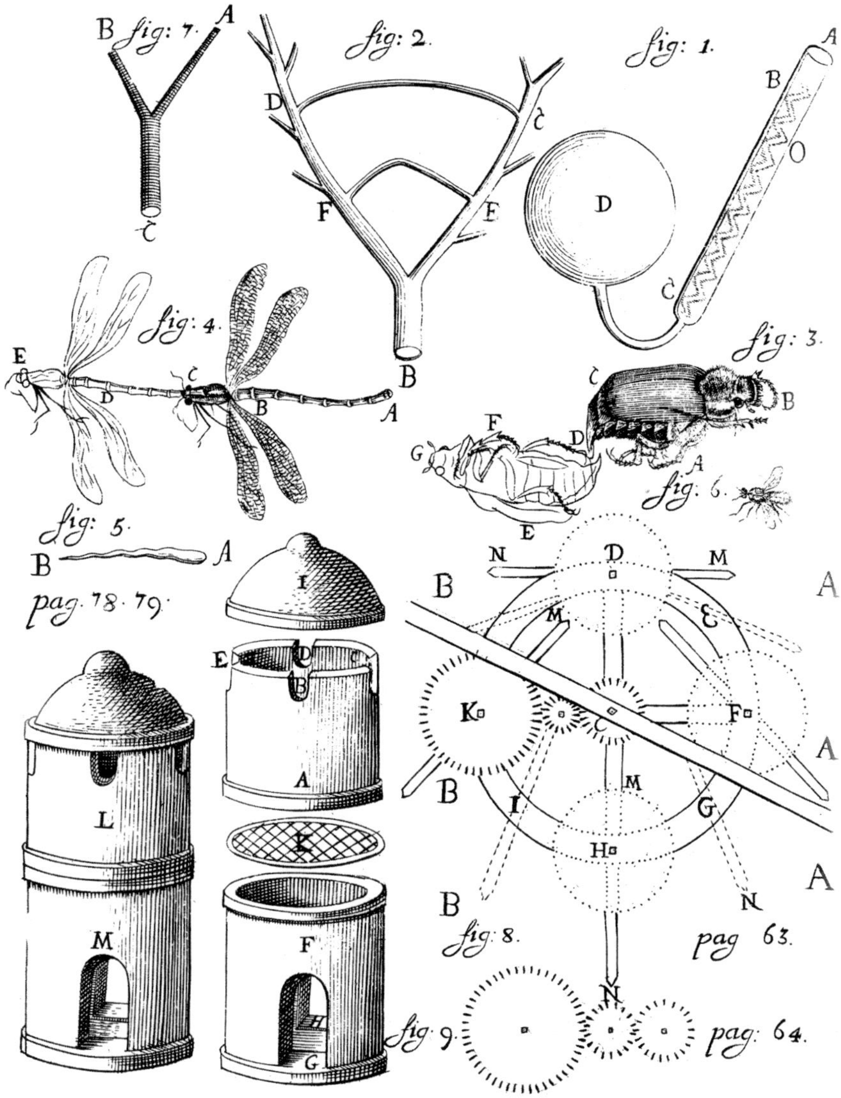

Figure 1.3 Depiction and exploded diagram (*lower left*) of Marc Antonio Cellio's furnace for preparing phosphorus, as printed in the illustrative plate of Robert Hooke's *Philosophical Collections*, no. 3 (1681).

present. In turn, phosphorus, made from sulphureous chalk and niter (as Hooke claimed of Balduin's preparations), demonstrated the making of light without heat when its chemical reagents were exposed to air's fermenting force. That was what Boyle found as he read in bed by the lights of his "aerial" and "icy" noctilucas. But whether properly self-shining or merely relaying photosensitive response like the prepared Bononian stone, phosphorescent effects *decay*. "It retains not its light long, nor its virtue above 5 or 6 years," so Hooke had written when reviewing Cellio's treatise in 1681. As with Beale's glowing pig parts, phosphorescent bodies lose their shine.[65]

That last fact was essential to Hooke's understanding of time. The repository of the mind, he had argued, was a museum of soul-made artifacts representing sensorial impressions. Like his contemporaries, Hooke possessed recipes detailing how art in a museum could be made to glow when rubbed with phosphorescent materials. But even as he identified an instant of human time with the duration required for making a single, bulky artifact called an idea, Hooke stressed how the soul's retrospective regard for its ever-growing stock of mnemonic records disclosed differences of integrity between them. Artificial phosphorus shines brightly at first, then decays. It may be restimulated, but its powers diminish with age. So too with human cognition: older memories are ever less capable of yielding back rich, lively return as the soul attempts to stimulate them. It is that discrepancy between new and old—between fresh chemical capacity to receive and relay light against a decayed or obsolescent preparation—that founds time itself. Built from experiments joining living heat, light, and fire; forged from mechanical clockwork and chemical technics; compounding artistic representations made to glow with artificial phosphorus and Hooke's lively transposition of such a collection into an imagined mind, time is a human product several times over.

"Unphilosophical and absurd"

Medical doctor and experimentalist Nehemiah Grew took a very different view of fossils than Hooke had done. Moisture can freeze on a windowpane into plantlike forms "by virtue of a *Nitro-Aerial* salt" in the water; so is it all the more likely, Grew claimed, that mixtures of salts can congeal underground, crystallizing into fossils' figural forms.[66] No residue or impression of former life is required. Armed with a theory of saline chemical principles descended from Johann Baptista van Helmont, Grew saw little reason "why the Salts of Plants, or Animal Bodies, washed down with Rains, and lodged under ground; should not there be disposed into such like figures, as well as above it? Probably, in some cases, much better, as in a colder place; and where therefore the Work not being done in a hurry, but more slowly, may be so much the more regular."[67]

Yet the process of cataloguing artifacts in the Royal Society's museum in the years around 1680 forced Grew into some uncomfortable taxonomic contortions. Etched as a rectangular tableau at the heart of his twentieth illustrative plate (fig. 1.4), a mounted slab of greenish brown stone pitched Grew's properly philosophical history of nature toward something like a natural history of human art. As he noted the stone's "pleasing variety both in colour and figure," Grew found its strong horizontal striations and vegetative boundaries highly suggestive. The stone resembled "a couple of *Rivers*. One crooked or very much winding to and fro; (as the *Thames* at

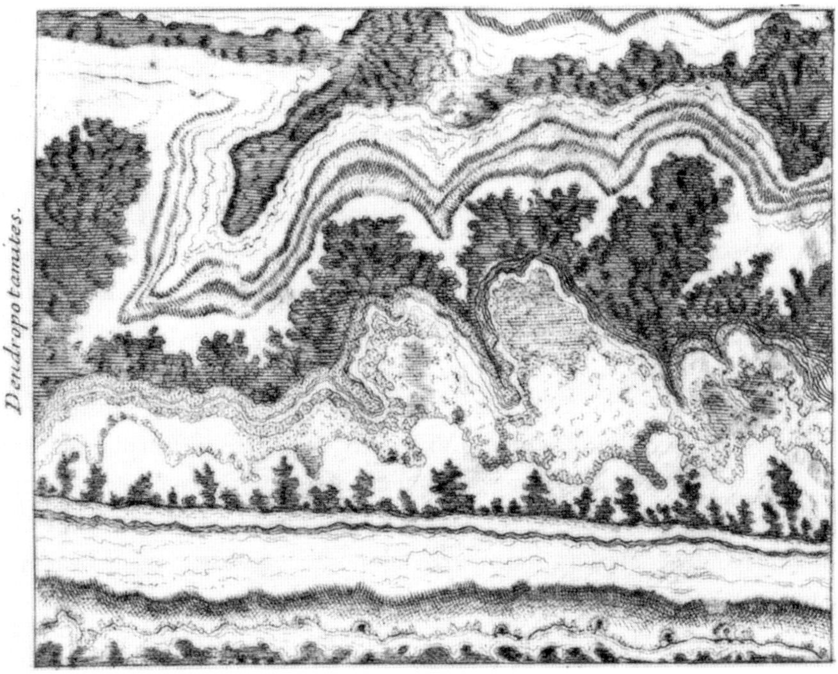

Figure 1.4 Etching of "Dendropotamites," tab. 20 in Nehemiah Grew, *Musaeum Regalis Societatis; or, A Catalogue & Description of the Natural and Artificial Rarities belonging to the Royal Society and preserved at Gresham Colledge* (1681).

Kingstone) and garbed all along with *Trees* upon the Bank. The other strait, with a *Footwalk* upon the Bank, and inclosed also with a little *Hedge-Row*."[68] In a certain sense, Grew's description was perfectly apt. Little is known of its donation, but this portion of the stone may well have been cut, framed, and given to the Royal Society precisely because it was taken to look like a riparian walkway. Moving with a classical tradition of the image made by chance, and crossed with a rising English appreciation for landscape views, Grew's slab may well have pleased precisely because it evoked a strange self-portrait: a bird's-eye view of Nature made by Nature herself.[69]

Grew also entertained a stronger possibility. True, he allowed it to be "unphilosophical and absurd" to propose Nature would imitate human art, but he faltered before a different stone slab, one not depicted in his catalogue and now lost.[70] More than "representing, as it were, a plain Field, inclosed with a HEDGE of TREES; some bigger, others less; all so lively, as if it had been the curious and elaborate Work of a *Painter*," this specimen suggested an intentional picture-making by Nature now imitating human techniques for imitating Nature.[71] The hedge figured into this slab appeared to Grew as though it had "been cast through a Glass (as *Kepler* shews the way sometimes of taking *Landskips*) upon a Tablet in a Dark Room."[72] Where Pliny had found the faces of the nine muses figured into the hearts of gemstones, so it could be within Nature's powers to cast images of her terrestrial works from the sunlit surface of the earth into the fabric of her rocky, subterranean bowels, just as humans had learned to project images by natural, optical propagations. Using a camera obscura, Nature might even fix those images in her own depths through a slower, steadier version of the chemical principles Grew saw producing the forms normally visible to human apprehension. "If *Coyns* are found, every day under ground," he reasons, "then why not sometimes also Pictures, and other Works, in time petrify'd?"[73]

In an experimental milieu wherein mechanical and chemical currents conjoined, Nature is seen making images of herself by some uncertain alloy of accident and intent. She uses means like that of the camera obscura and fixes fugitive forms in dark chemical spaces. I press on Grew's position in this manner to signal how the Royal Society's research into artificial phosphorus in the 1670s has been taken to anticipate photography. In the early 1880s, Austrian chemist and photo historian Josef Maria Eder (1855–1944) excavated the trans-European research into Bononian stone and artificial phosphorus as inaugurating "a new epoch in the history of the invention of photography."[74] Eder privileged the research of medical professor Johann Heinrich Schulze (1687–1744). "Attempting to reproduce the luminous stone of Balduin," Eder's Schulze happened upon a preparation of "nitric acid containing nitrate of silver for dissolving chalk. When he exposed this silver nitrous mixture of chalk and nitrous lime accidentally to light, he discovered that silver salts were sensitive to light." Once Schulze had established this key building block in the future medium's chemical edifice, Eder claims, photography beckoned. It unfolded in "a straight line from Schulze, via Wedgwood and Davy, to Talbot." While he concedes to Talbot and his negative-based camera technique the honor of being the inventor of modern photography, it was Schulze "a German, [who] is to be credited with the invention of photography" as such.[75]

The chapters that follow aim to disarm impulses such as these, along with the teleological procedure implicit to them whereby the interpreter trawls for components built in advance of photography's world-historical entry sometime after 1800. Instead of retracing some enchanted line back from Talbot's Lacock Abbey through Hooke's Gresham College, *Painting with Fire* argues that gestating practices with artificial phosphorus, their crossings with life, combustion, and clockwork mechanics compel us to reconsider fundamentally what Eder's later nineteenth-century moment could confidently call photography. To get to that point, three operations are salutary. First, we need to take cognizance of the ways in which making and thinking with chemical preparations that change conspicuously in time also afforded insight into the character of time itself. Hooke's final "Lecture of Light" speaks with that compulsion supremely as it alloys the precision chronometry dear to the experimentalist's heart with the ephemeral, decaying yield of chemical fire. Second, we should note how a range of chemical times occupied Restoration experimentalists. Labeling the second as the smallest instant of human time, Hooke calculates an outer limit of human intelligence at 3,155,760,000 ideas strung together in a linked chain. Adding one idea per second, a human body able to live for one hundred years could produce that much chemically glowing knowledge (although Hooke concedes the figure, in practice, to be around two billion due to sleep and other inconveniences).[76] Against such "instantaneous" speed, days were the units of measure proper to the sinking of aqua fortis (diluted nitric acid), aqua regia (a mixture of nitric and hydrochloric acids), and other chemicals into marble to yield "most lively Pictures, not only upon them, but passing *thorow* their whole substance," as translated from Athansius Kircher and published in the *Philosophical Transactions* in 1665.[77] Millennia, even eons, were the implicit temporal order of those extremely slow chemical operations envisaged by Grew and Hooke to explain the petrified "pictures" dug up from below the ground.

A third operation is a matter of method. These serried velocities of the evolving chemical object become perceptible to art-historical analysis when we allow ourselves to situate the privileged domain of painting within a matrix of experimental research bearing on combustion and life's speed.[78] A lesson gleaned as Restoration experimentalists apprehended Mayerne's notes on matters like King Charles's *Esther* and other blackened pictures within an expansive array of problems, temporally evolving chemical objects bleed into and spill beyond what would come to be called the fine arts. Resisting the impulse to compress their fixations into familiar talk about photography's "fossil ontology," a salient challenge will instead be to track the building of photography and other pictorial arts alongside (or, as) research into the ignition of fossil *fuels*.[79] But to understand how those materials and concerns would move from natural philosophers' recherché experiments into the subject matter, the material facture, and the theory at the very center of British picture-making, we need now to turn to a different Royal Society fellow: Sir Joshua Reynolds.

2

Joshua Reynolds's "Nice Chymistry" in the 1770s

In the first decades of the eighteenth century, Dutch polymath Lambert ten Kate engineered a collaboration with prominent painters seeking a mathematically grounded system for rendering human skin in oil paint. Ten Kate and his team assigned quantitative values to gray-scale tones. They calculated the percentages of tinting material required to achieve those values when approximating skin's manifold, intermingling hues. Then, suspending pigments in varying concentrations of binding media, collaborating painter Adriaen van der Werff appliquéd upward of twelve sheer layers to create flesh as a meticulously crafted veneer. "The 'skin' made up of paint layers," as have observed recent analysts of van der Werff's technique, "performs in an optical sense the same functions as real human skin."[1] Through ten Kate's mathematical harmonics, painted skin would *be* a skin: an isomorphic surrogate manifesting physical properties of the flesh represented.

Others in ten Kate's ken proceeded differently. Emigrating to England from Amsterdam in 1718, Jacob Christoph Le Blon (1667–1741) devised a method for printing in color. Four separate mezzotint plates—all presenting subtle variations on the same image—would be rocked, burnished, and loaded with red, yellow, blue, and black inks respectively (plates 4 and 5). Le Blon printed his plates one on top of another, apparently employing ten Kate's Pythagorean harmonics to establish percentages of black and white mixed with each primary tone to render nature's polychrome hues.[2] Philosophical grounding was important to Le Blon. In *Coloritto; or, The Harmony of Colouring in Painting* (1725), he aligns his project with Isaac Newton's theory of color. But what Le Blon would give back to philosophers who embraced his polychrome visualizations was substantial. Philosophers' standard intaglio printing techniques trapped ink in cut copper and pressed it into paper's white ground to give "a Design, but not a white Object."[3] Instead, Le Blon's superimposed chromatic separations attended to color's circumstantial interactions. Where van der Werff had depicted flesh by making a surrogate skin, Le Blon's technique renders the printmaker a Pygmalion to the studio-bust-Galatea on which he works.[4]

Painter and engraver William Hogarth (1697–1764) read the theoretical proclamations from ten Kate's circle through English translations prepared by Le Blon himself.[5] And he too took the stained, stony bust as exemplary prop as he offered his own technique for replicating the pigmentation of human skin. Master of flesh's monstrosities in his celebrated graphic art, Hogarth imagines skin in *The Analysis of Beauty* (1753) through tools and techniques imbibed in his artisanal training as apprentice to goldsmith Ellis Gamble.[6] Skin's epidermal "cuticula" is, he tells us, like the animal membranes employed by metalworkers when separating gold foil; its transparency is akin to the isinglass glue made from the air bladders of sturgeon and used in practical tasks of image transfer.[7] Hogarth also visualizes that subcutaneous, transparent matrix for holding liquids through a strange detail (fig. 2.1). Directing attention to the bust of a woman shown in profile—a mask concealing her face like a balaklava—he asks the viewer to invert optical values so that "the part blackest, you are to suppose . . . would be whitest." Thus imaginatively reversed, the print makes perceptible the "tender threads like network, fill'd with different color'd juices," which lie beneath the transparent epidermis and causes the skin's visible pigmentation. A suggestive reflection on the engraver's cut copper plate as it too holds and transfers the colored liquid called ink, Hogarth turns to the figure of sculpture stained to make his larger point, just as Le Blon had done.[8] "Let us now see how by art," he writes, "the like appearance may be made and penciled on the surface of an uniform coloured statue of wax or marble." In a Newtonian mood of his own, Hogarth's white is the mixture of all colors, not their privation. The reader is told how flesh tones can be achieved by sinking successive waves of red, yellow, blue, and purple pigments suspended in oil over a bust until it displays "a very fair, transparent and pearl-like complexion."[9]

In the 1660s, experimentalists at the early Royal Society had harvested recipes for chemical stains "causing a Picture, Drawn on a Surface, to Appear also in the Inmost Parts of the Stone." In the 1760s, carver Robert Chambers

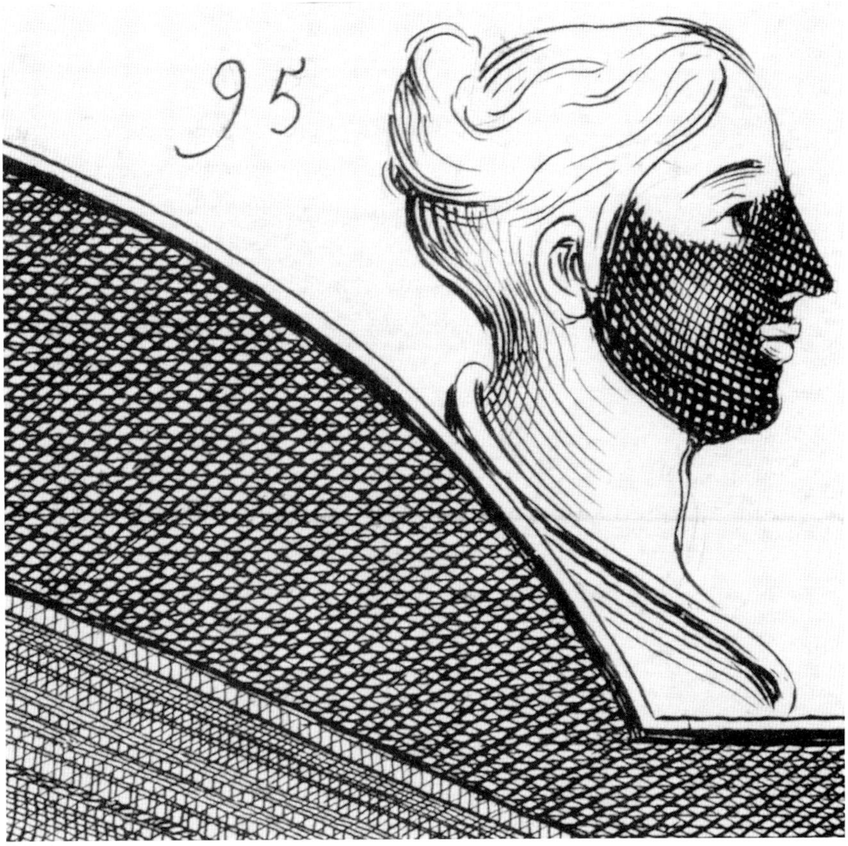

Figure 2.1 Plate 2, fig. 95, from William Hogarth, *Analysis of Beauty* (1753). Engraving.

would reclaim those techniques for the sculptor's ambit.[10] But in the 1750s, Hogarth saw the polychrome washing of sculpted busts as keenly instructive for painters. Since some pigments are made of "metal, some of earth, some of stone, and others of more perishable materials," Hogarth declares, the skin tones rendered by "the power of a skilful master, with all his rules of art" must necessarily be subject to uneven aging. "Time cannot operate on them otherwise," he argues, "than . . . that one changes darker, another lighter, one quite a different colour."[11]

Hogarth cut a robust figure through mid-eighteenth-century London's art world. Torchbearer for drawing from the life model at the St. Martin's Lane academy; champion of engraver's rights—secured by what is now known as Hogarth's Act of 1735—Hogarth was also a meticulous pictorial technician.[12] Among various attacks on connoisseurial pretension, he looked askance on the taste for patina and the proposition that age could improve the painter's art. "How is it possible," he asks, "that such different materials, ever variously changing (visibly after a certain time) should accidentally coincide with the artist's intention, and bring about the greater

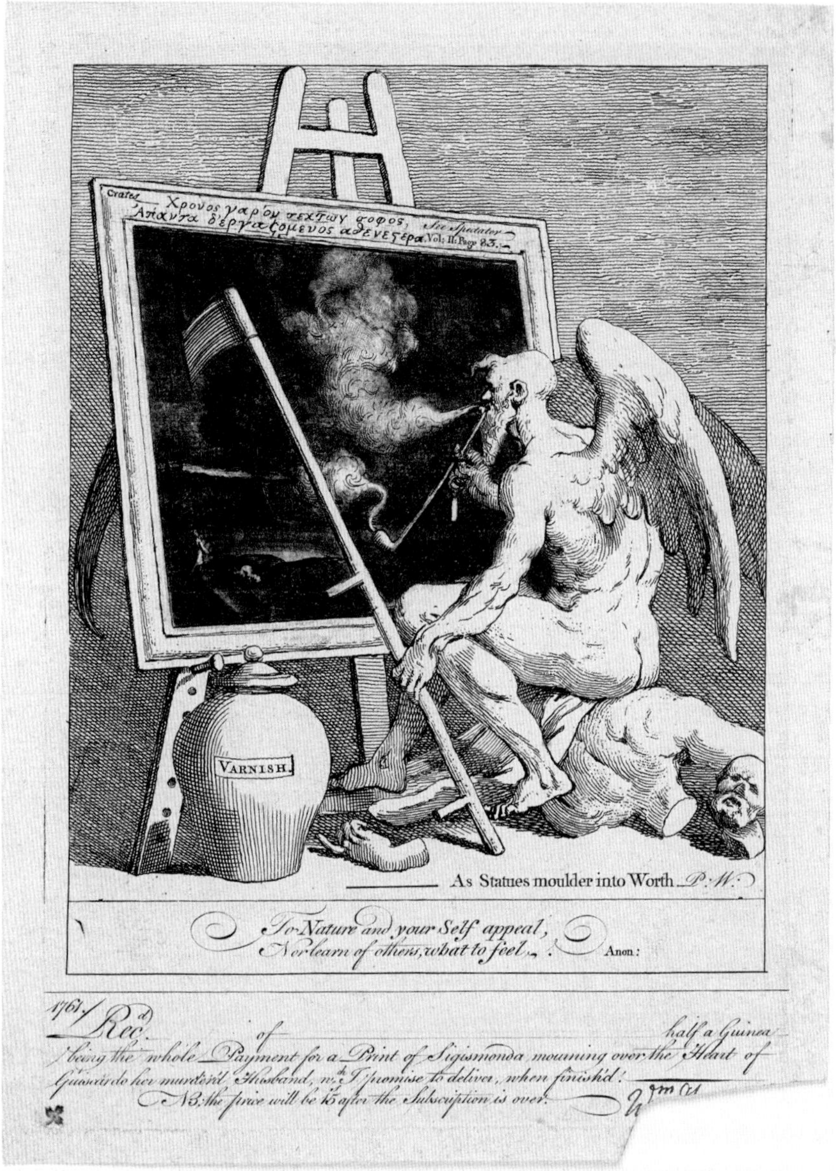

Figure 2.2 William Hogarth, *Time Smoking a Picture*, 1761. Etching/engraving. British Museum, London. © The Trustees of the British Museum/Art Resource, NY.

harmony of the piece, when is it manifestly contrary to their nature, for do we not see in most collections that time disunites, untunes, blackens, and by degrees destroys even the best preserved pictures."[13] *Time Smoking a Picture*, an etched and engraved subscription ticket from 1761 (fig. 2.2), is surely Hogarth's wittiest send-up to this proposition. Astride a headless marble sculpture, Father Time blows an inky, discoloring cloud of smoke onto a recently finished canvas.[14] Slicing his scythe through the canvas's skin, time is a destroyer of the mathematical harmonics underpinning artistic efficacy.

Of the fashionable truism that "time is a great improver of good pictures," Hogarth would conclude: "Nothing can be more absurd."[15]

On the face of it, Joshua Reynolds (1723–1792) might seem to have shared Hogarth's views. First president of Britain's Royal Academy of Arts, painter of over two thousand society portraits, and author of the famous *Discourses on Art*, Reynolds exemplified the "conservative programme in the true meaning of the word," in E. H. Gombrich's assessment.[16] A staunch defender of aristocratic order in an era of rapid social change, Reynolds and his academic program of grand-manner emulation have figured as bulwarks of "timeless" values. "The language of the *Discourses*," John Barrell has claimed, "repeatedly attributes value to what is fixed, settled, permanent, solid, as opposed to whatever is floating, fluctuating, fleeting, variable."[17] By the mid-1750s, Reynolds was consolidating a grip on the London art world that he would maintain for the next four decades, not by words but with pigments. The painter's case had been made supremely in his depiction of Augustus Keppel (1725–1786), scion of a leading Whig family and the naval captain with whom the painter sailed to Italy in 1749 (plate 6). Where rival portraitists labored over sitters' visible features, Reynolds cast Keppel in an Apollonian posture, thus aligning a contemporary officer with the prudent foresight of an ancient god.[18] Judged even by a hostile critic to be "a work of such truth and nobleness that it fixed universal attention," Reynolds's Keppel and the artistic project it augured might appear all the more convincingly allied with Hogarth against time when we note that both painters had been elected to serve in early 1757—alongside architect James "Athenian" Stuart (1713–1788) and chemist Peter Woulfe (ca. 1727–1803), among others—on a committee charged with testing the veracity of two hundred pounds of verdigris pigment delivered to London's newly established Society for the Encouragement of Arts, Manufactures and Commerce (henceforth, Society of Arts).[19] The isomorphic skins of ten Kate's circle here find their echo: high art would need to be highly stable.

Yet, as have discovered generations of conservators and collectors, Reynolds's "settled" images often began to deteriorate as *objects*—flaking, discoloring, visibly altering in time.[20] Analyzing his *Self-Portrait as a Deaf Man* (ca. 1775; fig. 2.3), conservators in the late 1990s unearthed a strange thicket of internally differentiated media and techniques. Against a darkened ground, the painter props his elbow to hear the picture's implied beholder on a ledge forged from pigment suspended in beeswax, spermaceti extract, and linseed oil. Cupped to magnify sound for Reynolds's impaired ears (mythically deafened by a cold caught while drawing in the Vatican), that proper left hand and crooked arm cast shadow onto a vermillion jacket fashioned from pigment mixed with walnut oil and beeswax; other areas of the paint film has been built up from bitumen, varnish, and resins.[21] More pernicious than the disease was its putative cure. When cleaning the

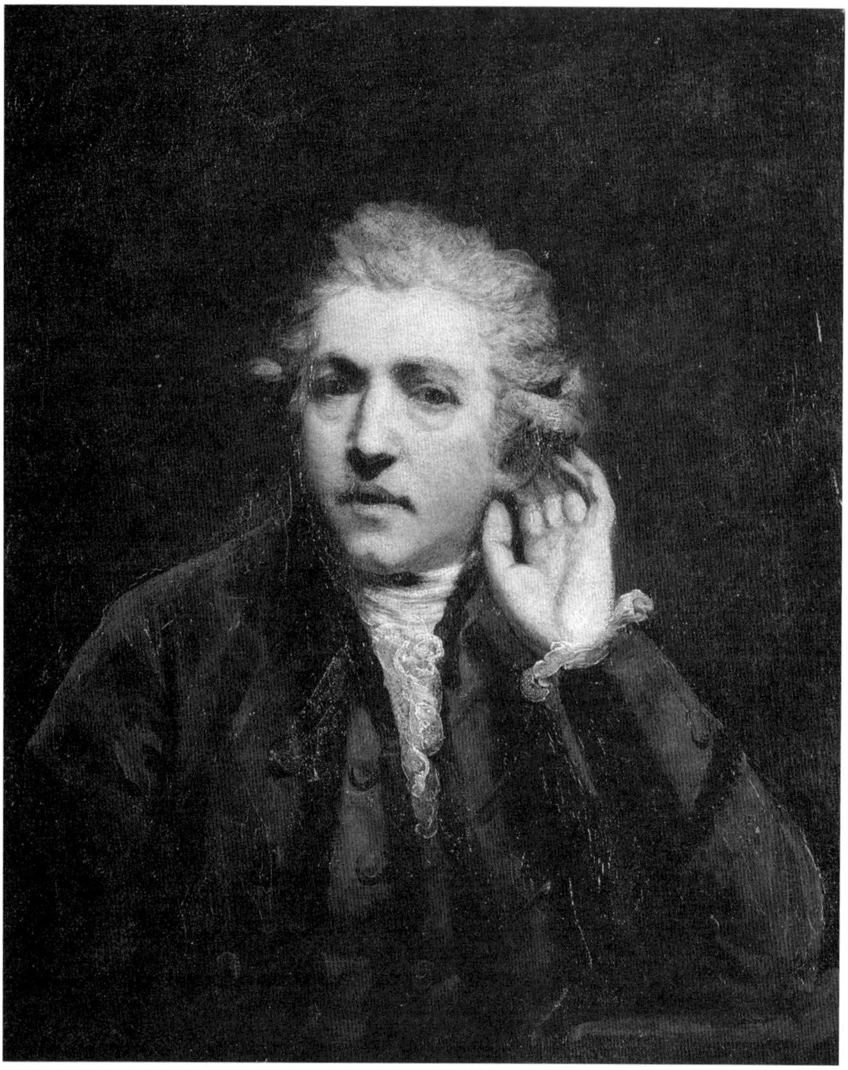

Figure 2.3 Joshua Reynolds, *Self-Portrait as a Deaf Man*, ca. 1775. Oil and other media on canvas. Tate Britain, London.

Tate Gallery's *Sir Watkin Williams-Wynn with His Mother* (ca. 1768–1769) in the 1940s, conservators found a puzzling morass of damage and attempted remedy.[22] "The worst area," they noted, "is in the upper part of the sky to the right, where a considerable archipelago of blue and blue-grey paint, mostly in the hatching, is new. . . . There are many repaints in the forehead and all over his head."[23] Excavation and would-be stabilization ceased once it became clear that further work "would certainly have shown up more of Reynolds' bad drawing."[24] This was hardly news. When treated at the National Gallery in 1859, the surface of Reynolds's *Three Ladies Adorning a Term of Hymen* (1773) was found "vine cracked in parts, tending to lift. Badly vehicle-cracked in numerous parts."[25]

Such instabilities had been known to period commentators as well. In the years following Reynolds's death in 1792, the fugitive character of his colors and painted surfaces would increasingly be linked to talk of "experiments" and chemical interests. "We have perpetually lamented," observed one obituary writer, "that what is technically called the *Vehicle* should have led him to chemic experiments, which, whatever brilliancy they may lend his colours for the present day, certainly will add to the fading powers of time upon the finest tints."[26] In a memoir published in 1819, painter Joseph Farington lamented Reynolds's risky shortcuts—his compulsive drive for "experiments in using his colours, although he had not acquired, in the earlier part of his life, sufficient chemical knowledge to enable him to judge of the result."[27] To sympathizers, Reynolds's flaking, fading works revealed the chasm separating his artistic aims from the capacities of mere pigments. Writing in the wake of the British Institution's 1813 landmark exhibition *Pictures by the Late Sir Joshua Reynolds*, when some one hundred forty pictures were put on retrospective display, painter Martin Archer Shee (1769–1850) took a conciliatory angle: "The ambition of Reynolds, was to produce fine colouring, not fine colours. His was the chastened glow—the subdued splendor—the 'deep toned brilliancy of the ancients;' which he so elegantly recommends in theory, and so successfully illustrates in practice."[28] Artistic imagination and ambition, not durable commercial products, were the proper subject on show. "The magnificent assemblage of his works so lately before the public," Shee contends, "did not indeed . . . excite the idea of 'a chemist's window.'"[29]

And yet, in life, Reynolds not only endorsed a chemical approach but set the Royal Academy's students on to the chrysopoetic path of the philosopher's stone. "By a nice chymistry, passing through his own mind," Reynolds would intone in 1774, detritus of art's history "shall be converted into pure gold."[30] The claim of this chapter is that, through Reynolds, a tradition of making and thinking with chemical materials that change conspicuously in time, previously nourished among experimental philosophers at the early Royal Society, enters into the very heart of British painting. That claim is not to deny the protracted history of productive engagements between painting and traditions of alchemy mythically embraced by Jan van Eyck in the early fifteenth century, traceable across the Silk Road and far back into Mediterranean antiquity.[31] Nor is it to assert that Reynolds's relation to chemistry was either unambiguously evident or peculiar to him; indeed, documentation of Reynolds's library, readings, and chemical contacts is just as fragmentary as his enterprise would prove influential.[32] Instead, I argue that where interests in chemical experiments with paint media and their capacity for change in time may have been courted by marginal figures in England's nascent, supposedly *retardataire* tradition of painting in oils, they were made standard to Anglo-American practice through the searing impression of what one historian has called the "vast moral function" of Reynolds's fabulously successful art.[33] Further, that equivocation between what the first president of the Royal Academy *actually did* by chemical means and *what he was taken to have done* as understood by acolytes is very much to the point. Ignited in paint, Reynolds's torch would be passed in relay to self-described chemical replicators in the years around 1790, whose work, in turn, would be recast as the chemical art of photography in the 1860s; so the subsequent chapters demonstrate.

To make that case, this chapter begins by situating Reynolds in his native milieu of greater Plymouth, a port city whose denizens and maritime concerns would long figure in his enterprise. Tracing an arc of chemical research forward from southwestern England to the 1750s Society of Arts at London's Charing Cross and into the painter's studio on Leicester Square in the 1770s, I build circumstantial as well as more concrete evidence to excavate chemical practices Reynolds apparently meant to conceal, placing them back into dialogue with a technical culture to which the painter has come to appear decidedly hostile.[34] Framed this way, Reynolds's public embrace of chemistry at the Royal Academy in 1774 and the heat of its rapid, contested response can open onto broader interpretation. As canvassed in the chapter's fourth section, Reynolds's chemistry allows return to a moment in the 1770s when the representation of time—no longer satisfied through the paternal personification of Hogarth's subscription ticket (see fig. 2.2)—had achieved particular importance. Expanding a theme of animation trenchantly identified in Reynolds's project, I conclude by tracing this unstable alloy of chemistry and ambitious painting through the proleptic temporality of fine art embodied in the painter's "infant portraits" of the 1770s. As with Robert Hooke's chemical comet, noted in the introduction, some patience is needed for these components to catalyze; the fireworks at the end make it worth the trouble.

Material Origins

Sunlight glints off a thumbnail, anchoring a splayed book into the lap of a female sitter who turns diagonally to the beholder (fig. 2.4). Descending from the left, illumination radiates outward from the vector of her white-clad bosom. It shimmers across the pleats of her satin dress, gathering into a calligraphic scrawl as folded cuff yields to scintillant beads of light playing off the gilded tooling of a book's spine at lower right. That luminous thumb is the sole flesh visible in a painting no longer identifiable with its sitter, Miss Irons, a noted beauty of mid-eighteenth-century Plymouth.[35] For, following a dispute over payment with the sitter's family, the portrait was defaced—better, *re*-faced—by Thomas Rennell (1716–1788), the artist commissioned to paint the picture in the first place. Upon his own rendering of Irons's visage, Rennell superimposed a *trompe l'oeil* depiction of a print.

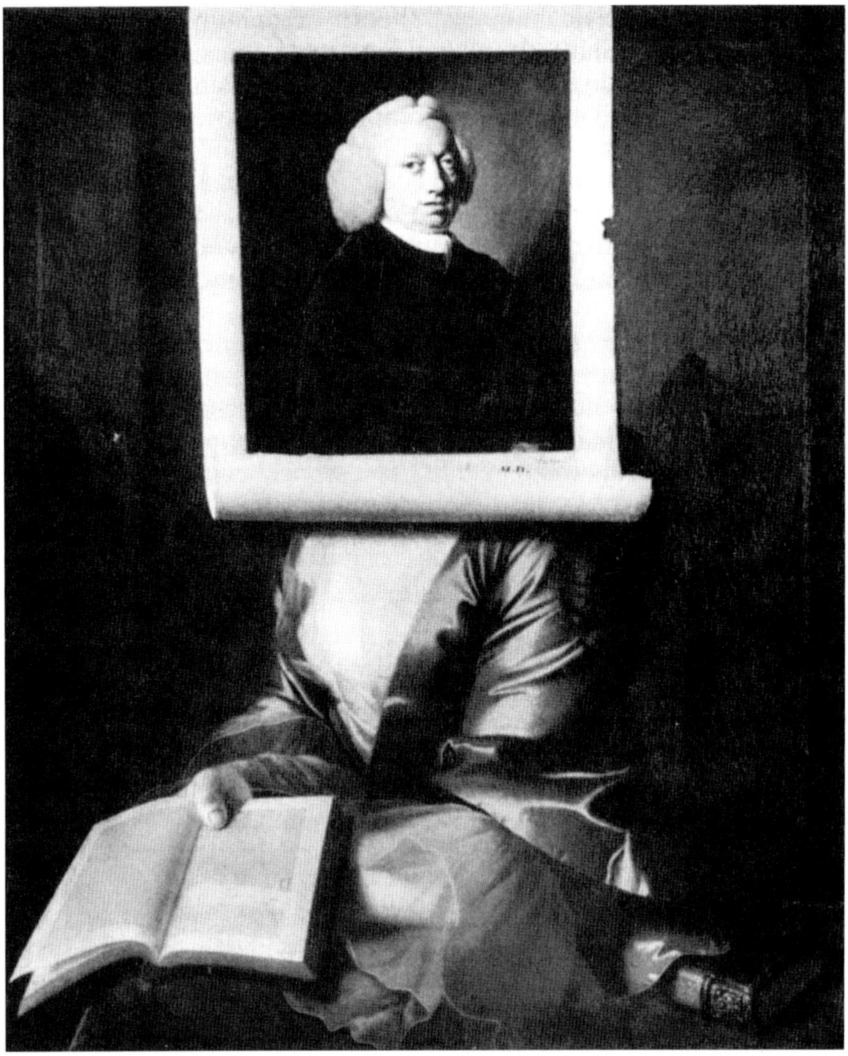

Figure 2.4 Thomas Rennell, *Miss Irons, with contemporary defacement*, ca. 1758–1788. Oil on canvas, 75 × 60 cm. Present location unknown.

With its curling, shadowy bottom casting obscurity onto the sitter's chest, the painted print shows the head and shoulders of a walleyed man staring out toward the beholder.

Jumbling bodily orientation, scale, and gender within the pictorial world of the portrait, this fictional print depicts a real engraving cut into a copper plate by Edward Fisher (1730–ca. 1785) after a *different* portrait by Rennell: a depiction of Dr. John Huxham, now in the collection of the Royal Society of London (plate 7). Elected a fellow of the Society in 1739, Huxham (1692–1768) enjoyed an illustrious international reputation for learned treatises like *Observationes de aëre et morbis epidemicis* (1752) and *An Essay on Fevers*, which had gone into five editions by 1767. An empiricist who refused the label "empiric," Huxham embraced the ancient Hippocratic corpus as the necessary cornerstone of learned medicine, a principle he extended to human endeavor in all its liberal branches.[36] "Of the utmost Advantage, and . . . universally acknowledged to be so," Huxham argued, emulating the ancients is particularly crucial to "whoever would excel in poetry, Sculpture, Statuary, &c . . . . as the most perfect Models, and most just Copies of Nature."[37]

How to craft generative life from canonical models without edging into that "delusive kind of industry," as Reynolds later call strictly imitative copying?[38] How to differentiate a broadening, tradition-enriching *emulatio* from the narrowing, deadening, and ultimately mechanical proposition of *imitatio*? Questions like those were central to the vexed logics of seriation vital to the early modern arts and sciences alike, where they bore fundamental aesthetic, epistemological, and political implication.[39] Such was heady stuff in the milieu of mid-eighteenth-century Plymouth that Huxham shared with both Rennell and the Irons family—and where the doctor was regarded as something of a charlatan.[40] The Irons portrait makes that point. As William Schupbach has convincingly argued, Rennell's disfiguration of his own work exemplifies "one of the traditional practices of painters to insult their bad customers in public by defacing their rejected commissions in ingeniously appropriate ways."[41] Since the family refused to pay Rennell for a portrait they felt did insufficient justice to Miss Irons's beauty, the painter blocked up the offending face with a vigorously unpleasant visage, thereby unmasking the sitter's true, duplicitous character.

Sitters, patron, and painter of this peculiar, doubled portrait were all active in midcentury Plymouth, as was Joshua Reynolds himself. Born in 1723 in Plympton (a village some five miles from Plymouth), Reynolds knew Devon and its contours well. In his juvenile commonplace book, he dutifully recorded the county's key features: "Devon in the Diocese of Exeter, 200 Miles in circumference, and 36310 Houses. The Air Sharp & healthfull. . . . It has divers excellent Harbours for the Navy-Royal, at Dartmouth Plymouth, etc. it contains 394 parishes, and 32 Towns. The Chief Town is Exeter, 138 miles from London, its Chief Seat is Biddiford."[42] Returning to greater Plymouth repeatedly throughout his adult life, Reynolds was elected alderman of Plympton in 1772, and mayor in 1773.[43] The Edgcumbes, the Parkers, and other politically connected Devonian families—along with the officer class of the major naval base in Plymouth—all played key roles in Reynolds's patronage and career advancement.[44]

Formative connections to the West Country's intellectual and technical cultures also ran in Reynolds's family. The painter's great-grandfather was Thomas Baker (ca. 1625–1690), a Somerset-born mathematician affiliated with Oxford's Wadham College, that epicenter of the physiological tradition examined in chapter 1. Relocating to Bideford, Baker subsequently published *The Geometrical Key; or, The Gate of Equations Unlock'd* (1684), an English/

Latin algebraic text that he dedicated to Seth Ward and Joseph Williamson, then president of the Royal Society.[45] When contemplating his son's career, the Oxford-educated Rev. Samuel Reynolds (who ran Plympton's grammar school until his death in 1746) aimed to mobilize those learned connections on his wife's side of the family. Reynolds senior thought either to apprentice Joshua as a painter or "of making him an Apothecary," thereby placing him in the traditional role of supplier of painters' pigments and other powerful drugs. Were the latter option chosen, the Rev. Reynolds would "make the proposal to my wifes kinsman Mr. Baker of Bideford."[46] Reynolds and his father also knew the medical doctor superimposed upon Rennell's sitter. Seeing the young Reynolds's drawing after the *Laocöon*, Dr. Huxham found his own principles of classical emulation exemplified so strongly that he was moved to prophecy. According to the Rev. Reynolds in 1742, Huxham had then declared "that he who drew that would be the first hand in England."[47]

Reynolds was even closer to the family through which Rennell's Irons/Huxham portrait descended: the Mudges of Plymouth.[48] The family's patriarch, Zachariah Mudge (1694–1769), had attended the grammar school operated by Reynolds's grandfather in Exeter, where he became friendly with the painter's father.[49] Building a glittering career in the Church of England, the elder Mudge was appointed vicar of St. Andrew's church in Plymouth in 1732, then prebendery of Exeter in 1736, while acting as a leading intellectual light to multiple Reynolds generations. Edmund Burke (1729–1797) traced an enduring imprint on the painter from Mudge, whose sermons he had brought back into publication at the outbreak of the French Revolution.[50] Writing with some ambivalence about his late friend, Burke identified Reynolds's aptitude for conceptual abstraction and pictorial generalization as "the source of every thing that can be called science. I believe his early acquaintance with Mr. Mudge of Exeter, a very learned and thinking man, and much inclined to philosophize in the spirit of the Platonists, disposed him to this habit."[51]

Science, as defined primarily in the 1755–1756 dictionary of Reynolds's close friend Dr. Samuel Johnson (1709–1784), was the divine knowledge professed by Zachariah Mudge. Johnson also offered other conceptions; science could be "Art attained by precepts, or built on principles."[52] The family's practical, principled disposition to science in these expanded senses was demonstrated amply by Zachariah's second and fourth sons. Patronized by George III and Spanish king Ferdinand VI, Thomas Mudge (ca. 1715–1794) was a leading, London-based maker of timekeeping instruments. In the 1760s, Mudge would be enlisted to adjudicate the distribution of the celebrated "longitude prize" to watchmaker John Harrison; Mudge's own improvements to Harrison's precision timekeeping devices were awarded further remuneration by the Board of Longitude in the 1790s.[53] Author of

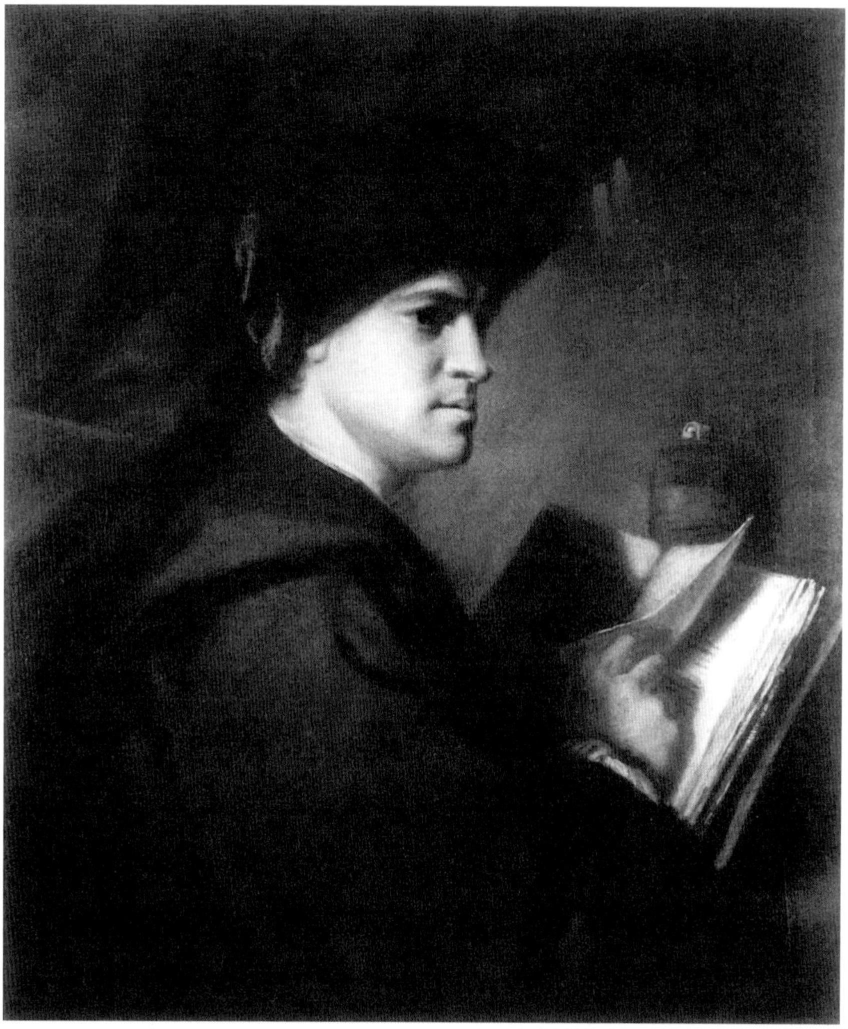

Figure 2.5 Joshua Reynolds, *Dr. John Mudge*, ca. 1752. Oil on canvas. Private collection.

Thoughts on the Means of Improving Watches and More Particularly Those of the Use of the Sea (1765), this "Beethoven among watchmakers" left London in 1771, relocating to the maritime center of Plymouth, where precision time-keeping and determination of longitude were vital concerns.[54] There, he joined his brother John Mudge.

Soon after returning from his Italian travels in 1752, Reynolds depicted his lifelong friend John Mudge (1721–1793) pausing pensively in a darkened interior (fig. 2.5). Mudge rests his hand on an opened book propped before him, an extended index finger extracting an illuminated curve from the page it lifts. Supported by a vertical volume, that lighted page empties into a shadowed gutter, splitting its face like the crease of Mudge's furrowed brow to underscore his reciprocal bookishness. Deeply inset eyes gazing away from the book and out of the pictorial frame to the right, Mudge's lantern-jawed visage turns in three-quarter view to the beholder. Shaded folds in his voluminous gown implicitly overhang the picture's bottom edge, breaching the proverbial fourth wall while simultaneously asserting the depth of the sitter's thought.[55]

An acclaimed medical doctor, John Mudge built his career in Plymouth, but was known more widely. After Reynolds had been proposed as a "Gentleman of learning, a lover of Philosophical enquiries," and elected fellow of the Royal Society of London in 1761, one of his major documented acts within the institution was endorsement of bids for fellowship—as he did for Mudge in early 1777.[56] Promptly elected, Mudge also won the institution's Copley Prize that year. The prize was awarded for Mudge's advances in crafting the parabolic mirrors required for reflecting telescopes, with which Isaac Newton had aimed to overcome chromatic aberrations, unmanageably long focal length, and other obstacles hindering telescopes operating by principles of optical refraction.[57] Outlining his prizewinning work, Mudge explained the remarkable metallurgical precision needed for the job. Because "errors in reflection are four times as great as in refraction," the metals demanded for a telescope after Newton's design have to be sufficiently hard, light, and free of pores as to yield nothing less than "perfection in the figure of the speculum."[58]

Casting perfection was no easy business. Mudge had initially failed in all attempts to mix serviceable alloys of copper and tin according to procedures set out by standard craftsmen's guides such as Dr. Robert Smith's *Compleat System of Opticks in Four Books* (1738). Smith's proportions of three parts copper to one and a quarter part tin invariably yielded metals perforated with microscopic pores, which compromised image-resolution on polishing.[59] It was only "in some measure by accident," as Mudge put it, that this challenge was overcome. Adding to the molten metal some scrap from "one of the bells of St. Andrew's" (the Plymouth church where his father had previously been vicar), Mudge cast a metallic alloy that "turned out perfectly free from pores, and in every respect as fine a metal as I ever saw."[60] By what principle had the problem been resolved? Extensive experiment and reflection on the matter revealed to Mudge that the significant heat needed to fuse the copper and tin produced calcination and, in turn, the problematic pores. By reheating the fused, metallic mass while mixing in an additional portion of tin, the serendipitous bell-fired metal could consistently be recreated and the grounds for a parabolic speculum laboriously wrought.[61]

If Plymouth connections thus brought Reynolds into the ken of the learned, male sitter defacing the Irons portrait as well as the family accomplished in multiple definitions of science through which it descended, the painter was on still more intimate terms with the network of Devonian artists to whom the strange portrait has variously been attributed. Until its cleaning in 1979, the Irons/Huxham picture was credited to Thomas Hud-

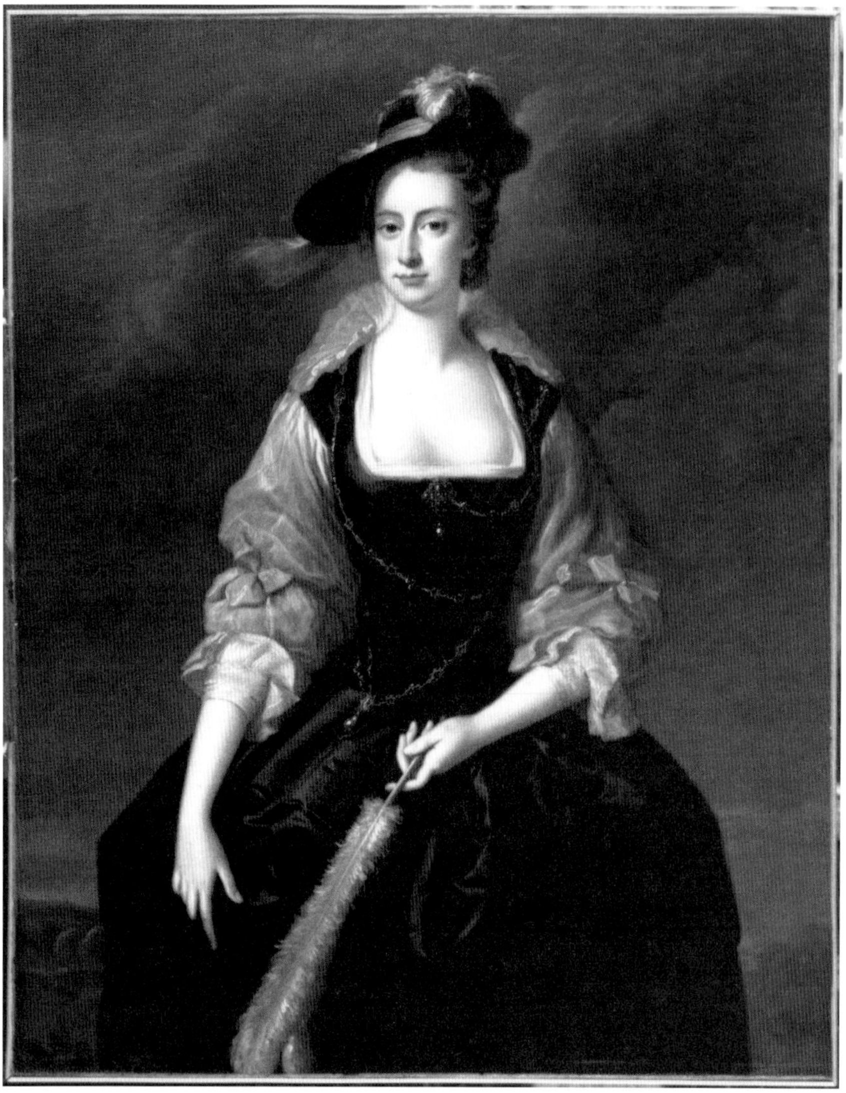

Figure 2.6 Thomas Hudson, *Lady Frances Courtenay*, ca. 1741. Oil on canvas. Courtesy of the Huntington Art Collections, San Marino, CA.

son (1701–1779), the painter to whom Reynolds was apprenticed in 1740.[62] Son-in-law and literal inheritor of the portrait practice of painter/theorist Jonathan Richardson (1667–1745), Hudson himself hailed from Exeter. When visiting from the capital to deliver some thirty portraits of leading citizens to the Devon town of Barnstable in 1740, Hudson had agreed terms with the Rev. Reynolds for the four-year apprenticeship of his seventeen-year-old son.

Exemplary of the approach to painting Hudson practiced during the era of Reynolds's apprenticeship is his *Lady Frances Courtenay* of the early 1740s (fig. 2.6). Turned slightly away from the beholder, Courtenay's head is an aggregate of stacked, geometrical patches, their joins burnished and smoothed with a brush's dry tip to bulge as gleaming brow ridges and rouged cheeks. Coy strokes gather the undulating pouches of chin into an immaculate, porcelain flesh, opalescent as it reflects light up from the sitter's bared throat and bosom. Thinly opulent—exposed yet impersonal—Courtenay is kitted out in archaizing dress designed to recall the Flemish tradition of Rubens and Van Dyck, which has been rendered in far freer, loosely scumbled brushwork courtesy of the studio assistants and drapery specialists to whom Hudson subcontracted pictorial labor. "The last of the conscienceless artists": so goes Ellis Waterhouse's colorful analysis.[63] Hudson's safe, inoffensive renderings appealed to what his most dedicated, modern interpreter has called "the most conservative element of British society." Yet his capacity for savvy reference to the Old Master works he collected (along with his shrewd cultivation of patronage from ruling elites at metropolitan core and Devon periphery) offer important precedents for the portraiture business later built so successfully by Reynolds.[64] Known to have undertaken several projects of pictorial restoration in the 1730s, Hudson also trained two of eighteenth-century Britain's most adventurous painters: Reynolds and Joseph Wright of Derby.

More frankly experimental in his pictorial facture was Thomas Rennell, the most convincing agent behind the Irons/Huxham portrait—as well as the painter with whom Reynolds likely lived after breaking his apprenticeship with Hudson and returning to Devon in 1743.[65] "The knowledge of Mr. Rennell was universal," avowed painter James Northcote (1746–1831). "He was fond of chemistry, to which he devoted a considerable portion of his time. Most of his colours, which he prepared himself, went through that operation: and he is said to have discovered the art of fixing those which are the most fading."[66] Little of Rennell's work survives, but the Irons/Huxham portrait speaks to those experimental proclivities and a bricoleur's ruthless wit. Such features are also evident in the earliest specimen of Reynolds's painting in oils, which too was literally tied to Devon's coast. While many early biographers would credit Plymouth-based power broker Richard, first Baron Edgcumbe, for launching Reynolds's artistic career by introducing him to Captain Keppel, a darker story traces the painter's earliest picture to the agency of Edgcumbe's wastrel son, also called Richard, an incorrigible gambler who died heirless at the age of forty-five.[67] "Sir Joshua Reynolds' first picture, Painted when under 12 years of age," per the caption of its 1822 mezzotint, this jowly portrait of the Rev. Thomas Smart with his skeletal, gloved hands resting in canine obedience on a fictive pulpit became an object of Victorian mythology (fig. 2.7).[68] "Coloured in a boat house at Cremyll beach under Mount Edgcumbe," so would claim antiquarian collector William Cotton in 1856, the likeness was painted by a juvenile Reynolds "on canvass which was part of a boat sail, and with the common paint used in shipwrights' painting sheds."[69] Florid though the fiction may well be here,

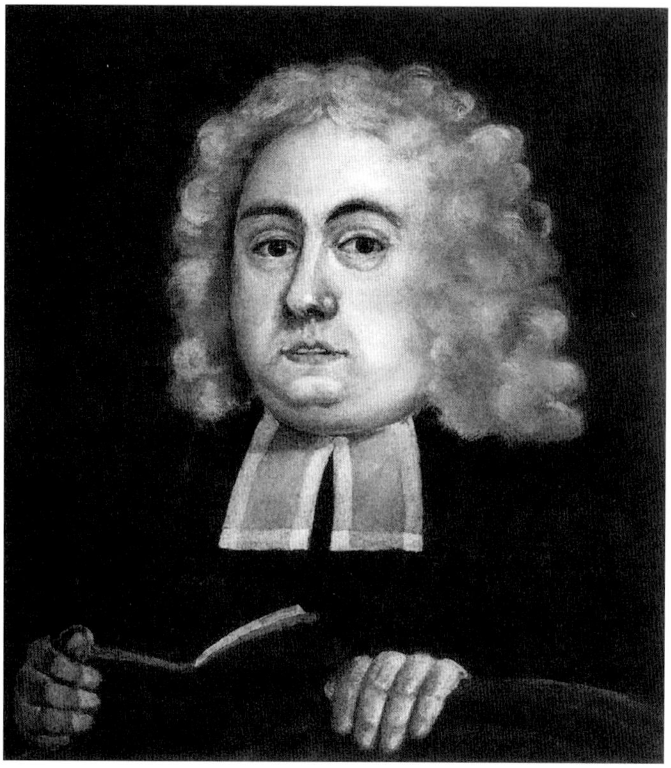

Figure 2.7 Joshua Reynolds, *The Rev. Thomas Smart*, ca. 1735. Oil on canvas. Private collection.

Reynolds's attraction to unconventional painting materials—as well as to the chemo-mechanical *nous* of Plymouth's technical community—would live on in his mature London practice.

From Plymouth to London (and Back Again)

Departing with Keppel from Plymouth in 1749, Reynolds voyaged to Italy and on to France for some three years of artistic study. Then unusual for British artists, those Continental travels would prove central to his subsequent career as a painter. Interpreters have long stressed how the sketches taken, observations made, and contacts forged in Italy would shape the grand-manner aspirations of Reynolds's approach to portraiture. On his return to London, they facilitated his entry to the Society of Dilettanti and other elite clubs; Italianate sites and conversations would later litter his *Discourses*. True, the metropolis to which Reynolds returned in 1753 was hardly an artistic rival to Rome or Paris. But an increasingly sophisticated matrix of venues and institutions was consolidating in 1750s London to support makers, dealers, and consumers of "the polite art" of painting. From the early 1740s, pictures by Hogarth, Thomas Gainsborough, Richard Wilson, and other living British artists could be seen at Thomas Coram's Foundling Hospital near Bloomsbury. At Vauxhall Gardens to the southwest, some fifty paintings of popular festivity by Francis Hayman and his studio had been installed in supper boxes, which catered to the reformed site's fashionable clientele. Beyond the Duke of Richmond's gallery of casts after ancient sculptures or the Raphael cartoons and other prized Old Masters from the royal collection visible at Hampton Court Palace, these new venues—alongside a bustling network of private artistic academies and a robust traffic in pictures sold by third-party dealers—nurtured a consolidating public for contemporary art.[70] If this "rise of a decentralized institutional system for the arts independent of the Court and the Government" moved in parallel with the emergence of a British consumer society, then a capstone event in that bourgeois aggregation was the show hosted for two weeks in the spring of 1760 by the Society of Arts: an event mythically known as Britain's first specially organized art exhibition.[71]

Reynolds would go on to join many clubs and learned institutions, but his involvement with the Society of Arts from 1756 merits close attention. Founded in 1754, the institution aimed to spur British art and industry against perceived French superiority. Drawing funds from subscribing members, it awarded monetary prizes to men and women for the manufacture in Britain of goods ranging from landscape paintings to sal ammoniac (a hard, white salt composed of ammonium chloride). The Society of Arts grew meteorically; it expanded from fewer than twenty members in 1755 to over two thousand a decade later. The institution routinely used what it called "Trials" or "Experiments" to establish whether submissions vying for cash prizes had successfully achieved published targets. In January 1760, a committee chaired by Scottish painter Allan Ramsay (1713–1782), then one of Reynolds's major professional rivals, met to assess a sample of copal varnish made from tree resin indigenous to the Americas. To win the prize, the varnish had to equal in quality rival preparations imported from Paris; it was to possess "great Hardness, perfect Transparency without disclouring any Painting it is laid over, being capable of the finest polish and not liable to crack."[72] Charged with evaluating submissions, Ramsay's team devised a regime of trials. Painting a design on a panel and cutting it down the middle, they would varnish one half using the candidate's submitted preparation; the other half would be varnished by the candidate. Brought together, the pieces were compared against effects of the Parisian standard.[73] More extreme than the fine-art trials for which Reynolds was enlisted were protocols set in April 1759. Then, sculptor Joseph Nollekens was locked in a room, challenged to recreate a prize-seeking model of *Lot and His Wife* as proof it was truly his own work.[74]

Experiments designed by the institution to adjudicate monetary awards also encouraged the submission of experimental materials and processes. Late in 1755, chemist Woulfe demonstrated "a Blue Colour which he believes is free from the Imperfections of common verdigrease."[75] In early 1758, Birmingham manufacturer John Gardner relayed that he had "invented an Oil Varnish that will not crack, is very transparent and softens and embellishes a piece of painting to Admiration."[76] Varnishes and Woulfe's concoctions would both subsequently appear in Reynolds's pictures. But surely the most interesting contact between the painter and the experimental techniques fostered by the Society of Arts came in 1760. Then, Reynolds was named by Johann Heinrich Müntz (1727–1798) as a prime candidate for adopting the encaustic process he had adapted from the Comte de Caylus—a technique claimed to be capable of fixing an image "permanently," as we will see in the next chapter.[77] Dedicating a treatise on the topic to none other than Reynolds's mythical childhood muse Dick Edgcumbe Jr. (who "saw the first Essays and Experiments in Encaustic"), Müntz sent his text to the Society of Arts with two landscape pictures prepared in encaustic technique, one "varnished with the white of an Egg; the other is a Symple dead-colouring fixed, as it came from the fire."[78] There, Reynolds was selected in June 1760 to serve on a subcommittee charged with assessing Müntz's submission.[79] To future conservators' horror, egged, wax-based paint-layers would soon materialize in Reynolds's practice.[80]

Endorsing a vision of the fine and industrial arts conjoined, the Society of Arts promoted painting, but it also acted as a metropolitan catalyst for the rapidly changing domain of chemistry. Where scholarship once asserted but meager support for chemistry, either at eighteenth-century Oxbridge or the Royal Society of London, that historiographical narrative of neglect is evolving.[81] Chemistry's position was certainly changing historically, especially following the establishment of William Cullen as the chair of chemistry at Edinburgh University in 1756.[82] Beginning in 1758, the Society of Arts itself had established a freestanding Committee on Chymistry that commissioned translations of texts including Pierre-Joseph Macquer's *Elémens de chymie-pratique* (1756) and Georg Ernst Stahl's *Fundamenta Chymiae dogmaticae et experimentalis* (1746–1747). It also assessed contributions and distributed prizes for British production in bulk of borax, bismuth, and other materials useful to the arts and industry.[83] Yorkshire-born polymath Robert Dossie (1717–1777) was a central figure in the early Society's chemical endeavors. From the spring of 1760, the minutes show him variously presenting with Benjamin Franklin on methods for purifying cloudy onyxes; setting inquiries into a green pigment manufactured by Woulfe; using chemical means to catch rats; and taking over the chairmanship of the Committee on Chymistry in November 1760.[84]

Dossie's *Handmaid to the Arts* (1758, expanded 1764) is an apt summa of the institution's unified vision for the fine and industrial arts.[85] Positioning chemistry as an Ur-science, Dossie details the materials and steps required for making pigments, painting vehicles, grounds, and varnishes, as well as outlining the virtuosic tradition's panoply of gilding, staining, japanning, bronzing, and associated processes.[86] Surveying the topsy-turvy world of modern manufactures and additives to artists' *materia practica*, Dossie declares, "It is not indeed to be wondered at, that persons who are wholly unaccostomed to chymical processes, should miscarry in the attempting some of these, which are of the nicest kind, and where the intention may be frustrated by so many minute accidents." Only by advancing command of "natural history, experimental philosophy, and chymistry" could British artists in the broadest sense challenge their vaunted French rivals.[87]

An articulate spokesman for that guiding function for chemistry, also active in the Society of Arts, was William Lewis (1708–1781), fellow of the Royal Society and Oxford-educated medical doctor.[88] Lecturing to London audiences through the 1730s and 1740s, Lewis positioned the chemist as reckoning with altogether more elusive quarry than the "determinate forces, subject to mechanic laws, and reducible to mathematical calculation," that were of concern to the prestigious tradition of mechanical philosophy promulgated at the Royal Society since the later seventeenth century.[89] "Chemistry," Lewis claims, "considers bodies as being composed of such a particular species of matter; dissoluble, liquefiable, vitrescible, combustible, fermentable, &c. impregnated with colour, smell, taste, &c. or consisting of dissimilar parts, which may be separated from one another, or transferred into other bodies. The properties of this kind are not subject to any known mechanism, and seem to be governed by laws of another order."[90] Were it to properly complement Isaac Newton's mathematicized physics, chemistry had to proceed as an autonomous field of natural knowledge concerned with discrete methods, forces, and materials.[91] This was no province for the armchair philosopher, a point visualized emphatically by the frontispiece to Lewis's *Commercium Philosophico-Technicum; or, The Philosophical Commerce of Arts* (1763; fig. 2.8). Furnaces for assaying, ovens for glassblowing, retorts for distilling, a palette for painting (which hangs from the left-hand edge of the central chimney): these instruments and more would be needed to track chemistry's slippery targets.

Nowhere is Lewis's practical, chemical contribution to the early Society of Arts more aptly grasped than by an artifact he sent to the Rev. Stephen Hales (1677–1761), a founding member of the institution and noted author of *Vegetable Staticks* (1727), which extended Newtonian chemical thinking into the circulation of fluids in plants.[92] In early 1760, Lewis wrote to Hales on a vexing problem: how to fabricate a concoction useful for marking ownership of sheep without leaving disfiguring stains in the wool. Lewis reported that his most successful preparation had been made by combining melted tallow with finely powdered charcoal and stirring the composition until it was entirely black. "It quickly fixes," Lewis explained, "does not spread and 'tis presumed will undergo little or no alteration from the sun or rain. It has also this advantage, that it will be completely discharged at the

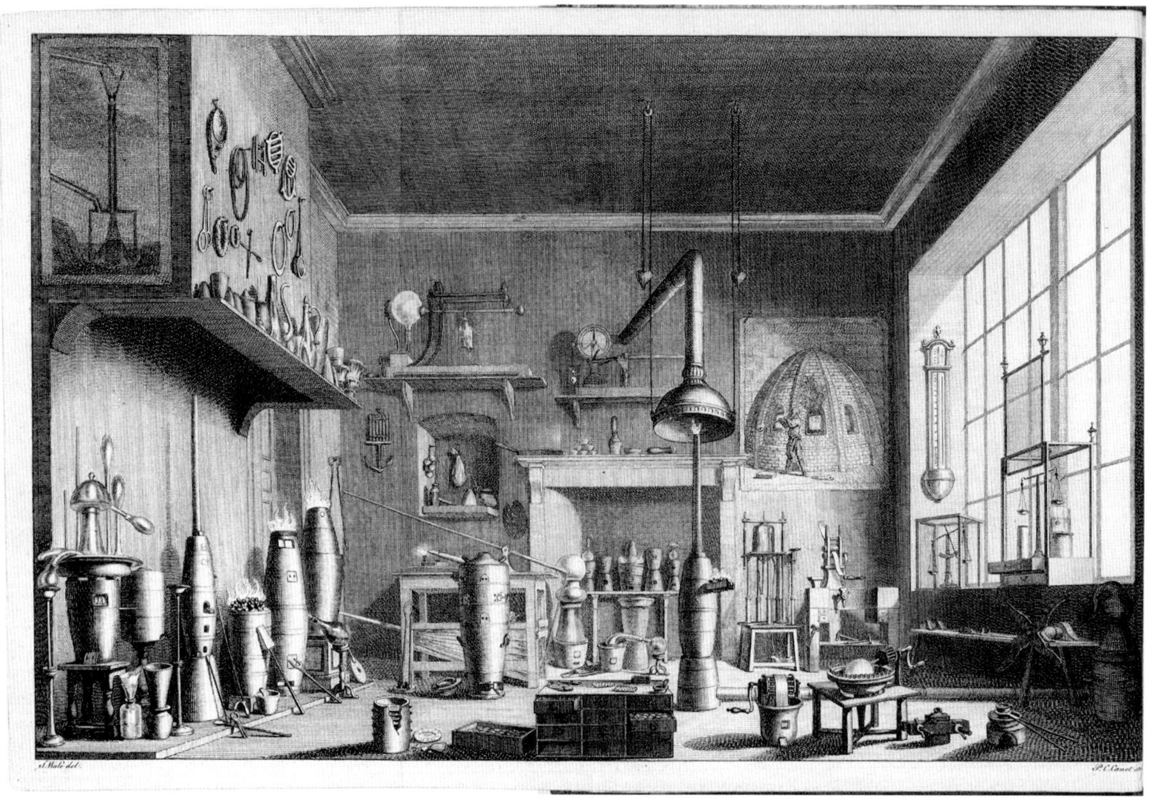

Figure 2.8 P. C. Canot's engraving of S. Wale's frontispiece to William Lewis, *Commercium Philosophico-Technicum; or, The Philosophical Commerce of Arts . . .* (1763).

same time that the wool is cleansed from its own grease."[93] Light-fast and insoluble in water, the blackened tallow would allow owners of sheep—that traditional backbone of Britain's textile industry—to identify their property without compromising its industrial yield since the marks could be washed cleaned with alkali-rich soap (fig. 2.9). A sample of Lewis's reversible marking preparation survives as a hatchet of matte black stain sunk into the wool's coarse weave. Alongside this "specimen of flannel marked with the composition by a stamping iron" appears a bare swatch, its weave buffed and rubbed where it "had been marked in the same manner and afterwards washed clean."[94] Evidence of practical chemistry made and reversed, the sample also enacts a telling geotemporal relay. Produced at Lewis's laboratory in Kingston-on-Thames, it was transported some two miles to Hales in Teddington by Alexander Chisholm, Lewis's assistant, who would, in the 1780s, act as amanuensis and chemical tutor at Josiah Wedgwood's Etruria.[95]

Influential stories tell us that, by the mid-1760s, Reynolds and other leading painters had abandoned the Society of Arts with its "improved spinning wheels or machines for slicing turnips" precisely to free themselves from the taint of all things mechanical and industrial.[96] Indeed, fol-

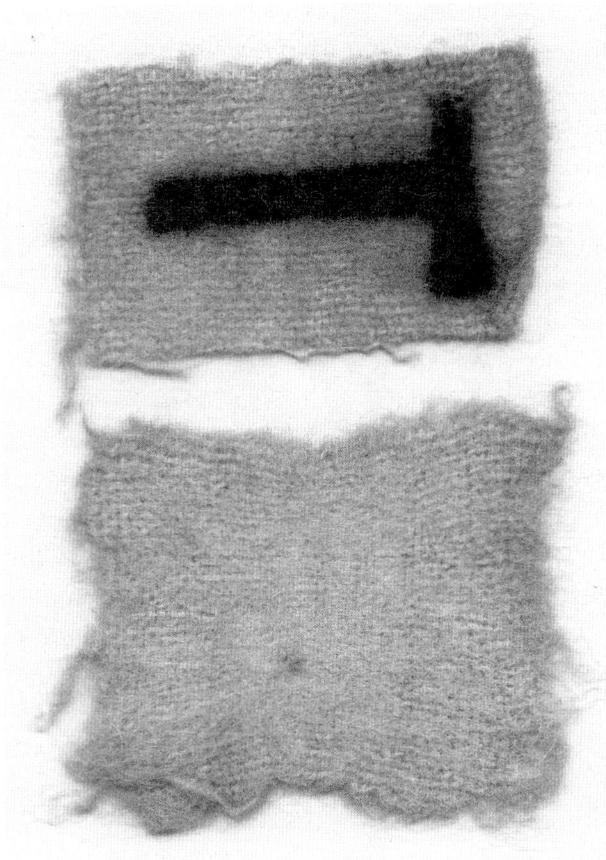

Figure 2.9 William Lewis's reversible sheep-marking preparation, as sent to Stephen Hales, February 16, 1760, Royal Society of Arts (RSA/PR/GE/110/25/1).

lowing their inaugural exhibition with the institution, a faction of Hogarth-led painters, sculptors, and architects intent on defining their practices in liberal terms broke from the Society of Arts to stage their own exhibition in 1760 as the Society of Artists of Great Britain.[97] The Royal Academy of Arts that would eventually emerge from this defection—with Reynolds knighted as its president in 1768—systematically eliminated instruction in paint-handling, pigment-making, and chemistry as it promoted an elevated vision of fine art.[98]

Yet some of the most compelling links connecting the emergent Royal Academy to that earlier, more ecumenical culture can be found running through Reynolds's own studio. James Northcote is a central figure here.[99] Son and brother of Plymouth watchmakers (and a former apprentice to an apothecary), Northcote was introduced by John Mudge to Reynolds; he would go on to apprentice with the painter between 1771 and 1776. Connections between technical culture in Plymouth and ambitious painting in London read across the pages of Northcote's contemporaneous correspondence. During his time in Reynolds's studio, Northcote wrote frequently to his elder brother, Samuel Northcote, variously requisitioning metal-casting,[100] fitting and shipping optical instruments,[101] and commissioning a telescope for Reynolds to use at his new country house in Richmond.[102] Northcote saw the expertise required for crafting precision optics as relevant to the production of experimental paint effects. Explaining to his brother how Reynolds "uses his colours with varnish of his own because the oils give the colours a dirty yellowness in time," the young apprentice lamented the procedure's consequence: "This <method> of his has an inconvenience full as bad which is that his pictures crack, sometimes before he had got them out of his hands." He then posed a query for their father: "I should be glad if my father would let me know if he thinks it is oweing to the varnish, and if so the reason that varnish should crack sooner than oil."[103] A month later, the apprentice cast a skeptical eye on the preparations his brother had recommended: "The camphire will not do for painting because it only keeps the Varnish moist a longer time but when it drys as it will in time the consequences will be the same."[104] Northcote thanked his brother for passing along a now-lost varnishing technique, which he promised to attempt when "I am perfect enough to try experiments but as yet I know no more of varnishes than that the colour is apt to change and the paint to crack."[105]

With its octagonal painting room, Chippendale furniture, and private gallery of Old Master paintings, Reynolds's fashionable town house at 47 Leicester Fields might seem just as far away from the cramped quarters of Plymouth Dock as the Royal Academy's diet of drawing after classical casts and life models did to the Society of Arts' polymathic program of cash premiums and experimental tests. Yet it is significant to note that Thomas Rennell—maker and remodeler of the Irons/Huxham portrait—continued to figure in Northcote's relay of intelligence between London and Plymouth. Transmitting a sequence of recipes to his brother from the studio in 1771, Northcote insisted on silence: "I would not have you mention to Rennell any thing of what I may have said concerning Sir Joshua, nor any defects which I may have remark'd to you in him nor to any body because he would not chuse to have it known and it would be ungenerous in me."[106] Nearly three decades after Rennell and Reynolds had cohabited amid its bastion of naval power, Plymouth's technical culture of materials and precision metal-casting literally lived on within the bounds of Reynolds's studio at Leicester Fields. And by the mid-1770s, that chemical knowledge had become a polarizing proposition at the Royal Academy itself.

Three Pyrotechnicians: Wright, Reynolds, Hone

From a low vantage point, a well-furnished laboratory opens, its immediate darkness punctured by two lights (plate 8). At left, flame blossoms from an oil lamp to describe the profile of a standing youth who directs visual attention. As the lamp's oily glow casts elongated human shadow onto a distant, partitioning wall, the youth's eyes and pointing forefinger guide his seated companion to the bearded experimenter at center right. Fallen to his knees, the older man extends a hand as if to silence his assistants' chatter. The juvenile gaze of the apprentices, the roving glances of the implied beholder: all may be distracted by gleaming implements, a mountain of manuscript scrawls, or time's implicit passage as registered by the clock and by the moon that cuts in through the arched window's leaded tracery at upper right. But the eyes of the kneeling adept are locked. They stare into a second light: a jet of blue-white gas erupting triangularly from the glass vessel in the pictorial foreground. This luminous blast has come unexpectedly. Per the title under which Joseph Wright of Derby (1734–1797) exhibited this picture in spring 1771, the subject is *The Alchymist, in Search of the Philosopher's Stone, Discovers Phosphorous, and prays for the Successful Conclusion of his operation, as was the custom of the Ancient Chymical Astrologers*.[107]

The claim that the inventions of phosphorus had come by happy accident was common currency in early 1770s Britain. In 1772 dissenting minister and natural philosopher Joseph Priestley (1733–1804) would direct the reader to a posthumous collection of Robert Hooke's writings for an account of the first preparation of Bononian stone by Vincenzo Cascariolo in the years around 1600, while narrating a broader history of self-shining materials. "It was an accident," Priestley affirms, that C. A. Balduin identified the light-giving properties of chalk distilled in aqua fortis during the 1670s "in his pursuit of the philosopher's stone."[108] Acknowledging the promiscuous character of phosphorus research in later 1660s German-speaking lands, noted in the previous chapter, some modern scholars have identified Wright's bald-headed alchemist neither as Balduin nor Cascariolo, but as Henning Brandt.[109] But contemporary viewers of Wright's painting were unmoved by such niceties. Critic Robert Baker took *The Alchymist* to task for "the disagreeable confusion of uninteresting objects" without regard for compositional principle, leaving the painter's foray into chemical history to one side.[110]

Chemistry, though, was par for the course. Connected to the Birmingham-based Lunar Society, an informal group of provincial intellectuals who gathered on full-moon nights to discuss matters relating to natural philosophy, Wright had built a reputation in 1760s London's emerging exhibition culture through *A Philosopher Giving a Lecture on the Orrery* (1764–1766), *An Experiment on a Bird in the Air Pump* (ca. 1767–1768), and other "candlelight" scenes of learned subjects.[111] Further, Wright exhibited *The Alchymist* in 1771 alongside two chiaroscuro views of fiery metalwork in blacksmiths' shops, and he showed his work with an institution that embraced chemical pedagogy. Unlike the rival Royal Academy under Reynolds, the Royal Incorporated Society of Artists of Great Britain (a splinter group formed after the Society of Arts' inaugural fine-art exhibition of 1760) taught chemistry and the making of artists' pigments from the early 1770s.[112]

Against Wright's scenes of forges and fires, the pictures Reynolds exhibited at the Royal Academy's spring show of 1771 (the head of a gentleman, a half-length portrait of an old man, a Venus, a nymph) looked positively limp.[113] Yet, where Wright was depicting chemical experiments, Reynolds was effectively performing them. Consider his early 1770s portrait of Vertue Jodrell, wife of classical scholar and baronet Richard Paul Jodrell (plate 9). With dark eyes cast downward, Jodrell turns at three-quarter view to the implied beholder, her head accentuated by an elaborate, twisted scarf shimmering with impasto highlights. Serpentine trails of pearl hang from her bound braid and proper left ear, echoing the crimson cable that cinches a screening curtain against the ovular frame of blue sky at right. A study in pictorial poise and balance, the portrait was also a flirtation with technical disaster. "Good heavens! Let us recapitulate in English," exclaimed painter Benjamin Robert Haydon in the 1840s: "The head painted in oil, then waxed, varnished, egged, varnished again with Wolff's [i.e., a varnish prepared by chemist Peter Woulfe], then waxed, sized, oiled, egged again, and them finally varnished with Wolff!!!"[114]

Haydon was then translating and publishing the notes on facture that Reynolds inscribed incrementally in the backs of his ledger books now in the collection of the Fitzwilliam Museum, Cambridge. Folio-size financial account books, these ledgers list sitters' names in alphabetical and chronological orders to record payments on paintings (fig. 2.10).[115] At the back of each volume, those pictures are selectively redescribed as recipes: as materials and procedures recorded in a mixture of pidgin Italian, Latin, and English (fig. 2.11).[116] Of Jodrell's painted head, Reynolds writes: "Oil. Cerata. Varnish with ova, poi varn. Con Wolf. Panni Cerra senza olio. Vernicata con ova poi con Wolf."[117] Registering the work of Woulfe, the egging of wax-bound pigments promoted by Müntz to the Society of Arts, Reynolds's terse, polylingual summa drove Haydon to raptures: "The surface Sir Joshua got was exquisite . . . and though the reward was worth the risk, in such extraordinary infatuation he must be a beacon."[118]

To these encrypted notes, Reynolds paired material artifacts that could exemplify his risky, "infatuated" paint preparations. One such device is a stretched canvas in the collection of the Royal Academy of Arts, now called *Studio Experiments in Colours and Media* (plate 10). Paint appears here in rich, variegated potential. Anchored to the bottom edge as an amber archipelago, pats of pigment spread upward in a floating mauve dot; a sashaying deposit of unctuous knife-work at right center; and a fat pucker of paint-media on the earthy ground at the right-hand edge. This canvas is one of

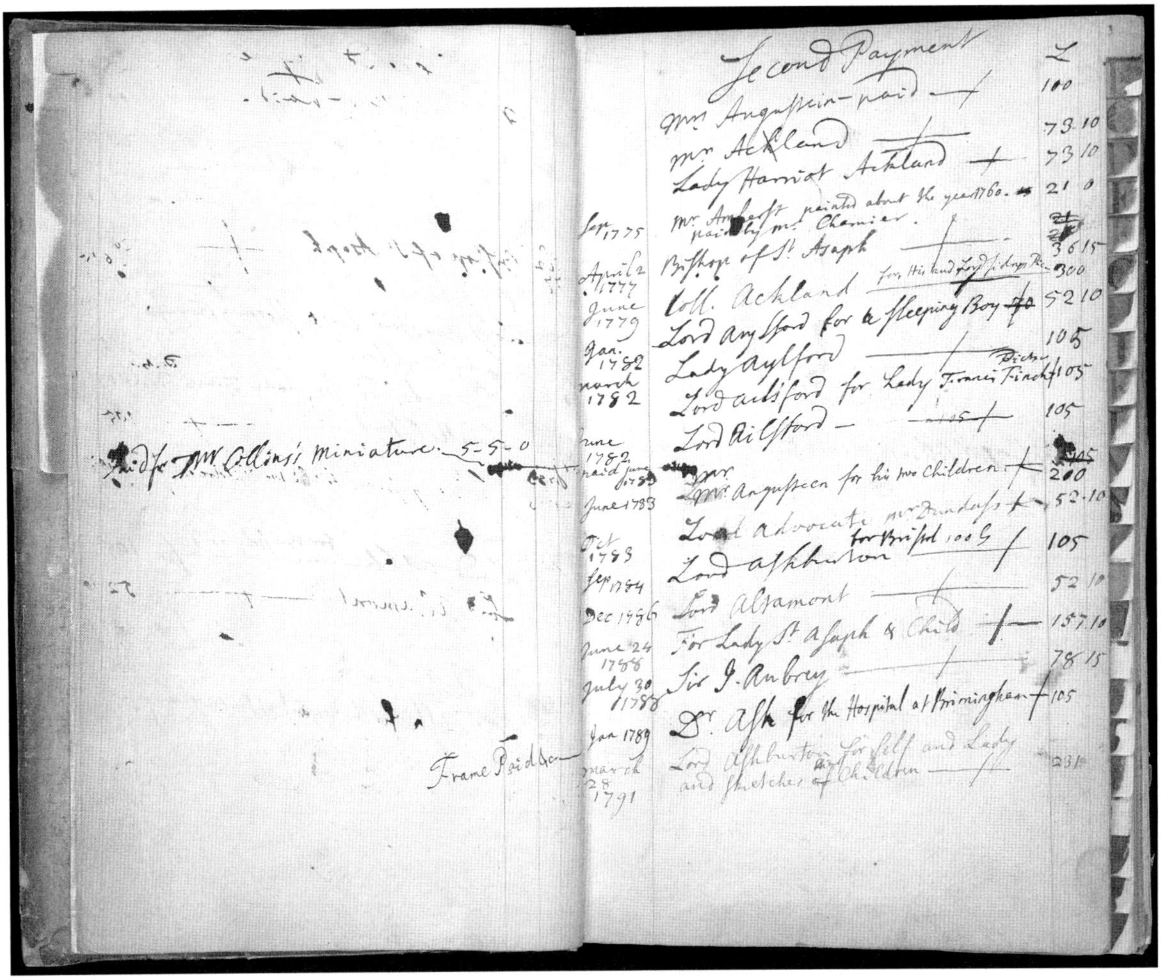

Figure 2.10 Joshua Reynolds's ledger book. Fitzwilliam Museum, Cambridge (MS 2.1916, f3r). © The Fitzwilliam Museum.

two "Experiments in Color" offered for sale to the Royal Academy in 1877 by George Barker Jr. (b. 1819), son of noted picture restorer George Barker Sr. (1794–1838), who had acquired the paintings at the auction of Sir Thomas Lawrence's effects in 1830 (fig. 2.12).[119] Although now illegible, Reynolds's dated annotations, written alongside these paint swatches, had been sufficiently discernible that Sir Charles Eastlake could publish patches such as this in the 1840s: "Gamboge with turpentine, March 6, 1772. Prepared gamboge with cera. Verditer, varnish alone. Gamboge with Venice turpentine, June 3, 1772."[120] By linking paint samples, texts, and dates together in intercepting circles, Reynolds had squared material evidence of his pigment preparations on the canvas support used through most of his vast production against the iterative, textual records kept in the backs of the ledger books. If less systematic than the meticulous documentation of experiment

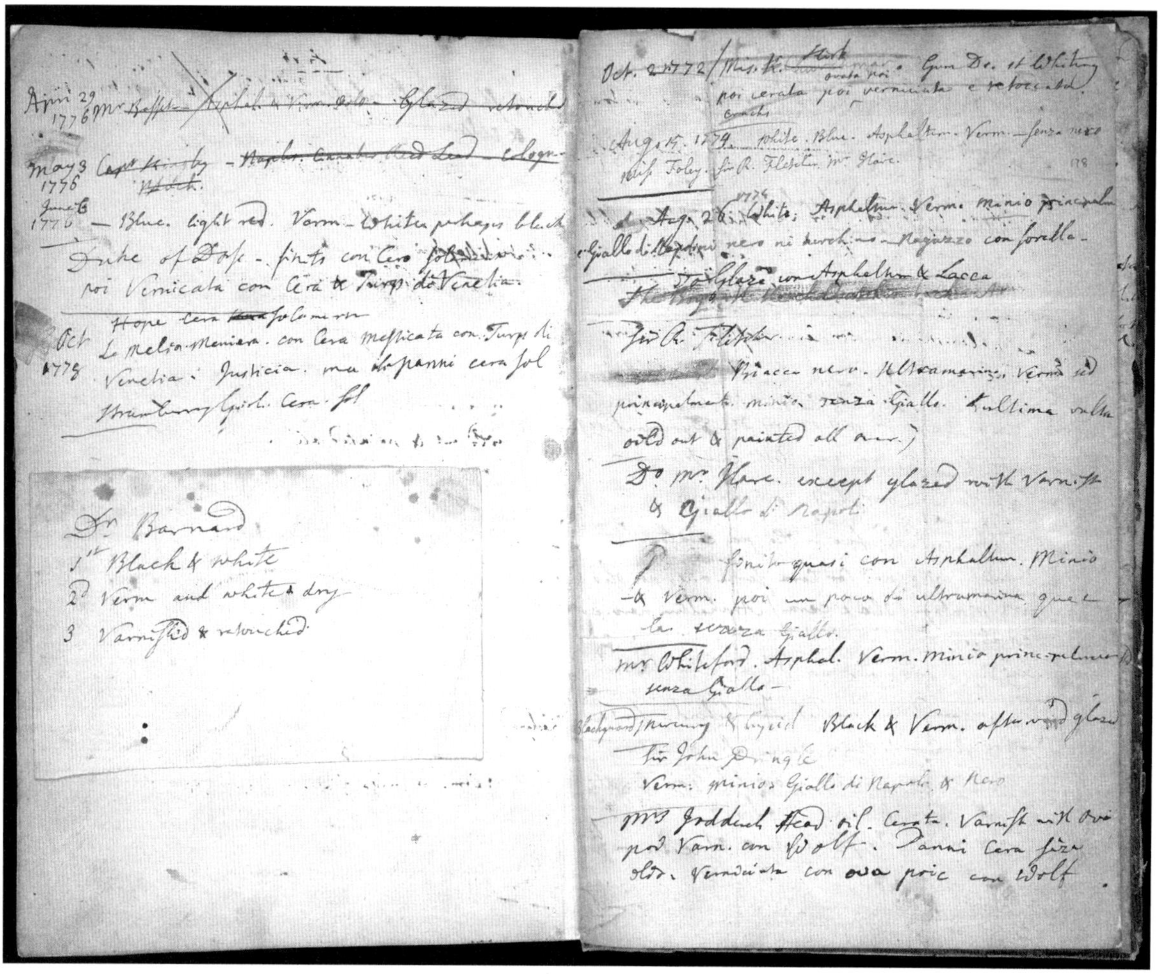

Figure 2.11 Joshua Reynolds's ledger book, showing recipes in back pages, ca. 1772–1778. Fitzwilliam Museum, Cambridge (MS 2.1916, vol. ff177$_{v}$–178$_{r}$). © The Fitzwilliam Museum.

contemporaneously devised by Josiah Wedgwood amid his innovations of the Staffordshire potteries, Reynolds nonetheless possessed means for preserving chemical combinations and their effects in time, long after the high-price pictures into which they were integrated had left his studio.[121]

Decades after his apprenticeship with Reynolds, Northcote would cast those experiments of the early 1770s under cloak of secrecy. The painter's "own preparations of colour were most carefully concealed from my sight and knowledge, and perpetually locked secure in his drawers; thus never to be seen or known by any one but himself."[122] But by then, the instability of Reynolds's pictures had become a standard article of critical comment. In 1763 connoisseur Horace Walpole (1717–1797) outlined an isomorphism between painting and sitter that emerged as a topos of Reynolds's reception. The colors of the painter's portrait of Henry Frederick, Duke of Cumberland (1745–1790), Walpole wagered, appeared "as much changed as the original is

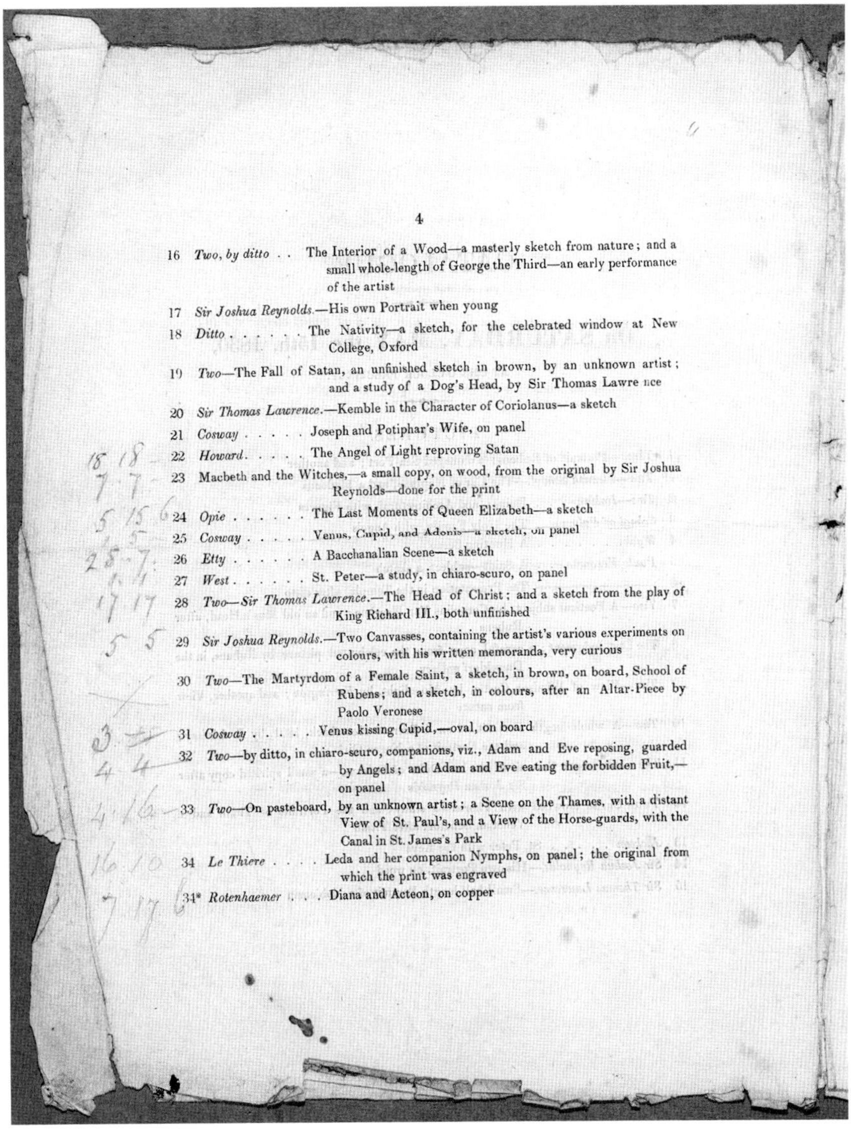

4

16 *Two, by ditto* . . The Interior of a Wood—a masterly sketch from nature; and a small whole-length of George the Third—an early performance of the artist

17 *Sir Joshua Reynolds.*—His own Portrait when young

18 *Ditto* . . . . . . The Nativity—a sketch, for the celebrated window at New College, Oxford

19 *Two*—The Fall of Satan, an unfinished sketch in brown, by an unknown artist; and a study of a Dog's Head, by Sir Thomas Lawrence

20 *Sir Thomas Lawrence.*—Kemble in the Character of Coriolanus—a sketch

21 *Cosway* . . . . . Joseph and Potiphar's Wife, on panel

22 *Howard*. . . . . The Angel of Light reproving Satan

23 Macbeth and the Witches,—a small copy, on wood, from the original by Sir Joshua Reynolds—done for the print

24 *Opie* . . . . . . The Last Moments of Queen Elizabeth—a sketch

25 *Cosway* . . . . . Venus, Cupid, and Adonis—a sketch, on panel

26 *Etty* . . . . . . A Bacchanalian Scene—a sketch

27 *West*. . . . . . St. Peter—a study, in chiaro-scuro, on panel

28 *Two—Sir Thomas Lawrence.*—The Head of Christ; and a sketch from the play of King Richard III., both unfinished

29 *Sir Joshua Reynolds.*—Two Canvasses, containing the artist's various experiments on colours, with his written memoranda, very curious

30 *Two*—The Martyrdom of a Female Saint, a sketch, in brown, on board, School of Rubens; and a sketch, in colours, after an Altar-Piece by Paolo Veronese

31 *Cosway* . . . . . Venus kissing Cupid,—oval, on board

32 *Two*—by ditto, in chiaro-scuro, companions, viz., Adam and Eve reposing, guarded by Angels; and Adam and Eve eating the forbidden Fruit,—on panel

33 *Two*—On pasteboard, by an unknown artist; a Scene on the Thames, with a distant View of St. Paul's, and a View of the Horse-guards, with the Canal in St. James's Park

34 *Le Thiere* . . . . Leda and her companion Nymphs, on panel; the original from which the print was engraved

34* *Rotenhaemer* . . . Diana and Acteon, on copper

Figure 2.12 Annotated copy of *A Catalogue of the Valuable Collection of Paintings by Ancient and Modern Masters of Sir Thomas Lawrence* (1830), 4. Huntington Library (40883).

to the proprietor."[123] A commentator in 1772 would parse Reynolds's practice into a mode of "extreme freedom" and a minute manner, apportioning their value this way: "In his bolder, better works, the colours are graceful rather than chaste; they have the ease of drawings, and mark how little attention was given by the artist to make them durable."[124] Of the twelve pictures the painter exhibited at the Royal Academy in the spring of 1774, three were particularly highlighted for their rapid change. Painted in 1770, Reynolds's full-length depiction of Sophia Aufrere feeding corn to chickens from a basket balanced on her hip was one such casualty. "The Colours of this Picture having been laid on some Time," noted a reviewer in London's *Public Advertiser*, "are exceedingly gone off."[125] A full-length, seated portrait of Walpole's niece Maria, Countess Waldegrave, gazing rapturously into the heavens had been "uncommonly elegant when first executed," apparently earlier that very year. Shown in the spring of 1774, the picture had "nearly lost the whole of its Force with the Beauties of its Colouring."[126] By 1777, a "poetical epistle" addressed to Reynolds envisioned those changes unfolding in real time: "as I thus enraptur'd stand / Before the wonders of your hand, / I see the lively tints decay, / The vivid colours melt away."[127]

Pictures made from materials that change in time need not be—and need not then have been—apprehended in chemical terms, of course.[128] But the imperative to think melting, wondrous evolutions at the heart of fine art and to understand all that as a chemical problematic—those mandates had been established by Reynolds himself in his sixth discourse. Read in December 1774 and published early the following year, "Discourse VI" set out to counter the claim that artistic creation was best advanced by relying on private cerebration and those ideas that flowed from the inspired, untutored gift of genius. Not only were such notions inimical to the emulative pedagogy to which the fledgling Royal Academy was committed, but they were positively hostile to the industrious assembly and transformation of artistic precedents exemplified by the Apollonian Keppel that had launched Reynolds's successful practice (see plate 6).

Auditors in 1774 would not have had to search far for voices advocating the kind of genius Reynolds sought to disarm. Author of the famous *Night-Thoughts*, Edward Young (bap. 1683–1765) had nourished a thriving georgic metaphor for the originality he claimed as genius's identifying characteristic. "An *Original* may be said to be of a *vegetable* nature," Young contended in *Conjectures on Original Composition* (1759). "It rises spontaneously from the vital root of genius; it *grows*, it is not *made*."[129] Reynolds's privileged species of emulative art was plotted as antipode to Young's vegetal fertility in telling terms. "*Imitations*," Young observes, "are often a sort of *manufacture* wrought up by those *mechanics*, *art* and *labour*, out of pre-existent materials not their own."[130] Published in the year in which Reynolds's sixth discourse was delivered, Alexander Gerard's *Essay on Genius* extended that georgic terrain further. Gerard too took a dim view of "servile imitation, or . . . plodding industry"; he extols instead the vegetative bounty that flows from imagination under tempering guidance of judgment in a mind of genius. "As a rich soil produces not only the largest quantity of grain, but also the greatest profusion of such weeds as tend to choak it," Gerard claims, "so a fertile imagination . . . produces many trifling, false, and improper thoughts, which, if they be not immediately examined by reason, and speedily rejected, will over-run and obstruct the truth or the beauty."[131] True genius's mental faculties are exquisitely self-correcting. The fancy's "mechanical and instinctive" yield of ideas linked by association (an action one interpreter has called "a direct analogue of the Newtonian mechanism of the physical world") is controlled by the faculty of judgment.[132] "Like a skilled refiner," Gerard puts it as he shifts "managerial" metaphors into a metallurgical register, judgment sorts through the jumbled ores naturally produced by the imaginative faculties and "purifies the gold . . . from the dross with which it is intermingled."[133]

Reynolds quoted from Gerard directly in preparatory notes for his discourses.[134] Citing Gerard's interpolation of Jean le Rond d'Alembert, the painter affirms in metallurgical mood how the inventive mind "loves to rush forward without control and without rule, to produce indiscriminately the monstrous and the sublime, and to carry down in rapid stream gold and mud mingled together by the impetuosity of its course."[135] Elsewhere in the same papers, he too appeals to Gerard's georgic concerns: "When I recommend . . . <enriching> & manuring the mind with other mens thoughts I suppose the Artist to know his Art so as to know what to choose and what to reject."[136] But, as Reynolds could well have known from his days among the spinning wheels and turnip-slicers at the Society of Arts, the refining of precious metals and the manuring of soils shared a common agent: the catalytic chemical transformations "Discourse VI" places at the center of grand-manner art.[137]

In the discourse's published form, Reynolds turns to those chemical arts, seeking to conceptualize what he grants could appear a paradoxical making of artistic novelty from practices of imitation. Contesting the tendency of "those who are unacquainted with the *cause* of any thing extraordinary, to be astonished at the *effect*, and to consider it as a kind of magic," he enjoins the Academy's students to rebuff talk of untutored genius. "The mind," he affirms, taking up the georgic metaphor, "is but a barren soil . . . soon exhausted . . . unless it be continually fertilised and enriched with foreign matter."[138] Rather than private cerebration, the artist is to canvass the great painting schools of Europe, to select and carefully mix existing artistic materials, and thereby to derive composites with novel properties.[139] "The fire of the artist's own genius operating upon these materials which have been thus diligently collected," he proclaims, "will enable him to make new combinations, perhaps, superior to what had ever before been in the possession of the Art."[140] Resonating with the emphasis on intention that had underpinned Hogarth's conception of art, Reynolds's painter is to use reasoned fire to transmute base materials: "He will pick up from dunghills what by a nice chymistry, passing through his own mind, shall be converted into pure gold; and, under the rudeness of Gothic essays, he will find original, rational, and even sublime inventions."[141] With the sulphureous manure that fortifies the soil of the mind—or the heated piles of dung crucial to the manufacture of premodern pigments (including the verdigris that Reynolds and Hogarth had been charged to assess in 1757)—the artist sublimes what he finds to make something new, altogether different and inestimably precious.[142] Emulative fire is the philosopher's stone of academic art.

Yet, as Wright of Derby had done when depicting the unintended creation of phosphorus through the arts of fire some three years earlier, Reynolds *also* models artistic invention as chemical accident made by fire. The creation of novelty from miscellaneous, artistic dross can equally happen, Reynolds claims, as in the chance mixture by which various "metals . . . are said to have been melted and run together at the burning of Corinth."[143] Reynolds likely knew that story from Pliny the Elder's *Natural History*, which identifies Corinthian bronze as "a compound that was produced by accident, when Corinth was burned at the time of its capture."[144] Sacked in 146 BC by Roman troops under Lucius Mummius, Corinth's conflagration and its surprising metallurgical yield were familiar myths of the ancient world.[145] "At the fall of Ilium," so claims confused buffoon Trimalchio in Petronius's *Satyricon*, "Hannibal, a trickster and a great knave, collected all the sculptures, bronze, gold, and silver, into a single pile, and set light to them. They melted into one amalgam of bronze [*unum aera miscellanea*]."[146] By Reynolds's telling, fire had liquidated Corinth's standing stock of metals to form a "new and till then unknown metal . . . equal in value to any of those that had contributed to its composition. And though a curious refiner may come with his crucibles, analyse and separate its various component parts, yet Corinthian brass would still hold its rank amongst the most beautiful and valuable of metals."[147] Artistic invention, then, is an intentional, teachable science founded on studiously assembled precedents and fired under trained, mental chemistry toward the production of lasting value. But, as his friend John Mudge would soon claim of his prizewinning metallic alloys, innovation equally unfolds by fire's happy accidents as the artist collaborates with and profits from its unexpected combinations.

The sixth discouse, from December 1774, was not the last word Reynolds would have on the subject of "serendipity," that compound of accident and sagacity mythically minted by Horace Walpole in 1754.[148] Reynolds's thirteenth discourse of 1786 would recommend that architects should "take advantage *sometimes* of that to which I am sure the Painter ought always to have his eyes open, I mean to the use of accidents; to follow when they lead, and to improve them, rather than always to trust to a regular plan."[149] Yet, in the immediate wake of his 1774 lecture, Reynolds and his cloven chemistry would garner direct, critical exposition—also by means of fire. Less than five months after "Discourse VI" was read, Irish painter Nathaniel Hone (1718–1784) submitted *The Pictorial Conjuror, Displaying the Whole Art of Optical Deception* to the Royal Academy's 1775 summer exhibition (plate 11). Perched on the rim of a terrestrial globe, an owl gazes out from inky darkness at upper right. Turning from this winged companion and the profile of St. Paul's Cathedral visible beyond the massy columns at left, a bearded figure sits cloaked in crimson housecoat and fur-lined vest, a hexagram pendant dangling from his neck. Identifiable to contemporary viewers as George White (one of Reynolds's favored models in the early 1770s), this would-be conjuror grasps a page of the spread volume at lower right as a young child looks on, crossed arms folded into the old man's lap. The action of the picture is organized around the elegant flick of illuminated wrist with which the conjuror wields his wand. Down that slim, diagonal span the beholder's eyes are directed, drawn in by an arc on the floor inscribed with zodiacal symbols, to the site of combustion at extreme lower left. Fire leaps forth there, under order from the conjuror's wooden wand. It consumes a cascade of copperplate prints after old master pictures.

In 1771 Walpole had anticipated the subject of "Discourse VI" as he defended Reynolds's use of "borrowed attitudes from ancient masters" through a conceit drawn from the sister arts.[150] "When a single posture is imitated from an historic picture and applied to a portrait in a different dress and with new attributes," Walpole explained, "this is not plagiarism, but quotation: and a quotation from a great author, with a novel application of the sense . . . may have more merit than the original."[151] Hone disagreed. If a minor, modern literature has catalogued the complex array of thefts with which Hone's satire saddles Reynolds, period viewers were no less savvy.[152] Commenting in London's *Public Advertiser* in May 1775, "A Lover of Wit" declared the picture not a lampoon but a compliment to the president of the Royal Academy: "The Conjuror . . . which represents Sir Joshua Reynolds, is supposed to be attracting by his Magic Wand, the various Excellencies which are dispersed in the various Works of other Masters."[153] Displaying a keen reading of "Discourse VI," this wit claims, Hone had rendered concrete the "Fire of Genius with which his Works are animated, and which is considered (as the President has observed) by those who see not the Means by which Art is accomplished as miraculous, the Effect of supernatural Powers, Witchcraft, Conjuration, or whatever you will please to call it."[154] Allowing that Hone was frustrated in his commercial interests by Reynolds's domination of the London art market, *The Conjuror* stands as a complex metapictorial argument. Not only does it allegorize the emulative strategies Reynolds had recommended to the Academy some five months earlier, but it visualizes the vulgar error of mistaking artistic skill for magic then critiqued by the president. Joseph Farington outlined a more direct interpretation in the early nineteenth century: "The principal figure in the composition was supposed to be a wizard who had discovered by his skill in the black art there proofs of Sir Joshua's plagiarism."[155] Indeed, initially accepted for exhibition, Hone's picture was publicly rejected, once identified as an attack slandering Reynolds and Angelica Kauffman both.[156]

Whether satirizing a misapprehension of Reynolds's model of grand-manner emulation or exposing the cynical, money-grubbing ends of his pictorial chrysopoeia, Hone's painting materializes the complex network of chemical ideas gathered around Reynolds by the mid-1770s.[157] Like Wright's *Alchymist*, Hone represents his Reynoldsian conjuror as an elder adept at work with arts of fire in a sinister space surrounded by the apparatus of inscrutable learning and a youthful entourage. Moreover, as Johann Zoffany had done in 1770 when depicting actor David Garrick as Abel Drugger in Ben Jonson's *The Alchymist* (fig. 2.13), Hone places his practitioner in a darkened interior surrounded by voluminous writings, globes, and nocturnal winged creatures. All of these elements would have been familiar to Reynolds, who acquired Zoffany's portrait of his friend Garrick in an alchemical role reprised for decades on the London stage; *The Alchymist* was actually showing at Drury Lane in December 1774 when "Discourse VI" was delivered.[158] That period viewers could also connect the alchemical penumbra of Hone's painting to Reynolds's unstable paint experiments is evident in a ludic lament from the following decade. "It is a pity," one wag claimed in 1785, "the pencil of Sir *Joshua Reynolds* does not possess an equal power of *magic* with that of *Hone's Conjuror*!—he might else do something to remove a *spell* which will be fatal to his reputation ages hence."[159] In this view, Hone's critique of Reynolds's silver-tongued theft from the artistic tradition would endure long after "the objects of the ridicule are lost in that general confusion of tints, which the progress of a few years only, occasion in Sir Joshua's works!"[160]

Through their interwoven pyrotechnics, Wright, Reynolds, and Hone show how, by the summer of 1775, the practice and theory of painting embraced by the president of the Royal Academy had come to be linked publicly with elements of an older, vexed alchemical tradition. In his eccentric studio facture, Reynolds's *may* have mobilized the metallurgico-chemical technics of the Mudges, the Northcotes, fellow members of the early Society of Arts, and his erstwhile chum Thomas Rennell. But in his embrace of a "nice chymistry"—operations working both by steady reason and happy accident—Reynolds had arrived at a decidedly different position on the chemical change of artistic materials than the standing view voiced by his former collaborator William Hogarth. Far from lamenting metal pigments' visible change in time as a Hogarthian ruin of art, Reynolds had made metals' accidental evolutions an integral component of artistic invention itself. How should we understand this figure/ground reversal and the stakes it entailed?

"Embrio-passions"

Assessing Joseph Wright's depiction of the accidental making of phosphorus in 1771, critic Robert Baker saw a jumble (see plate 8). The mix of papers,

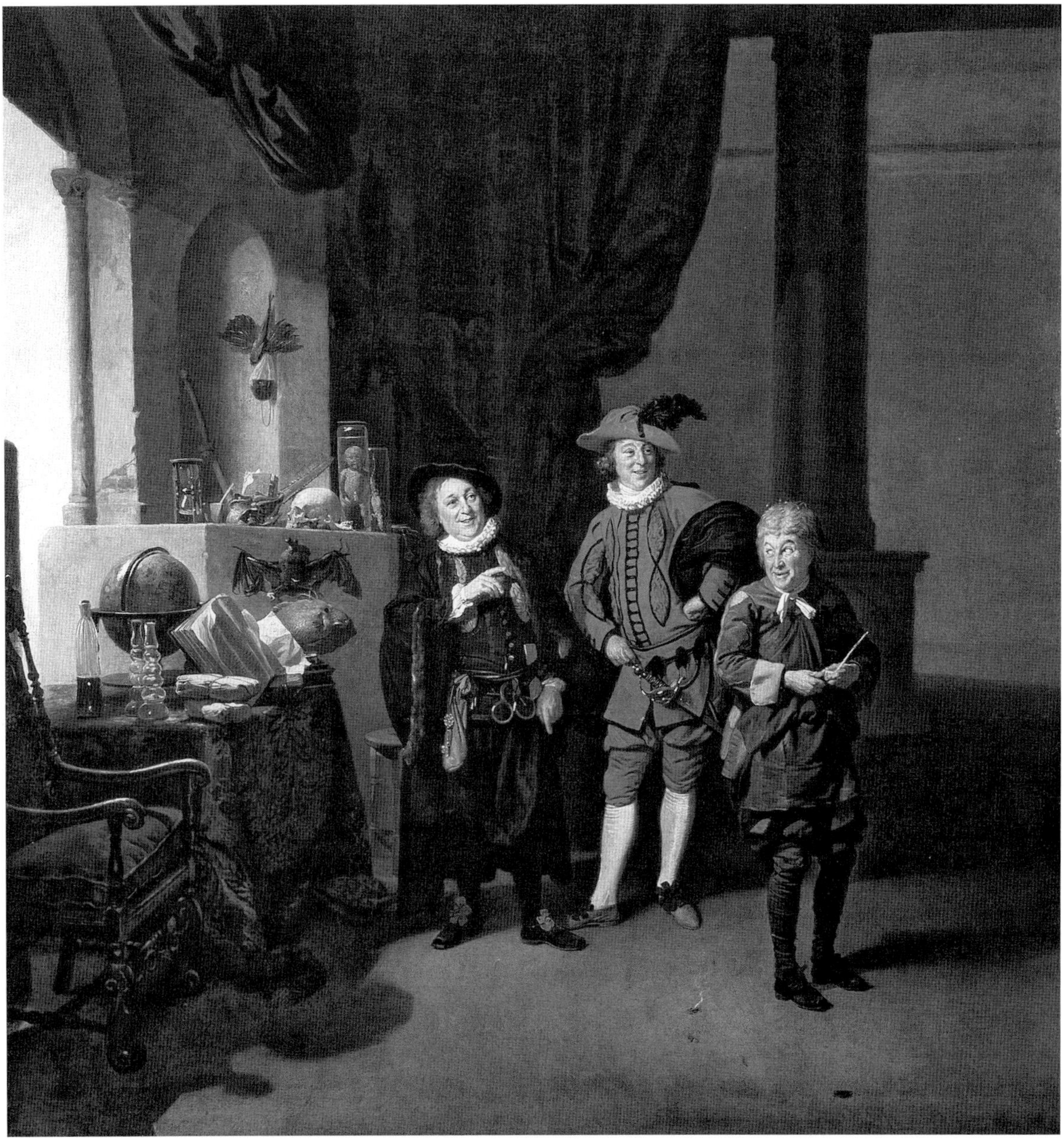

Figure 2.13 Johann Zoffany, *David Garrick with Edmund Burton and John Palmer in "The Alchymist,"* 1770. Oil on canvas. Private collection. Bridgeman Images.

glass, and fire in a chemical laboratory need not be hostile to the painter's art, Baker allowed. Reciting an amplification of Aristotle standard to period aesthetic thought, the critic underscored how, in the hands of a skilled painter, "a great number of uninteresting objects . . . may be disposed, as for the whole (the *tout-ensemble*, as the French call it) to be not only not disagreeable, but even pleasing to the eye."[161] Wright's want of compositional gift would surely have been heightened in Baker's eyes had he known

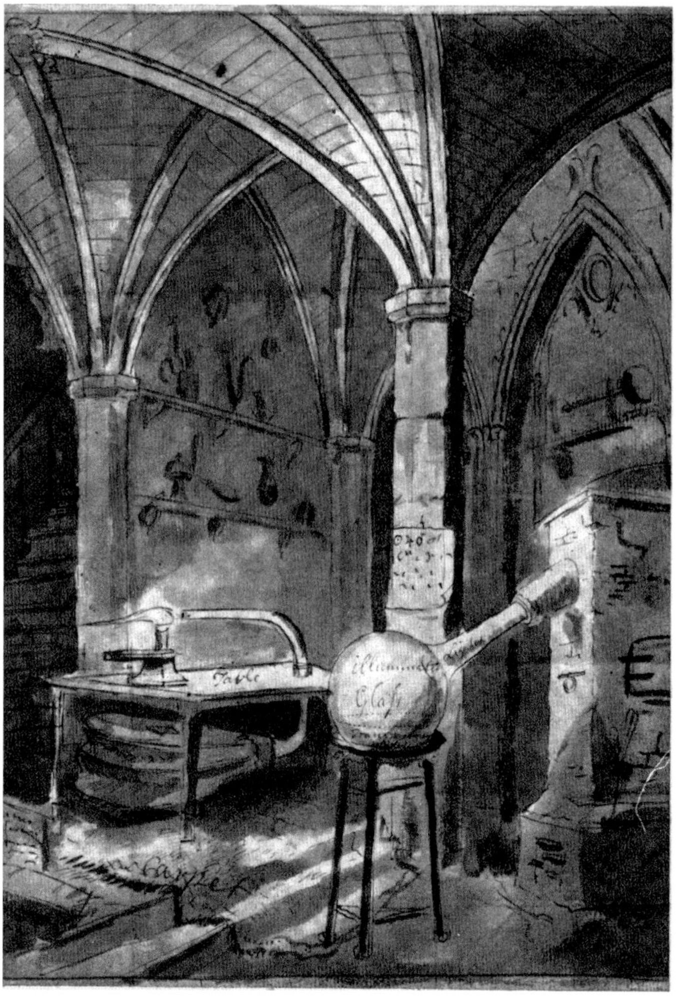

Figure 2.14 Peter Perez Burdett, *Study for the Alchymist*, February 1771. Derby Museum and Art Gallery.

just how much help the painter had actually received (fig. 2.14). Early in 1771, chemist Matthew Turner (d. 1789) and printmaker Peter Perez Burdett (ca. 1734–1793) in Liverpool had sent Wright a sketch and description plotting a pared architecture of chemical experiment. Denuded of human form, the sketch delivers the *mise-en-scene* needed for the dramatic moment when "the Chemist . . . turning his head <towards the Glass> upon his Assistants exclaiming upon the first appearance of the Phenomenon <of the luminous exhalation from the Retort>."[162] Striking in this light are the features added by Wright to make his oil painting a jumble not only of composition, but of time itself. The face of the full moon slipping in at upper right jostles with a horoscope's stellar projection inscribed on manuscript leaf behind the jet of self-shining gas. That gas casts the bulky shadow of a clock as it gives the beholder a mechanical hour of twenty minutes before midnight splayed at vertical center on a Gothic pier. Wright's mingling of modern industry and "age-old craft," his willful anachronisms and meditations on time's fragility informed by the *vanitas* tradition: all have been well noted among the chiaroscuro subjects he also exhibited in 1771.[163] But that heterochronic impulse also underpins *The Alchymist*, which clothes a century-old experimental event in an uncertain cloak cut from a deep time out of mind.

Wright's competing times within pictorial fiction are curious; Reynolds's experiments with materials that could, in time, lacerate pictorial integrity seem to move with altogether more elemental force. Reynolds's "persistence in following practices which he knew perfectly well would seriously shorten the life of his pictures," so claims M. Kirby Talley in a classic modern assessment, "can only be described as perverse."[164] So, *why would* Reynolds have made pictures that changed chemically in time? What explains his ushering of an experimental tradition of work with temporally evolving chemical objects into British painting? Maybe this is a story that turns on deskilling and alienation. After all, the collapse of traditional guild corporations like the Painter-Stainers' Company, coupled with the abandonment of large-scale artistic patronage by the early Hanoverian court, had left the business of painting in the Britain of Reynolds's youth facing an uncertain future. "With no restrictions from guild, academy or court," Rica Jones contends, "artists were free to set themselves up and bid for work at all levels with or without formal training."[165] Yet deregulation as such would not explain why Reynolds would embrace a poetics of chemical change whereas Hogarth (who had far less conventional pictorial training) would excoriate it. Might Reynolds's approach, then, tell a story about an incipient modernity in an age of "consumer revolution" through the embrace of synthetics?[166] Poet William Mason (1724–1797) told a tale like that. Appealing to "the pocket-book of old Beale [i.e., Charles Beale]," Mason contrasted Peter Lely's time-tested, Restoration-era pigments with Reynolds's preferred smalt and other industrially enhanced substitutes.[167] "The art of enamelling, and that of making china," Mason claimed, were freeing painting from regional constraint and would only continue to improve: "A pottery in Staffordshire is able to produce a blue equal to that of Nankin china, I see no reason why such smalt should not equal ultramarine in point of durability."[168] But that Reynolds had ingeniously built progression into paintings for an age of progress has been given most audacious expression at art's join with political economy. Deftly assessing a competitive marketplace, Neil De Marchi and Hans J. Van Miegroet have contended, Reynolds treated flashy effects of experimental pigments and media that dried quickly but might not last as a worthwhile gamble. "Even if only 10 or 20 percent of Reynolds's pictures cracked badly, lost paint, or faded," De Marchi and Van Miegroet propose, "would-be buyers were necessarily entering into a wager when purchasing a picture by him."[169] Calculated risk incarnate, Reynolds's potentially unstable pictures could thus be useful to John Frederick Sackville, third duke of Dorset, and other collectors vying for the top end of

London's luxury market precisely because they offered the opportunity to demonstrate elite *sangfroid* should their high-stakes wagers not pay off.[170] Less than depictions of contemporary individuals, we might conclude, Reynolds's changing pictures portray the systemic conditions of a British "society now living to an increasing degree by speculation and by credit: that is to say, by men's expectations of one another's capacity for future action and performance."[171] Reynolds's pictures would be supreme exemplars of that imperative Michel Foucault diagnosed at the heart of liberalism: "Live dangerously."[172]

Maybe; but, guided by Talley's talk of perversity, I want to push still further into the elemental. For a pressure of time lapped up as mother's milk by Reynolds, Wright, and other products of Thomas Hudson's studio from the seventeenth-century French tradition had increased sharply in the years around 1770.[173] Certainly, theorists in early eighteenth-century Britain's emergent aesthetic tradition had recognized a need for abiding time's unity. The painter, so declared Anthony Ashley Cooper, third earl of Shaftesbury (1671–1713), is "debar'd the taking advantage from any other Action than what is immediately present, and belonging to that single Instant he describes."[174] In the 1720s, Jonathan Richardson (Hudson's father-in-law, a painter/theorist key to Reynolds's artistic identity) had brooked no compromise. The painter must "*represent but One Instant of Time, no Action must be chose which cannot be suppos'd to be doing in that Instant*."[175] Such adjurations paled against the strictures on time's pictorial unity demanded in the wake of the 1750s reaction against the Rococo.[176] Writing in the journal that would soon publish Reynolds's first forays into art theory, Samuel Johnson ranged over possible topics for pictorial representation in 1759, only to dismiss them. Neither the death of Hector nor the demise of Hercules were fit subjects for painting in Johnson's estimate since "these fill the mind, but will not compose a picture, because they cannot be united in a single moment."[177] Johnson's insistence that the painter give "an action not successive but instantaneous" would find its most celebrated exposition with Gotthold Ephraim Lessing (1729–1781), publishing on the Greco-Roman sculpture that had so impressed Dr. Huxham when replicated by a juvenile Reynolds. Breaking apart *ut pictura poesis* tropes, Lessing's *Laoköon* (1766) drew a firm line between poetry's temporal sequence and the spatial syntax proper to the plastic arts. "Painting, by virtue of its symbols or means of imitation, which it can combine in space only," Lessing had declared, "must renounce the element of time entirely."[178]

Mythically, the problem of time in painting had been brought to Reynolds's very doorstep in an altogether more forcible manner in the early 1770s—a passage he himself declared a "revolution in the art" of history painting.[179] In addition to its scrutiny of the handling of time's instantaneity, academic convention demanded that modern events could enter into

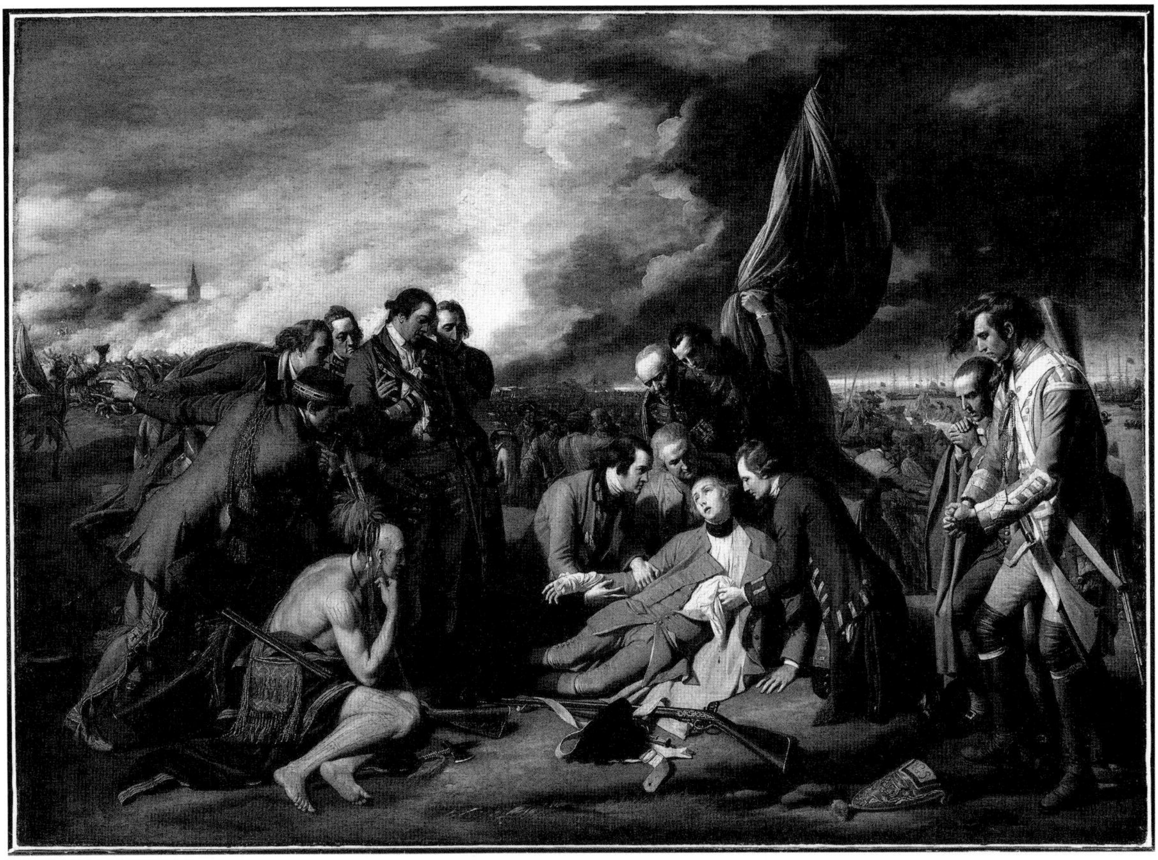

Figure 2.15 Benjamin West, *The Death of General Wolfe*, 1770–1771. Oil on canvas. National Gallery of Canada, Ottawa (gift of the 2nd Duke of Westminster to the Canadian War Memorials, 1918; transfer from the Canadian War Memorials, 1921).

the elevated genre of history painting only if their actors were cast into the timeless garb of classical antiquity. Anglo-American painter Benjamin West (1738–1820) had broken that convention in *The Death of General Wolfe* (1770; fig. 2.15), a heroic rendering of death at the colonial frontier, which he exhibited contemporaneously to Wright's *Alchymist* in spring 1771. Drawing upon standing, theatrical precedent, whereby "remoteness of setting is as conducive to grandeur as distance in time," per Edgar Wind's venerable formulation, West had licensed for painting "what the classical rules of the grand style had forbidden: the representation of contemporary events without disguise."[180] Depicted at the very moment when he gives up the ghost on Quebec's Plains of Abraham, West's General James Wolfe would die with his military boots on.

Far from being an innovation, David Solkin has argued, West's modern-dress history painting in the 1770s would have looked "not just vulgar but *retardataire*," as it replayed for the new Royal Academy moves ventured by painters for frankly commercial publics in the 1760s.[181] Undaunted, West imagined his as a declaration of independence of modern time from ancient convention. "The event intended to be commemorated," he proclaimed to biographer John Galt decades later, "took place on the 13th of September, 1758, in a region of the world unknown to the Greeks and Romans, at a period of time when no such nations, nor heroes in their costume, any longer existed."[182] Not only had a new ambition for history painting been inaugurated in West's telling, but a new relationship of the artist to historical time had been forged as well. Is the painter not "by the energy of his own mind, and by his attachment to his art," so West would ask in his inaugural (and lone published) discourse of 1792, "that *watchman* who observes the great incidents of his time, and rescues them from oblivion?"[183] It was Reynolds himself who had opposed West's temporal watch in the early 1770s. Appealing to the precarious hold of the fine arts upon British taste, Reynolds had allegedly urged West to revise his depiction of Wolfe, to "adopt the classic costume of antiquity, as much more becoming the inherent greatness of . . . [his] subject than the modern garb of war."[184] As told by West's recuperative biographer, Reynolds's ultimate volte-face was as much a modernizing victory won by the second president of the Royal Academy over the first as the conquest by an active practitioner of history painting against one who merely promoted the elevated genre in theory. But in his own way, Reynolds too was intensely engaged with the problem of time in painting in the early 1770s. Shifting attention from West's watchman back to the fabricators of watches in his Plymouth ken, it becomes possible to see Reynolds's "infant pictures"—his strange depictions of historical and contemporary figures as infants, which became an object of sustained attention in the early 1770s—as meditations on the conditions and temporal possibilities of painting under chemistry's regime.

*

Two children stand at the edge of a wood (plate 12). A russet canopy of deciduous foliage spans the left-hand corner of the canvas, its spreading leaves impervious to the gusty clouds' diagonal striations. Henry Spencer poses with an ostrich-plumed cap cocked across the crown of his head, hose-bound legs crossed in ivory accent to his scarlet get-up. He gazes cherubically toward the implied beholder as his sister, Charlotte Spencer, inclines forward. Her index finger extended to the heel of his palm, she tilts her head and fixes her eyes beyond the picture plane, as if summoning the palmist's powers of divination. Commissioned by George Spencer, fourth duke of Marlborough, this picture (now known as *The Young Fortune Teller*) was exhibited by Reynolds as *A Young Nobleman and His Sister* in 1775, where its chiromantic meditation would be challenged directly by Hone's *Conjuror*.[185] A complex skein of quotations from Van Dyck and Caravaggio that would have given grist to Hone's mill, the double portrait's representation of future prediction might equally ask after its material life.[186] For contrary to its saccharine imagery, the painting is a stew of volatility. "Nero, Cinabro, minio, e Azzuro" all applied "thick," Reynolds claims in his ledgers, stipulating that the pigments were mixed first with oil, then with wax without oil.[187] Encountering these extraordinary procedures, one modern conservator wrote to the picture's present owner in outright astonishment: "I was wondering, if you have any evidence that this is indeed the case and how the painting has survived consequently!"[188] Prompting meditation on the eventual actions and appearance of the future ruling class it depicts, it is as if the double portrait also doubles to divine its own, projected material life.[189]

At the moment in the early 1770s when evidence of his chemical engagements was becoming most intricately knotted, portrayal of infants and children emerged as a concerted preoccupation for Reynolds. No fewer than one hundred fifty sessions with child models were scheduled in the painter's sitter books between 1771 and 1773, building the backbone of his forays into the genre of the "fancy picture," where portrayal of specific individuals could cede to imaginative play.[190] While the painter's depictions of children would become a leitmotif of his Victorian reception, contemporaries in the eighteenth century's "age of the child" also attributed a signal import to these experimental, iterative works.[191] Walpole cited Reynolds's depiction of three-year-old John Crewe, cast in the manner of "Holbein's swaggering and colossal" Henry VIII, as he defended the painter against charges of simple theft.[192] As with the Renaissance putto's figuration of "all those uncontrollable sensations and irrational physical and mental alterations that constantly arise in the body unbidden," Reynolds's infants embodied suggestion.[193] Gesturing to portraits of four-year-old Sarah Price (now at Hatfield House) and the lost *Infant Jupiter* shown at the Royal Academy in spring 1774, Walpole asks "when was infantine loveliness, or embrio-passions, touched, with sweeter truth?"[194]

I want to build on Walpole's embryonic account of the infant portraits through the insights of some of Reynolds's most perceptive modern interpreters. When depicting a young child cautiously clutching a sword as the conquering commander Hannibal, Edgar Wind once noted, Reynolds had created a conspicuous discrepancy. The gap between young model and the historical warrior activates what Wind describes as a particular role for the beholder: "The presentiment of things to come which fills the young Hannibal's mind and, in the imagination of the spectator, is transferred from the historical exemplar to the actual boy in his role. In other words, the tension is due to the visible manipulation of the temporal sequence in the picture."[195] Staging an aesthetics of anticipation that Wind aligns with Burke's celebrated analysis of the sublime, pictorial diminution of historical Hannibal becomes the means through which the beholder can access sublimity by imaginatively projecting the child's visible features into heroic futurity. "The

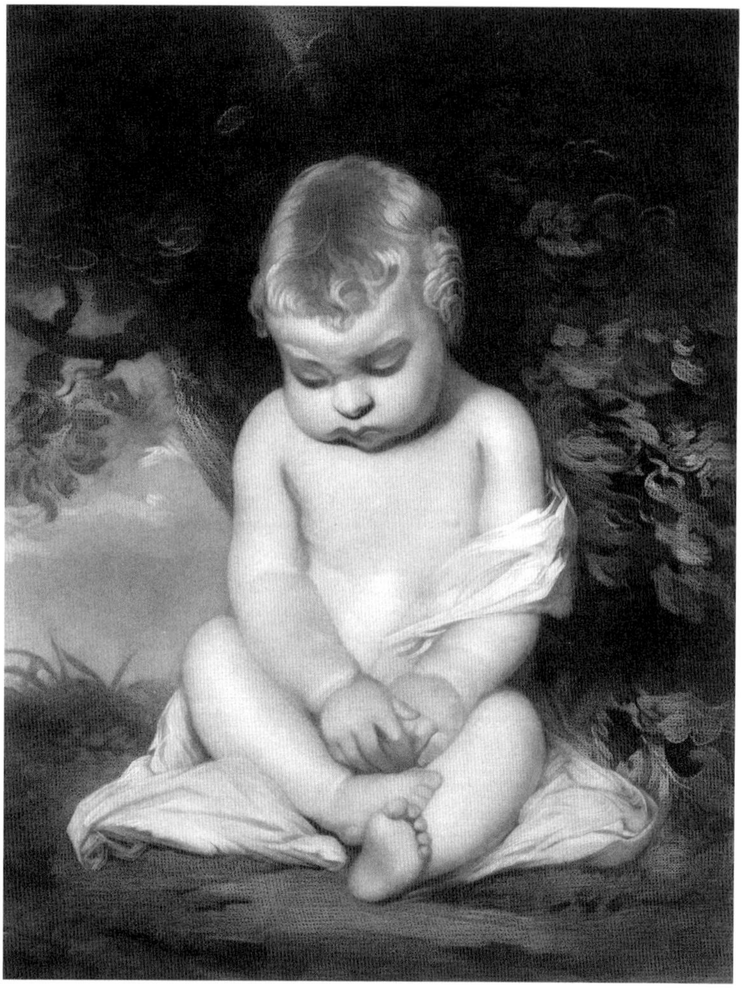

Figure 2.16 Georg Zobel after Joshua Reynolds, *The Infant Johnson*, 1858. Mezzotint. British Museum, London (1859,1210.1054). © The Trustees of the British Museum/Art Resource, NY.

mystery of heroism can be represented only," Wind proposes, "by showing the hero *in statu nascendi*, as a promise, but not yet a fulfillment."[196] Reynolds secures pictorial grandeur from a literally minor subject by depicting potential and enlisting the beholder to imagine its realization in time.

Less a dynamic between painting and beholder, Ronald Paulson has read the infant paintings as exemplifying a tension between a portraitist's attention to particulars and Reynolds's drive for abstraction, which Burke had cast as a source of science as such. Eyes downcast, ballooned cheeks pressing stretched lips, tiny left foot extended, *Infant Johnson* (1781–1782; fig. 2.16) gives the rotund poet Samuel Johnson in nascent essence: "The particular seed [is] related in a teleological sense to the man as ideal."[197] Paulson locates the infant pictures' ambition in a winnowing away of time's accidental accretions until the picture plane manifests an atavistic core. The infant portrait format allows Reynolds access "to the ideal form he is seeking for each individual without . . . the particularities and oddities and twitches picked up over the years. The child is the general truth from which the man will increasingly, with experience and age, particularize himself."[198] By this account, Reynolds's picture plane becomes a gestational space awaiting only time's unfurling to deliver identifiable sitters in all their concrete particularity—a conceit that operates in formal as well as conceptual terms. "In many canvases the picture space is constructed like a womb," Paulson proposes, "with the compact figure of the foetus floating within it."[199]

Nowhere is this pictorial womb and its "embrio-passions" more in play than in *The Infant Academy*, now at Kenwood House (plate 13). A painter's meditation on the act of painting, Reynolds's picture shows a child seated on a low stool before an easel, brush at the ready in his right hand. A loaded palette visible through the aperture of his elevated arm and bare chest, the infant painter waits. He pauses beneath the heavy weight of fluted architectural columns amid fading light at the horizon for an assistant to adjust the ostrich plume in the mob cap of his coy, infant model. Turning his back on the sculpted head at lower left—even on the flowers' effusion of color, by which Le Blon and Hogarth had purported to give such busts animation—Reynolds's painter sits poised before his darkened, oval canvas. The rhythmic impastos of that support's glinting nails both insist on its ovular form and alert the beholder to its material heft. Already within that egglike space, a figure has begun to grow. Seen in profile emerging from the darkness, this embryonic being partly resembles the model on which the infant painter gazes; its tousled hair and puffed cheeks also recall the features of the painter himself.

Seneca's famous adage whereby depiction should be judged like the resemblance of a father to a son bespeaks a hoary tradition for conceiving artistic representation in terms of sexual reproduction.[200] Rather than imagining the gestating figure on the darkened ground as the product of some union between sitter and painter, however, Reynolds's chemical ambit in the mid-1770s affords a different possibility; it compels us to see how fruitful growth and generative change in time could be sounded beyond sexual means. Such ideas had wide reach. In early January 1776, London's *Morning Post and Daily Advertiser* would relay striking news from France. A "doctor of chemistry . . . a native of Persia, a man of profound and continual meditation in all the different secrets of nature," had brought a shocking claim to Paris's Faubourg St. Antoine. This mysterious Persian adept (subsequently identified as one Kerim Soubas ab Bassera from the area of Isfahan in what is now Iran) avowed that, in a span of six hours, he could "make real living children, without knowledge of women, and by the art of chemistry alone."[201] Word of this startling assertion had been brought to the police and ultimately to the royal court at Versailles. Sequestered there under lock and key, charged with providing demonstrative evidence, the doctor finally emerged. He presented "to his Majesty, a most beautiful male child with all the natural faculties that accompany children by the usual way . . . it differed in nothing from other infants, but the astonishing manner by which it had been produced."[202] The correspondent goes on to warn London's women lest this chemical power of asexual replication should cross the Channel, enabling British men to "reproduce themselves without their aid."

Befitting the eclectic emulation Reynolds had endorsed in "Discourse VI," references to painters as diverse as Rembrandt and Boucher (along with Rubens, Van Dyck, and Carle van Loo) have been traced through *The Infant Academy*.[203] And like many of Reynolds's infant portraits, the picture has been highly susceptible to change in time through its experimental facture. "In no other area of his practice," so Martin Postle observes of Reynolds's infant-heavy fancy pictures, "is the analogy of the 'apothecary' . . . so apt."[204] Photographic details of the painting taken by Harold Plenderleith in the 1930s reveal the contusions, buckling, and rupture of the paint film, which would be subject to successive campaigns of restoration roughly once per decade between 1957 and 1997 (fig. 2.17).[205] Whatever socioeconomic motives Reynolds and his patrons might have had for viewing chemical change as a site for gain of value and not (or not only) loss, the tale of the Persian chemist opens a different possibility. For the claim that life could be created by means of art alone—that chemical actions could forswear sexual reproduction and yield vitality by their own terms—was central to a longer chymical tradition. Drawing from currents running deep into the ancient world, sixteenth-century alchemist Paracelsus and his followers "transferred the apex of human ingenuity from the fabrication of synthetic gold to the making of an artificial man."[206] Frequently rendered by the dung-heated processes of putrefaction Reynolds would recommend to chrysopoetic painters, the "homunculus" brought to life by chemical process became a supreme statement of human art's powers. "By fusing the traditions of artificial generation, alchemical debate, and an unorthodox Catholicism," historian of science William R. Newman argues, "Paracelsus and his epigones managed to create an image of the alchemist as a *magus coelestis* . . . [who] held the keys of art and nature."[207] True art worked by chemistry to make not gold, but life itself.

Connection between Paracelus and Soubas ab Bassera was drawn explicitly in the 1770s London press, but homunculi had also been made familiar in mid-eighteenth-century British culture through the remarkable success of *The Life and Opinions of Tristram Shandy, Gentleman* (1759–1767) by Reynolds's friend Laurence Sterne.[208] A witty intervention into medical debate between ovists and preformationists, Sterne's homunculus consists, as Tristram narrates, "as we do, of skin, hair, fat, flesh, veins, arteries, ligaments, nerves, cartileges, bones, marrow, brains, glands, genitals, humours, and articulations."[209] A visceral homunculus of that kind had been depicted in chromatic brilliance by Jacques Fabien Gautier D'Agoty (1717–1785) who

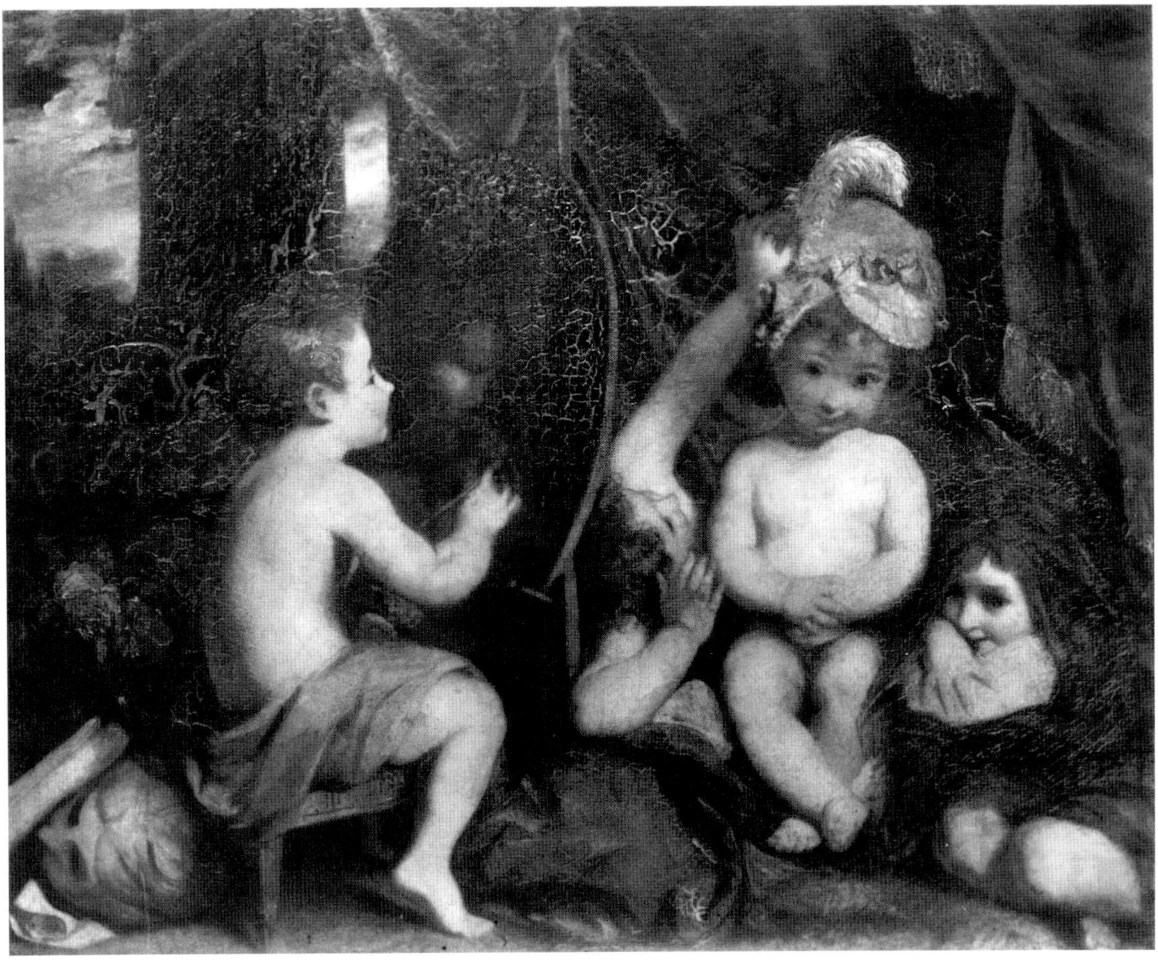

Figure 2.17 Harold Plenderleith, photograph of the surface of Joshua Reynolds, *The Infant Academy*, ca. 1930s.

successfully extended the French color-printing *privilège* obtained by Le Blon and realized its function as an instrument of philosophical publication (plate 14).[210] In Sterne's novel, this homunculus is conveyed from father to mother in the sexual act, an act inseparably linked through Tristram's narration to the winding and maintenance of clocks.[211] Where Sterne would sexualize time, binding talk of clock-winding to thoughts of spermatic homunculi ejaculated in copulation, Reynolds might be seen to inject the materials of clock-making—the camphor, the spermaceti, and other materials shared between Plymouth watchmakers and his London studio—into painting.[212] Abjuring the role of artist as watchman announced by West, Reynolds enters painting instead into a chemical time of asexual propagation whereby the picture *qua* homunculus takes on a life of its own. By the fire of the artist's mind, noble metal and dross (i.e., Rembrandt and Boucher) are chemically alloyed to make the new life emerging in the womb-space of the darkened canvas. Such a view also finds unexpected echo in Hone's *Conjuror* (see plate 11) if the besotted child is taken not as an object of the magician's deluding instruction but as the product of his chemical *making*.[213] Neither the ruin of art, as Hogarth had it, nor only an emulation of the image of capital, Reynolds's embryonic pictorial surfaces would reclaim the alchemist's dream of the homunculus. Emblematized by the infants they depict, these temporally evolving objects create life by chemistry's artificial fire in the very center of British academic painting.[214]

Now, given the restricted evidence we possess of Reynolds's library and the limitations thus imposed on expansion of such alchemical interpretation, the matter might also be apprehended by turning back to the testy relations between action and accident, the fine and the industrial arts and their join at the fire-driven problem of emulation, which runs down the spine of this chapter. Published some two years before Reynolds's death, Immanuel Kant's *Kritik der Urteilskraft* (1790) would make an infamous distinction between the very natural-philosophical and fine-art achievements this chapter has sought to reclaim within Reynolds's practice. Teachable concepts underpin Newton's science, Kant argues; they can be studied and acquired systematically. Different conditions obtain in Homer's poetry.[215] Realigning a seam built into "Discourse VI," Kant is here formulating a crucial theoretical cornerstone of the autonomy of fine art as it speaks to the highest cognitive faculties in their freedom, presenting with purposiveness while remaining irreducible to concepts or purposes. That splitting of natural philosophy from the fine arts brings Kant back around to the question of reusing artistic precedents, which Reynolds had tackled via chemical means in 1774. True fine art must operate by rules, Kant allows; but those rules cannot be formulas or recipes after the manner of Reynolds's ledger-book notations. "Rather, the rule must be abstracted from what the artist has done, i.e., from the product, which others may use to test their own talent, letting it serve them as their model [*Muster*], not to be *copied* [*Nachmachung*], but to be *imitated* [*Nachahmung*]. How this is possible is difficult to explain."[216]

The way Kant solves this dilemma—by appealing to the natural gift of genius to set rules within fine art—marks a key difference from Reynolds, auguring developments familiar to stories about Romanticism accented in the next chapter. Nonetheless, Kant's genuine work of fine art and Reynolds's infant portraits share a defining, proleptic temporality. Neither can be known immediately.[217] It is not enough for a candidate-specimen of the fine arts to be original, Kant notes, "since nonsense too can be original."[218] Instead, the true work of genius is only revealed in time as it comes to be recognized by future ages as a rule-giving guide: "The products of genius must also be models, i.e., they must be *exemplary*; hence, though they do not themselves arise through imitation, still they must serve others for this, i.e., as a standard or rule by which to judge."[219] It is not enough for a true entry to the fine arts to be supremely beautiful or engaging to the mind's free play; rather, the work of genius must be "an example . . . not to be imitated [*der Nachahmung*], but to be followed [*der Nachfolge*] by another genius."[220]

This is the time of the infant portraits: the time of fine art itself. Projected materially and figuratively into time, the "infantine" work discloses its true character and originality as it matures. An autonomous creation with a life of its own (and not just the superficial skins of vitality privileged by ten Kate's circle and Hogarth), the work must be followed to see where it will lead, what lessons it will teach as its implications and ramifications unfold as time's test discloses its true character. Up from the experimental, industrial vision of the Society of Arts, Reynolds had arrived at a temporality of fine art by a route redolent with the aims and practices of a chemical tradition Kant himself would dismiss as unduly empirical.[221] And whether or not Reynolds could have had conscious access either to chemical homunculi or to Kant, preservation of the life of his chemically unstable pictures by chemical means would be the expressed ambition of Reynolds's torchbearing followers in the 1790s. It is to them that a history of Enlightenment Britain's temporal evolving chemical objects must now turn.

3

"Rend'rd Imortal"

The Work of Art in an Age of Chemical Reproduction

In the biographies of Joshua Reynolds he began publishing some four decades after completing his apprenticeship, James Northcote retold a tale highlighting Samuel Johnson's ignorance of visual art (see fig. 2.16). From the pages of Hester Lynch Piozzi's *Anecdotes of the Late Samuel Johnson, L.L.D.* (1786), Northcote plucked an episode set at the Streatham haunt of wealthy brewer Henry Thrale in the 1770s. There, in the company of Reynolds and other luminaries, Johnson had expressed pique at seeing "so much mind as the science of painting requires, laid out upon such perishable materials."[1] Why, Johnson asks Reynolds, did painters not prefer copper to canvas as a pictorial support? As the painter attempts to deflect the question by gesturing to the difficultly in sourcing sufficiently large sheets of metal, Johnson cuts him short: "What foppish obstacles are these! . . . Here is Thrale who has a thousand ton of copper; you may paint it all round if you will, I suppose. It will serve to brew in afterwards: will it not, Sir?"[2]

Preserving the painter's science with the metal that would house the brewer's fermentations, Johnson took a keen interest in chemistry. To readers of *The Gentleman's Magazine* in 1739, he had explained how Herman Boerhaave (1668–1738) brought a chaotic chemical landscape "to Order and made that clear and easy, which was before to the last degree difficult and obscure."[3] An amateur adept who actively promoted the career of chemist

Robert Dossie, Johnson could impress on James Boswell the role of powdered bones in the smelting of iron.[4] Boswell found the floor of Johnson's library "strewn with manuscript leaves . . . I observed an apparatus for chymical experiments, of which Johnson was all his life very fond."[5] But where Hester Piozzi had rendered Johnson's fixation on metal supports as evidence of distaste for painting born from his limited ocular capacities, Northcote fashioned it as a blindness to the art's proper chemistry.[6] Contrary to the "white and raw look" found in pictures by Reynolds and other painterly masters when fresh off the easel, Northcote vaunted patina or the "the mellow and rich hue which we now see in them, and which time alone must have given to them, adding much to their excellence."[7] By promoting copper over canvas, Johnson overlooked those beneficial effects that chemical evolutions unfolding in time could have upon paintings.[8] And besides, Northcote averred, "the duration of a picture does not depend on the strength or durability of the canvas on which is it painted. The canvas can be renewed as often as it may be found necessary, and the colours will in time become nearly as hard and as durable as enamel."[9] Poor cleaning practices, not supports, were to blame for pictorial destruction.

But that oil pictures could be—were *being*—destroyed was increasingly difficult for Enlightenment observers to deny. Excavations begun in 1738 and 1748 of brilliantly colored paintings at the ancient Roman cities of Herculaneum and Pompeii prompted a peculiar species of melancholy for leading connoisseurs of European oil pictures. Among the *philosophes*, as Andrew McClellan has observed, "it was widely believed that the great paintings of the Renaissance were in the process of deteriorating (this was one of the reasons for making students in Rome copy the Old Masters in oil). The century witnessed growing interest in new and not so new methods of safeguarding art against the elements."[10] A leading voice in this conversation had been antiquarian Anne Claude Philippe de Tubières-Grimoard de Pestels de Lévis, Comte de Caylus (1692–1765). In the early 1750s, Caylus began a campaign of experiments to reconstruct modes of painting with wax used in classical antiquity. He parsed laconic references to encaustic painting left by Pliny and other ancient authors; he plied them against evidence of mural paintings smuggled out of the cities unearthed beneath Vesuvius's shadow. Thwarting the Bourbon ban on scholarly reproduction, fragmentary scenes of Herculanean murals depicting Theseus towering over the slain Minotaur nervously etched by Jérôme-Charles Bellicard had gone through numerous French and English editions by the late 1750s (fig. 3.1).[11] By then, Caylus had assembled an informal research team including his protégé, painter Joseph-Marie Vien (1716–1809), to recreate encaustic preparations. The triumph of this enterprise was Vien's *Head of Minerva* (now at the Hermitage, St. Petersburg), which caused a sensation when first exhibited at Paris's Académie des inscriptions et belles lettres in 1754.[12]

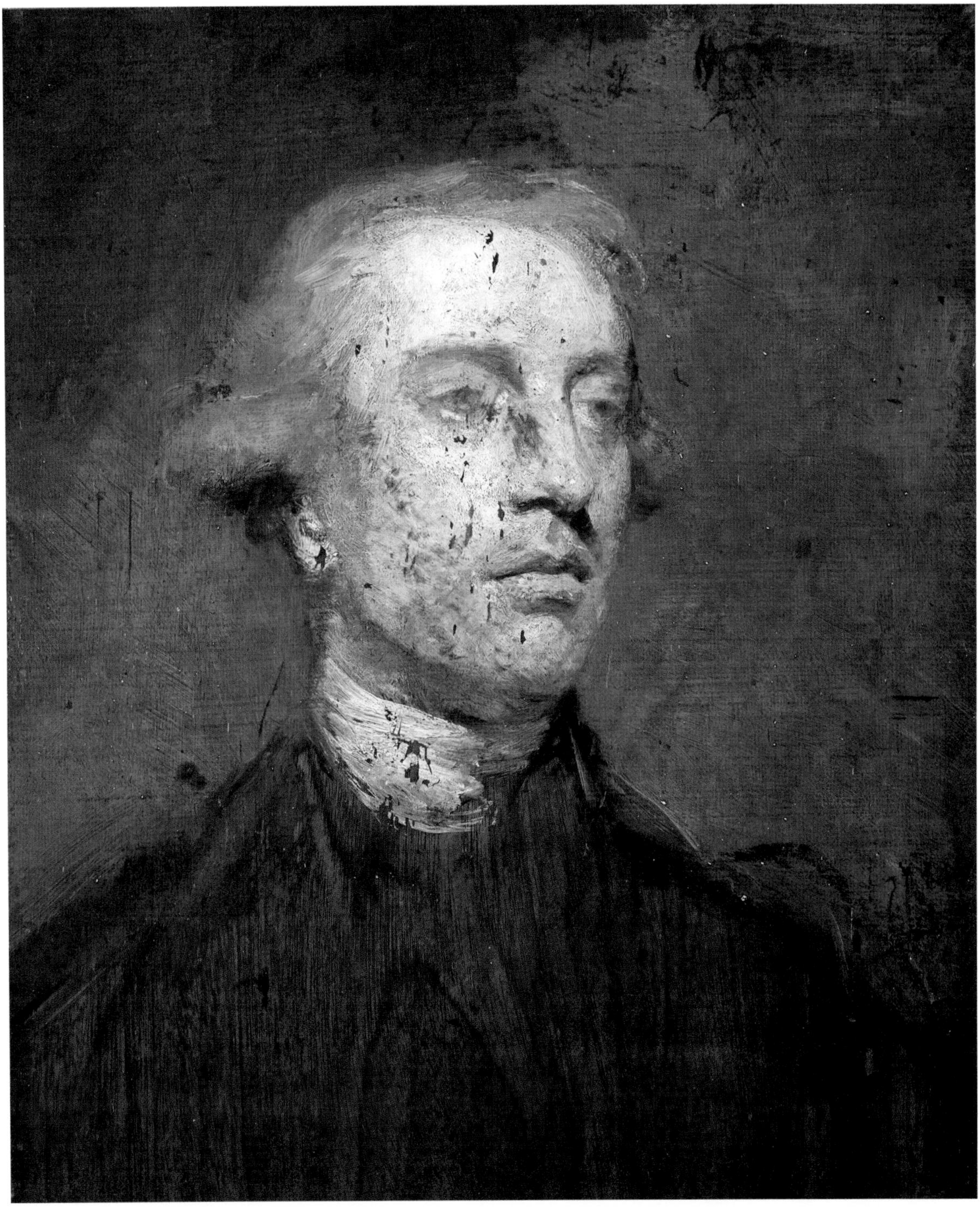

Plate 1 Joshua Reynolds, portrait of James Coutts, ca. 1771. Oil on mahogany panel. Scottish National Gallery, Edinburgh.

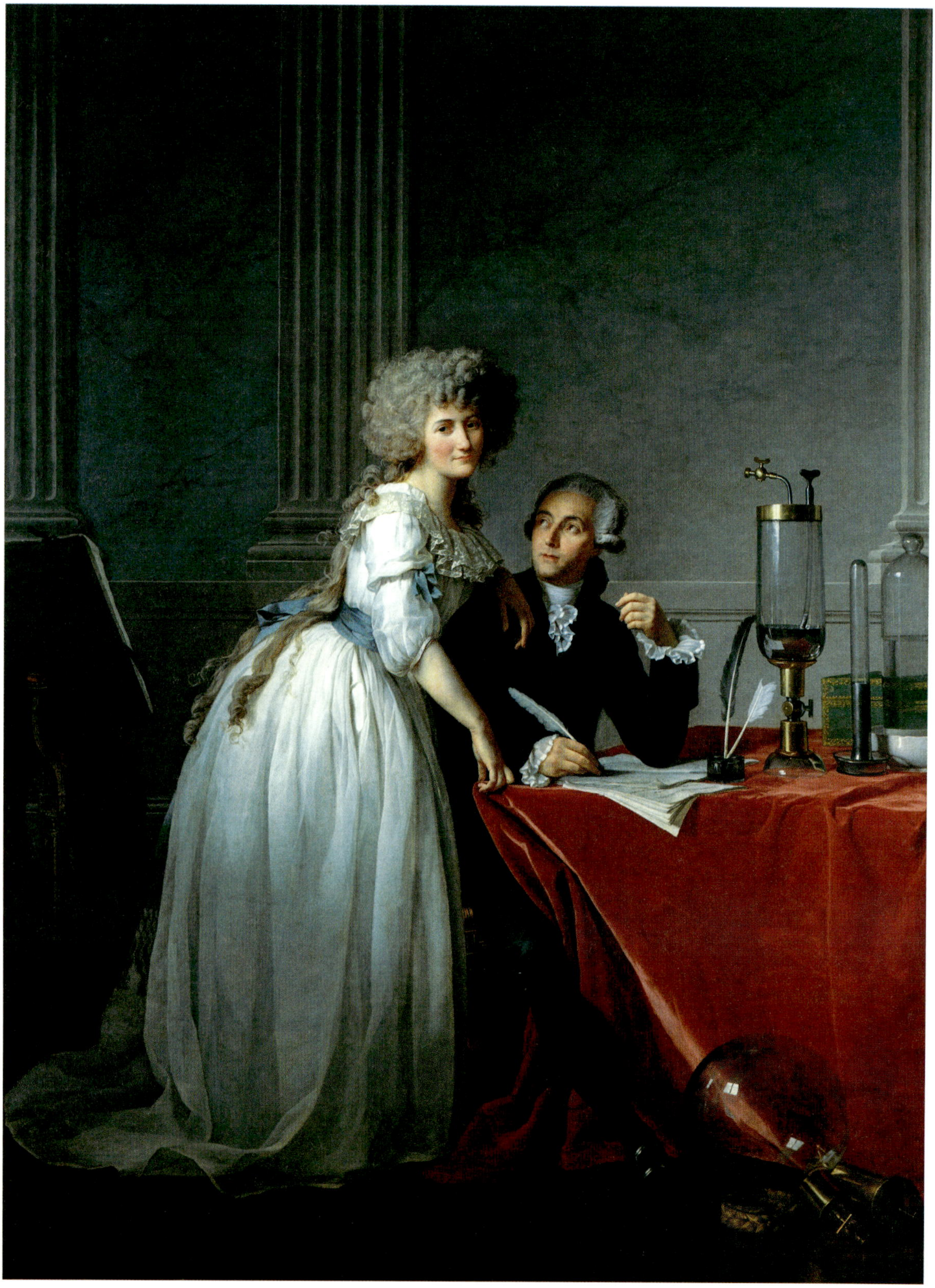

Plate 2 Jacques-Louis David, *Antoine Laurent Lavoisier (1743–1794) and His Wife (Marie Anne Pierrette Paulze, 1758–1836)*, 1788. Oil on canvas. Metropolitan Museum of Art, New York. Image © The Metropolitan Museum of Art. Art Resource, NY.

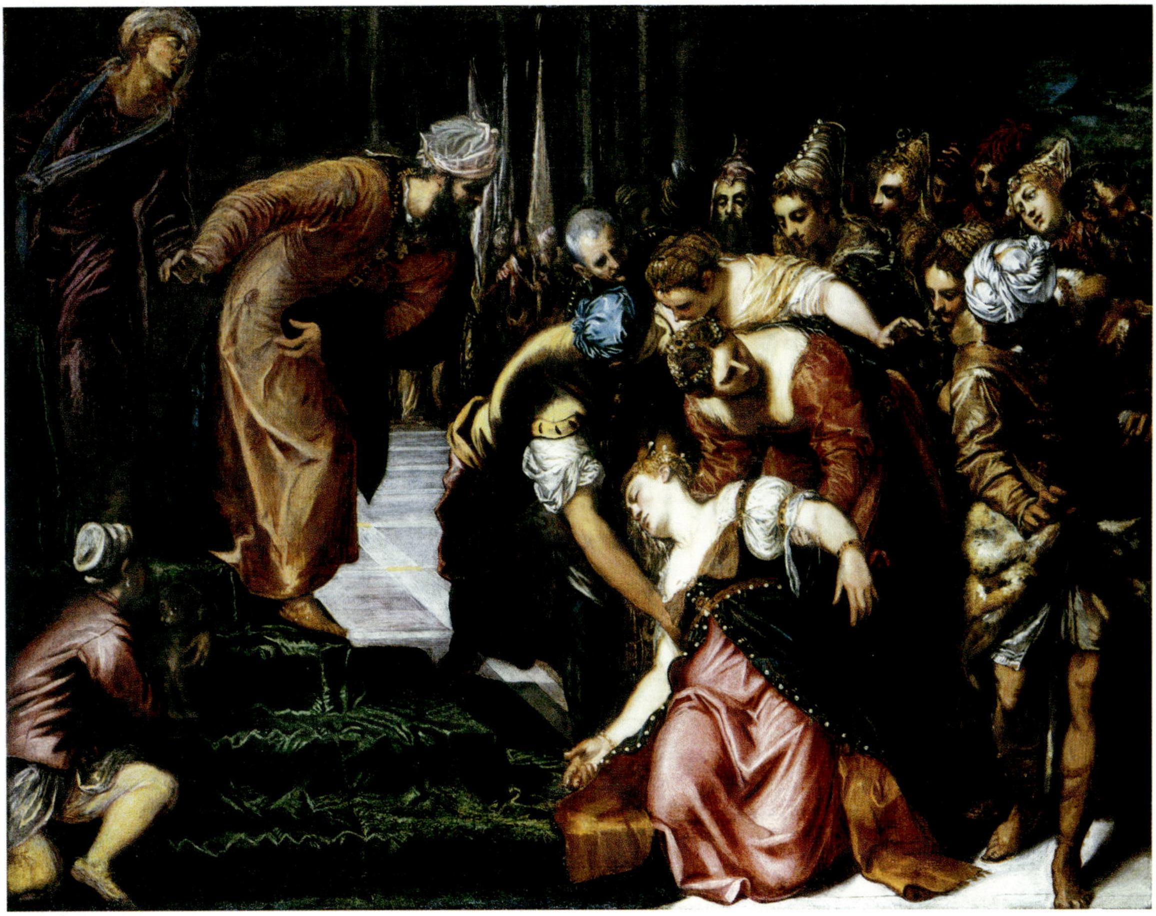

Plate 3 Tintoretto, *Esther before Ahasuerus*, ca. 1546–1547. Oil on canvas. Royal Collection Trust. © Her Majesty Queen Elizabeth II, 2018. Bridgeman Images.

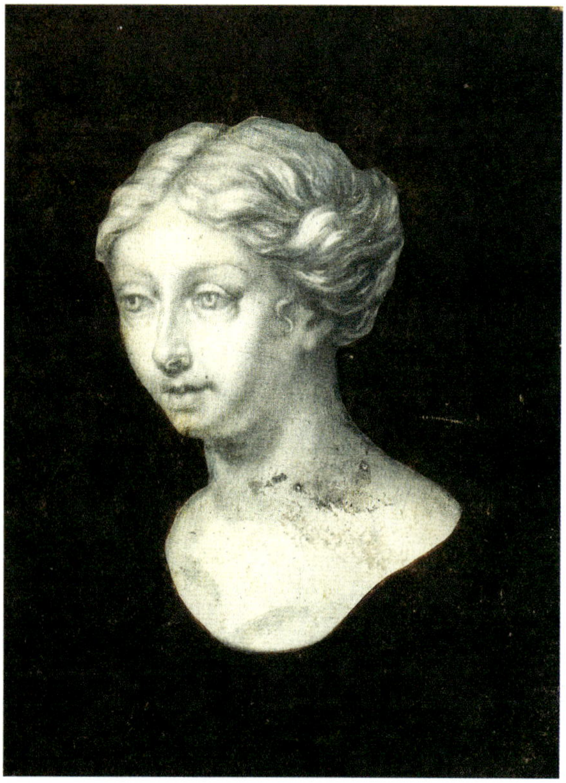

Plate 4 J. C. Le Blon, plate 2 from *Coloritto; or, The Harmony of Colouring in Painting* (1725). © Victoria and Albert Museum, London.

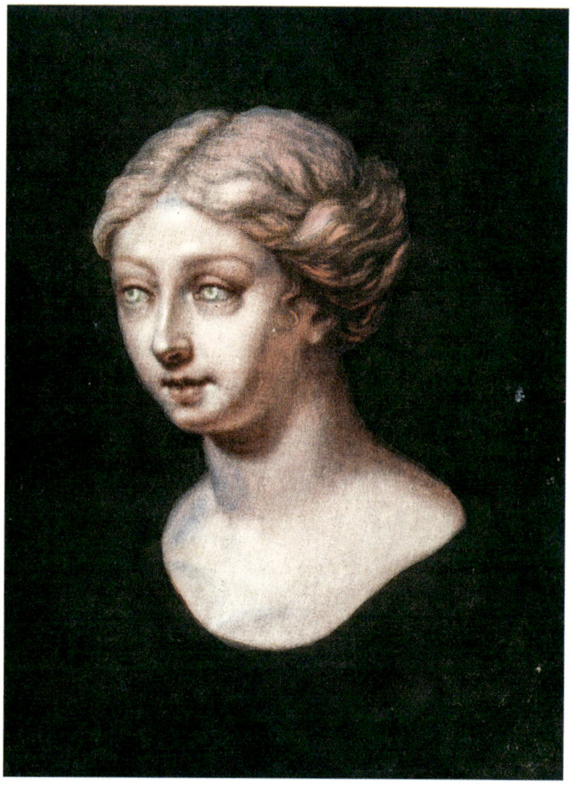

Plate 5 J. C. Le Blon, plate 3 from *Coloritto; or, The Harmony of Colouring in Painting* (1725). © Victoria and Albert Museum, London.

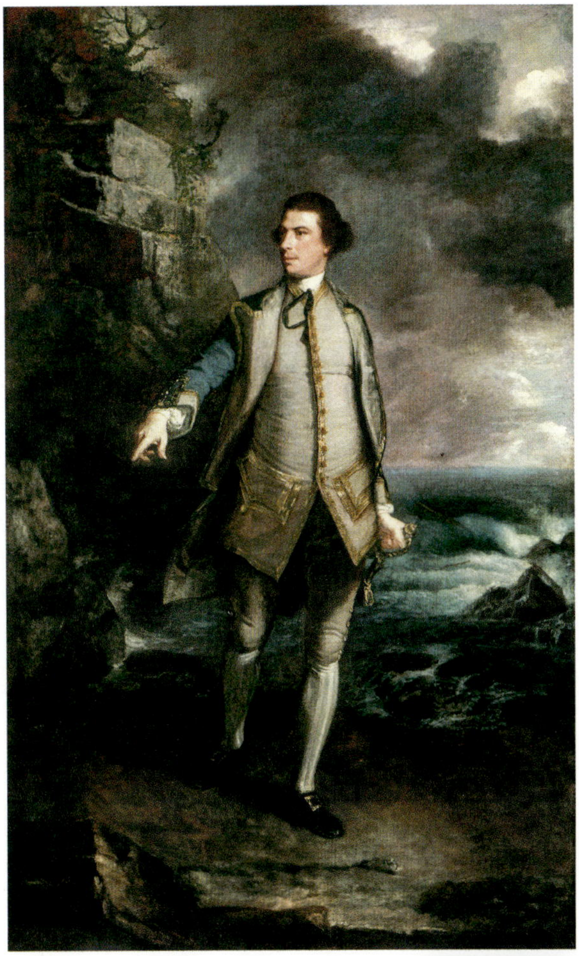

Plate 6 Joshua Reynolds, *Capt. Augustus Keppel*, 1752–1753. Oil on canvas. National Maritime Museum, Greenwich.

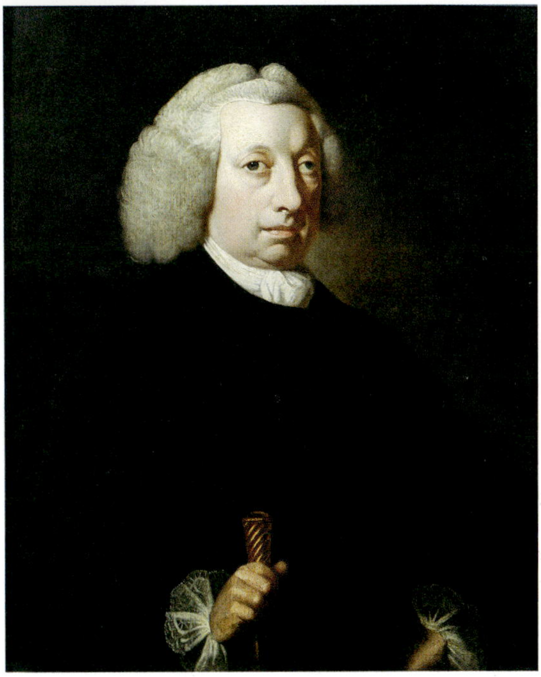

Plate 7 Thomas Rennell, *Dr. John Huxham*, ca. 1760. Oil on canvas. Royal Society of London.

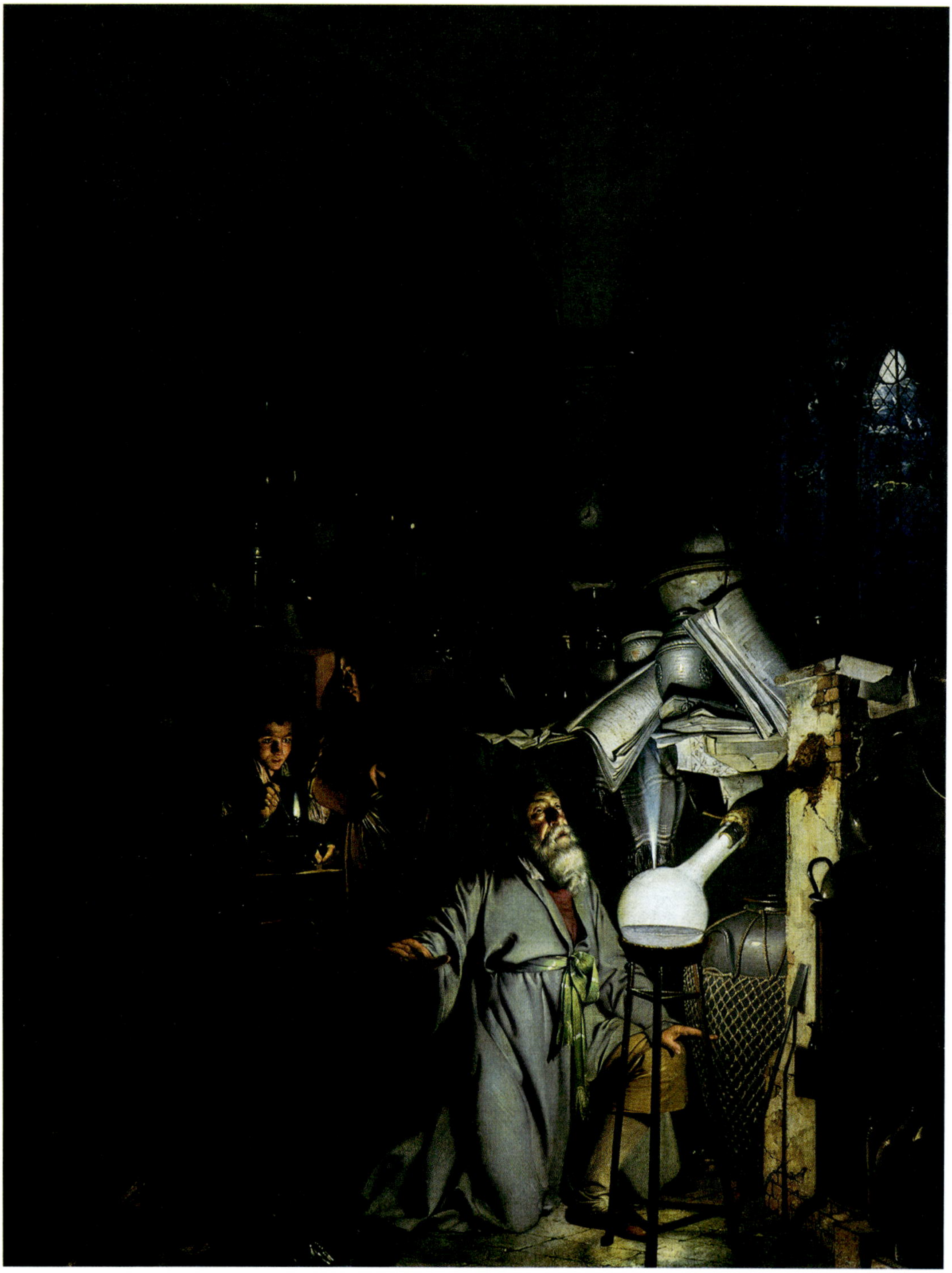

Plate 8 Joseph Wright of Derby, *The Alchymist, in Search of the Philosopher's Stone, Discovers Phosphorous, and prays for the Successful Conclusion of his operation, as was the custom of the Ancient Chymical Astrologers*, ca. 1771. Oil on canvas. Derby Museum and Art Gallery, UK. Bridgeman Images.

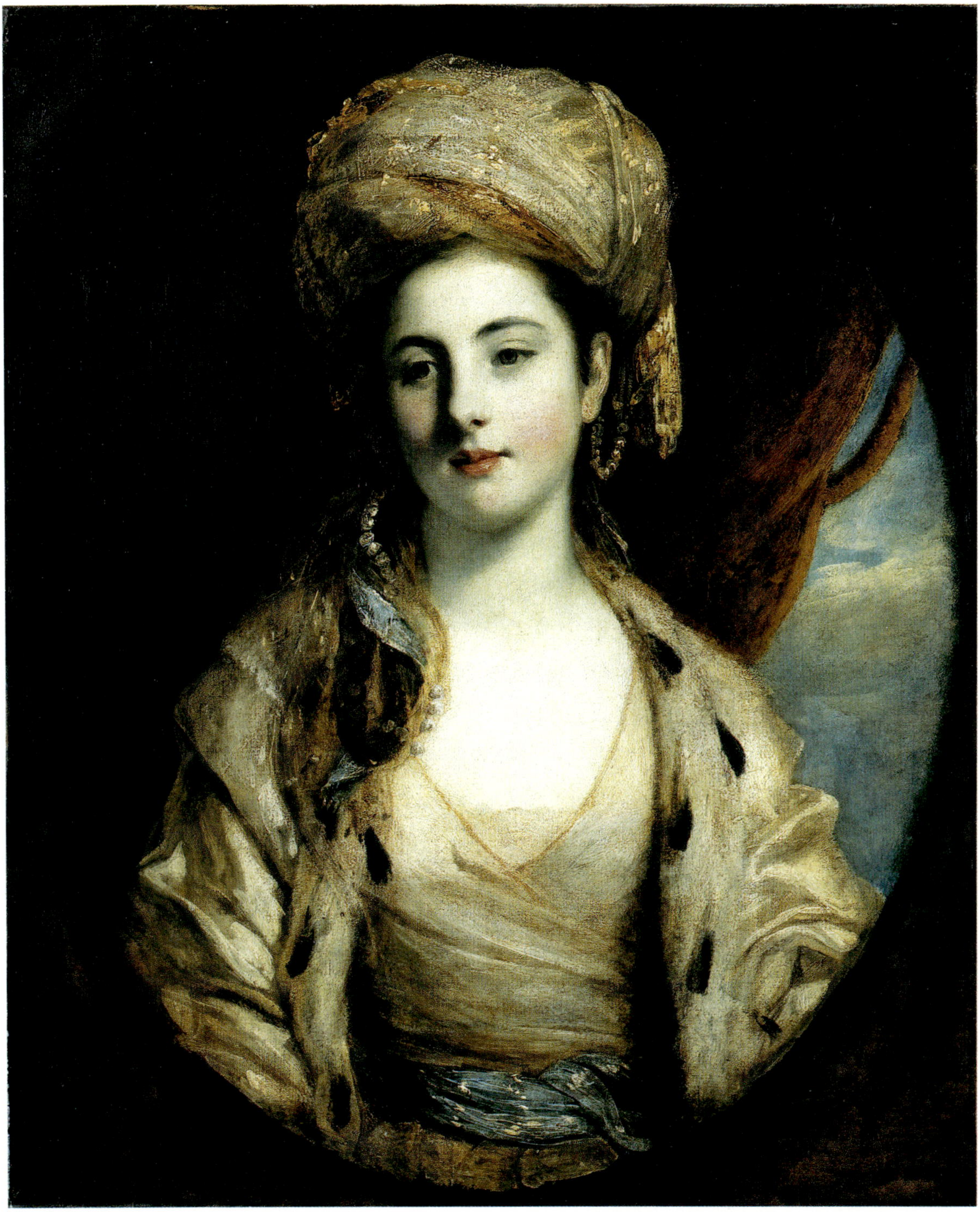

Plate 9 Joshua Reynolds, *Mrs. Richard Paul Jodrell*, ca. 1774–1776. Oil and other media on canvas, 30½ × 25¼ inches (77.5 × 64.1 cm). Bequest of Eleanor Clay Ford, Detroit Institute of Arts. Bridgeman Images.

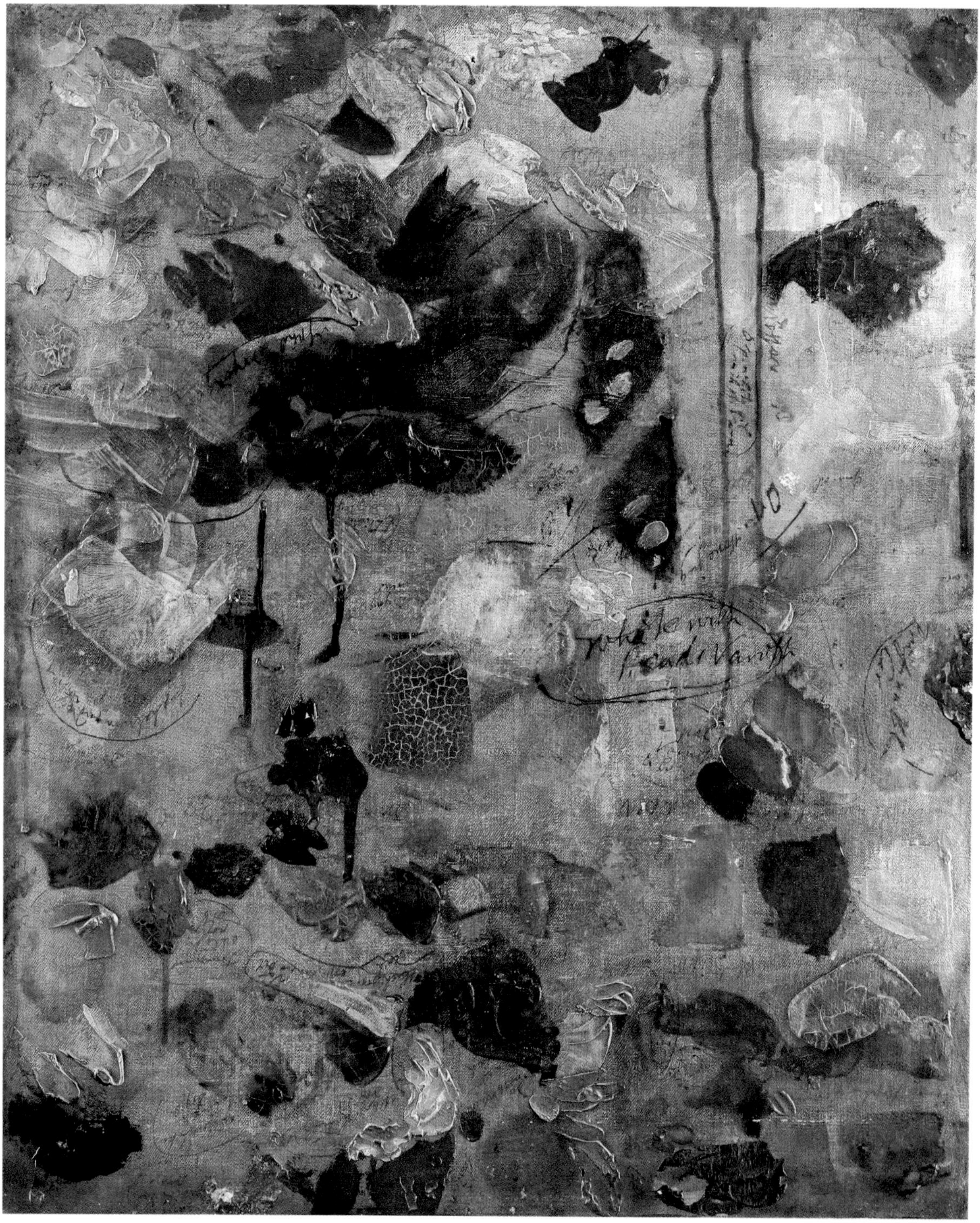

Plate 10 Joshua Reynolds, *Studio Experiments in Colours and Media*, 1770s[?]. Various media on canvas, 609 × 509 × 13 mm. Royal Academy of Arts, London.

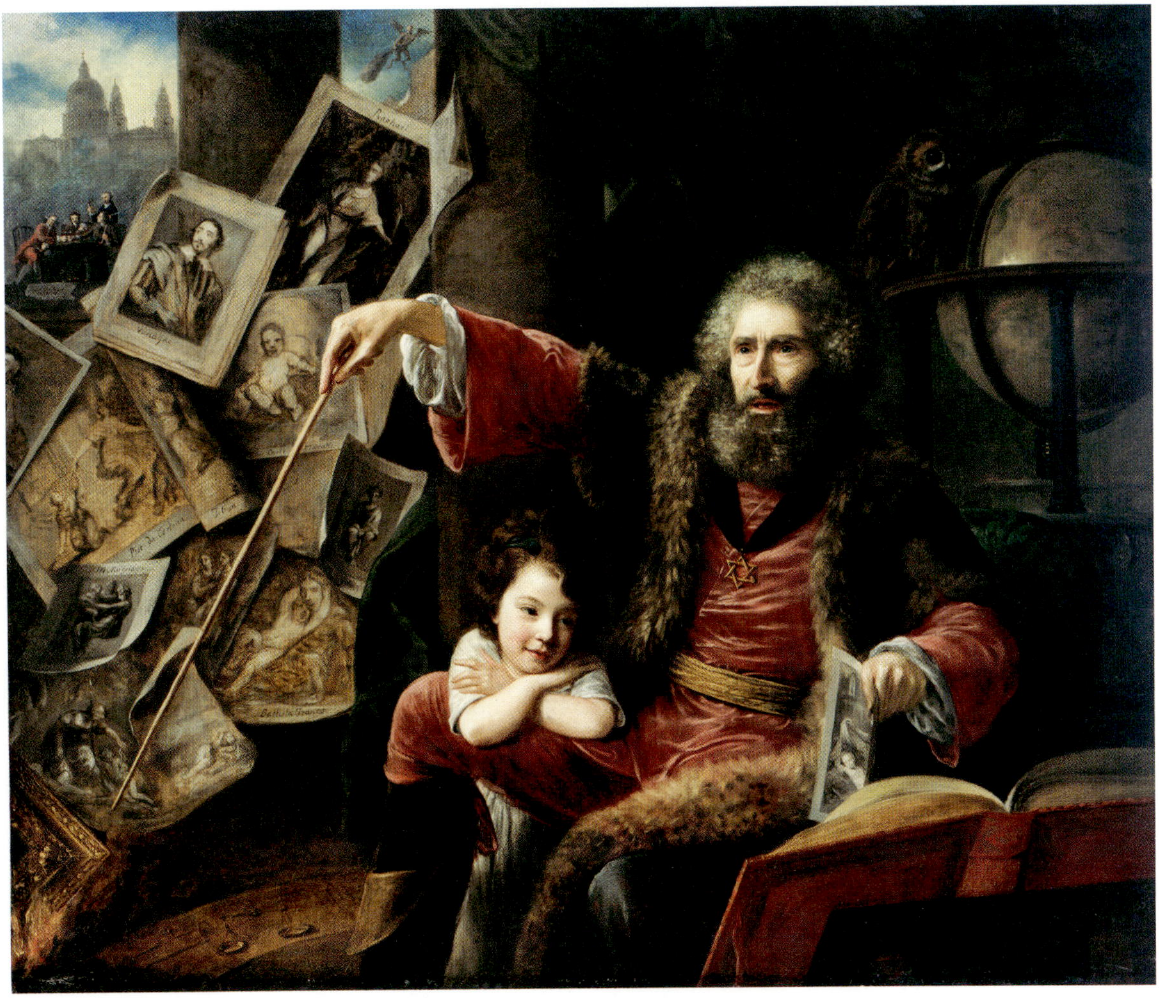

Plate 11 Nathanial Hone, *The Conjuror*, 1775.
Oil on canvas. National Gallery of Ireland, Dublin.
Photo © National Gallery of Ireland.

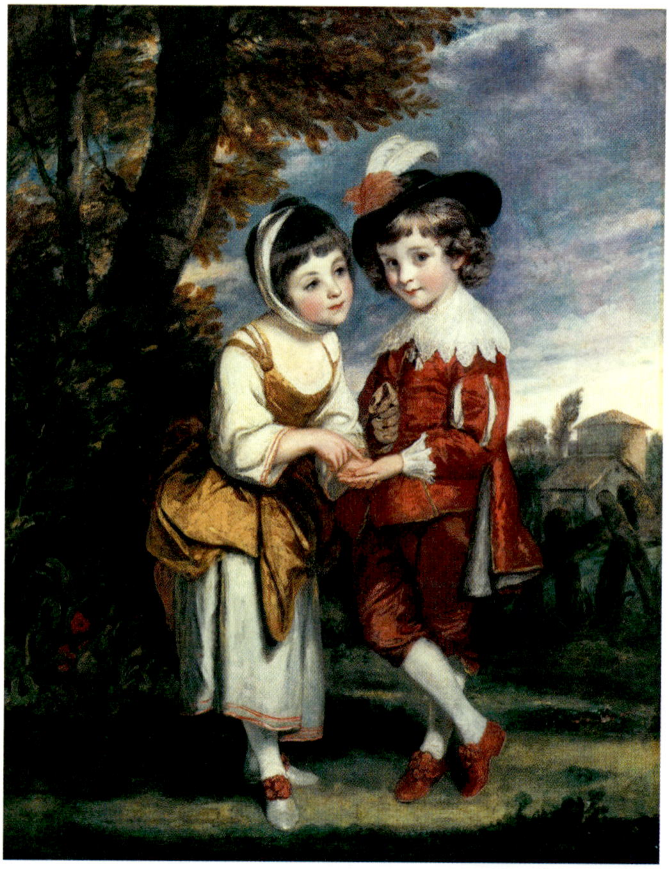

Plate 12 Joshua Reynolds, *The Young Fortune Teller: Lord Henry Spencer and Lady Charlotte Spencer*, ca. 1774–1775. Oil and other media on canvas, 56¼ × 44¾ inches (143 × 114 cm). Courtesy of the Huntington Art Collections, San Marino, California.

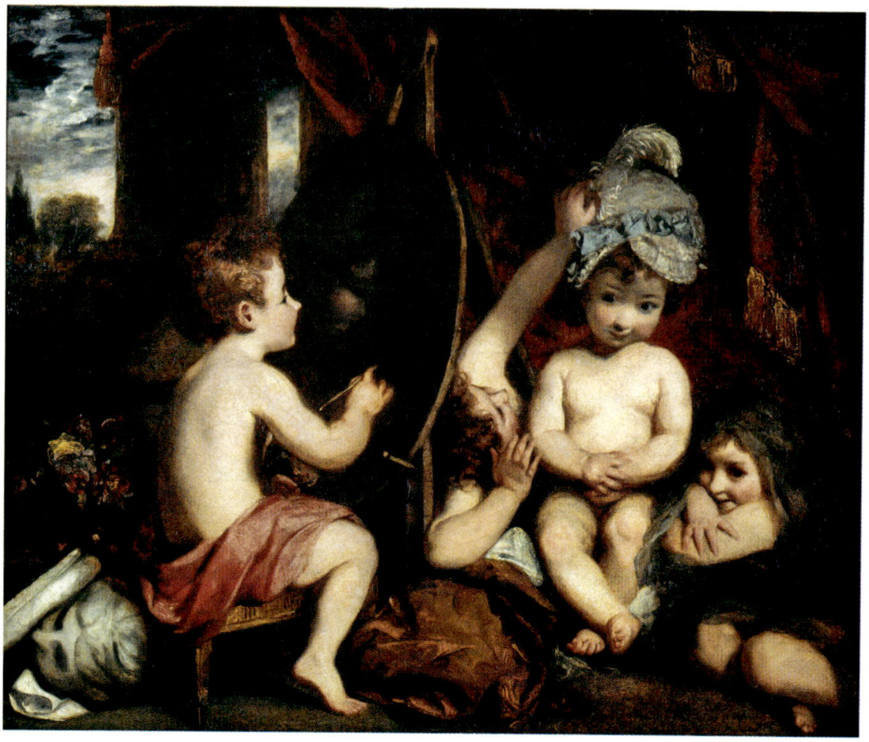

Plate 13 Joshua Reynolds, *The Infant Academy*, 1781–1782. Oil on canvas, 45 × 56 inches (114 × 142 cm). The Iveagh Bequest, Kenwood House, London. © Historic England. Bridgeman Images.

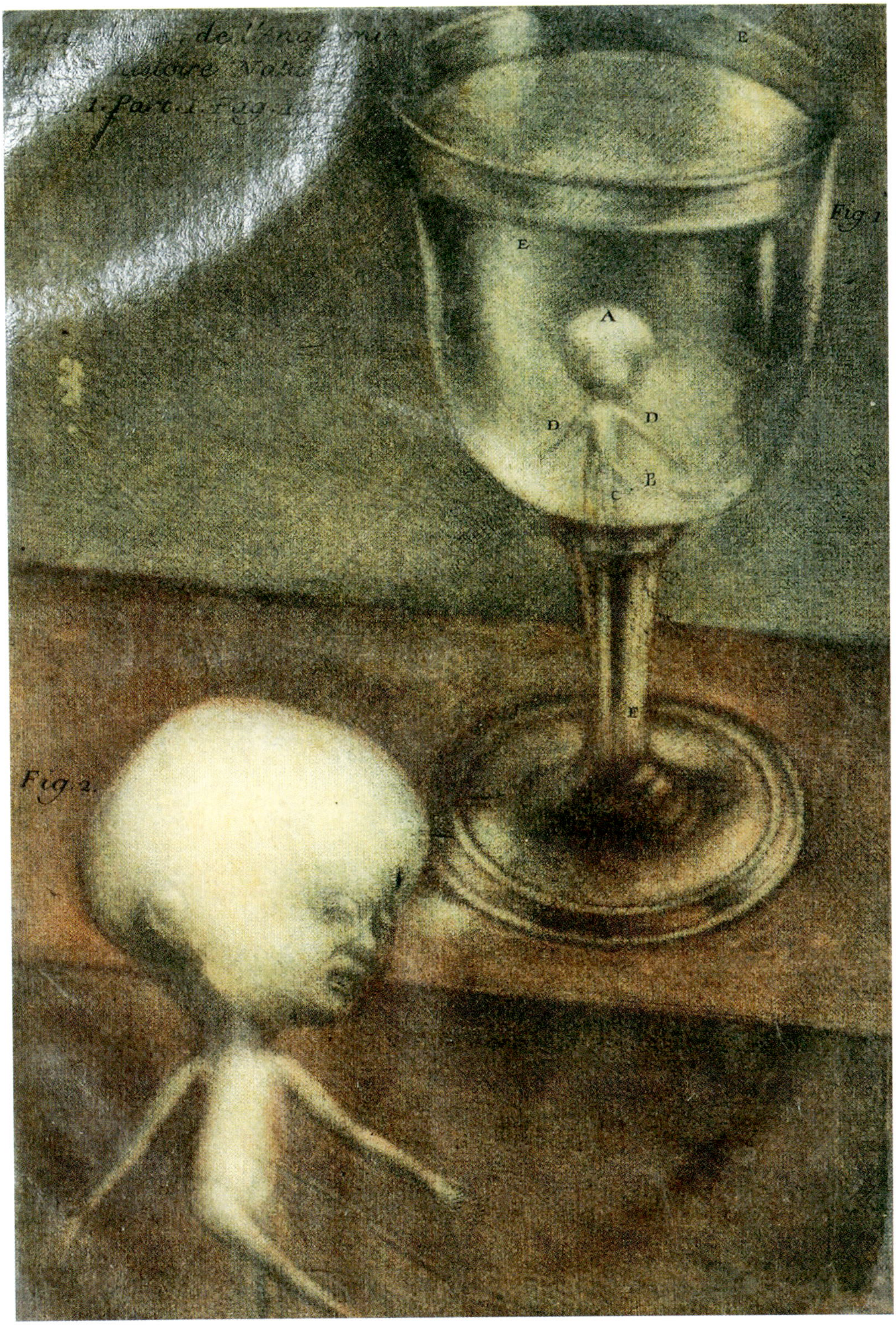

Plate 14 Jacques Fabien Gautier D'Agoty, color mezzotint of an embryo in a vial and optically magnified, as printed in volume 1 of *Observations sur l'histoire naturelle, sur la physique et sur la peinture* (1752).

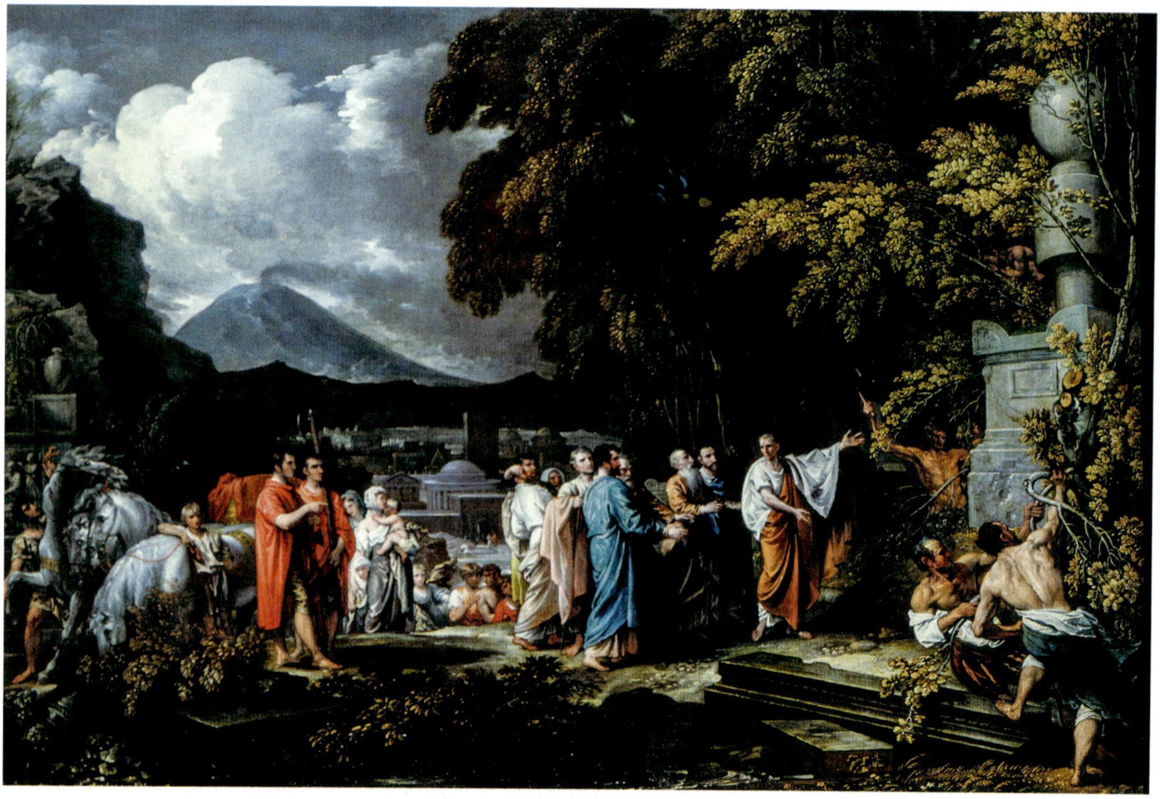

Plate 15 Benjamin West, *Cicero Discovering the Tomb of Archimedes*, 1796–1797. Oil on canvas. Private collection.

Plate 16 James Gillray, *TITIANUS REDIVIVUS; or, The Seven-Wise-Men consulting the new Venetian Oracle,—a Scene in ye Academic Grove. No 1*, 1797. Hand-colored etching and aquatint. British Museum, London. © The Trustees of the British Museum. Art Resource, NY.

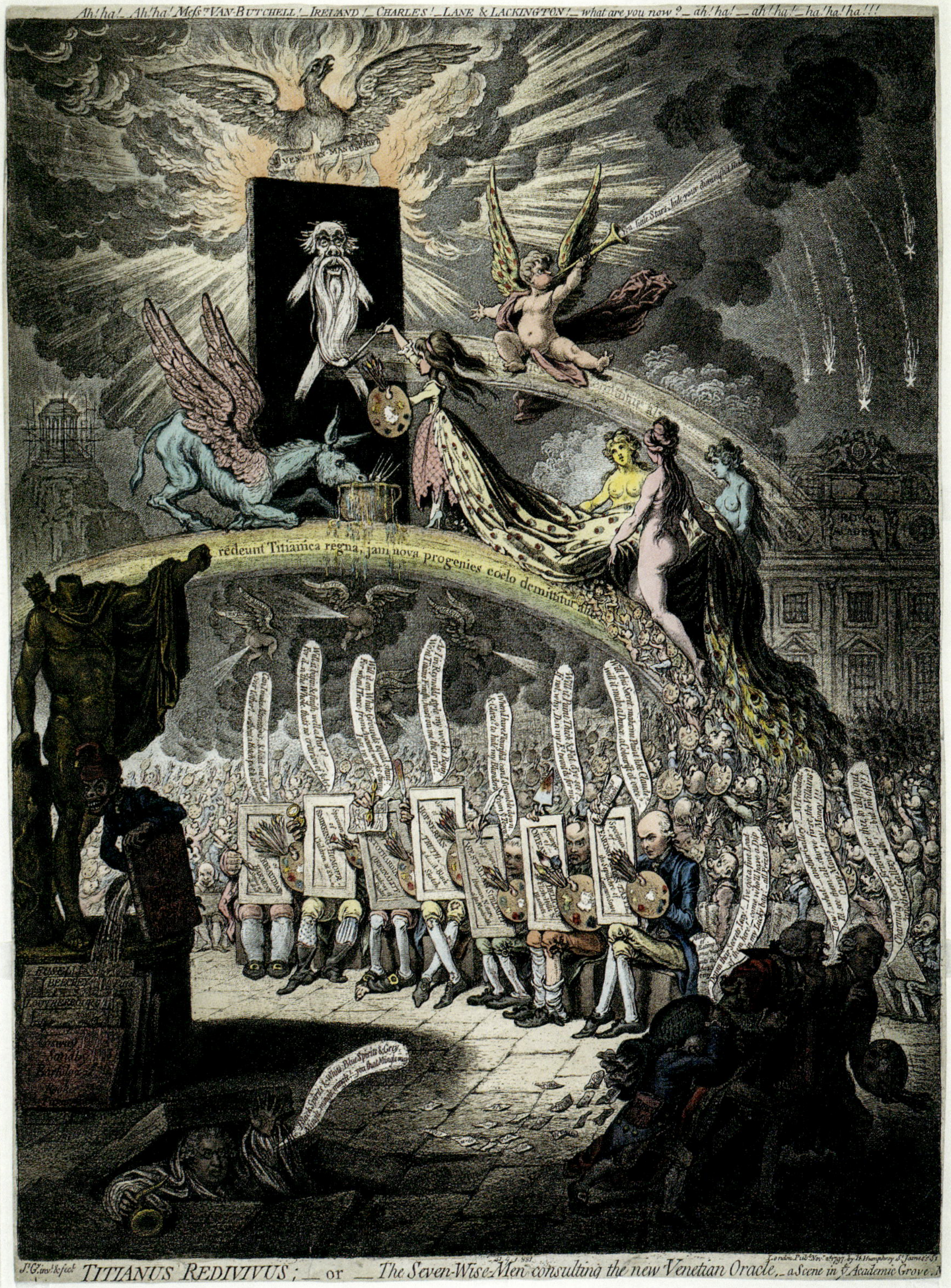
Ah! ha! _ Ah! ha! Mess.rs VAN-BUTCHELL ! _ IRELAND ! _ CHARLES ! _ LANE & LACKINGTON ! _ what are you now ? _ ah! ha! _ ah! ha! _ ha! ha! ha!!!
redeunt Titianica regna, jam nova progenies coelo demittitur
J.s G. inv.t & fec.t TITIANUS REDIVIVUS ; _ or _ The Seven-Wise-Men consulting the new Venetian Oracle, _ a Scene in y.e Academic Grove.
London Pub.d Nov.r 2.d 1797 by H. Humphrey S.t James's St

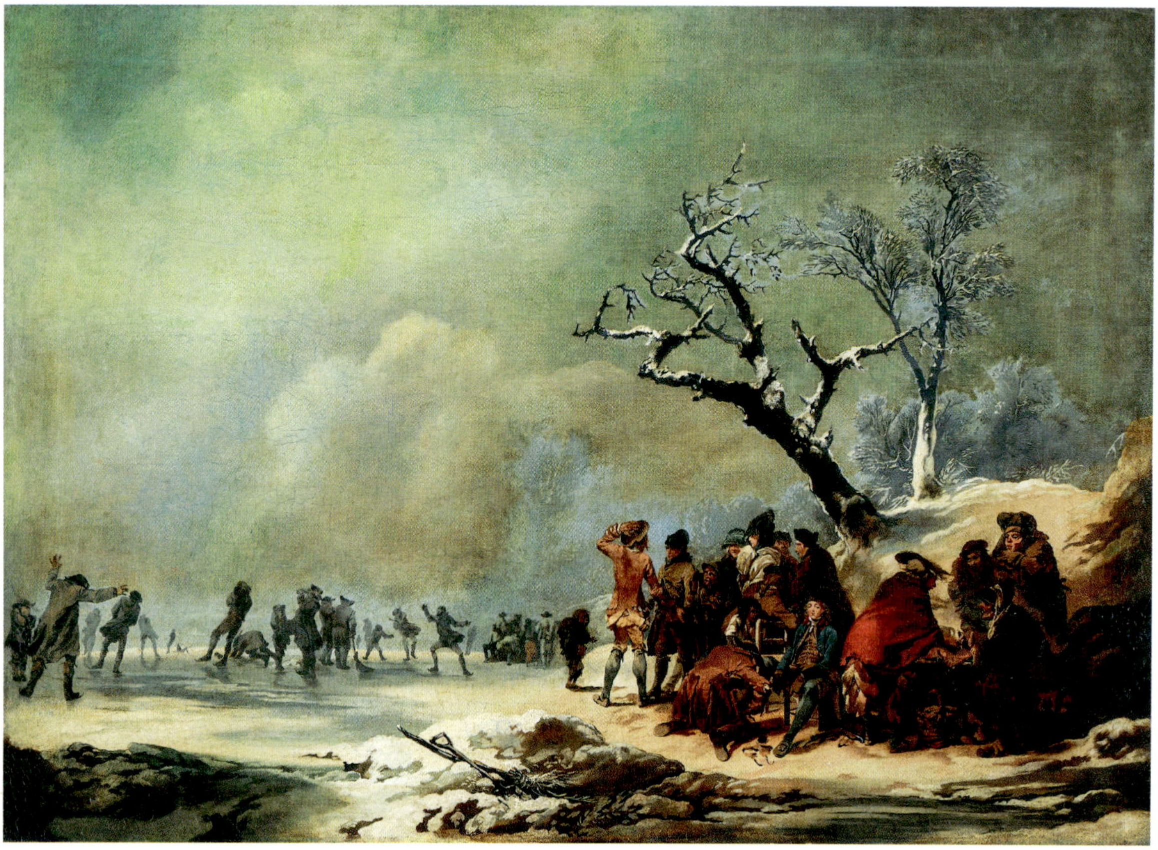

Plate 17 Joseph Booth, "polygraphic" print after Philippe de Loutherbourg's *Winter*, 1780–1790. Mechanical oil painting on canvas. British Museum, London. © The Trustees of the British Museum. Art Resource, NY.

Plate 18 Robert Waring Darwin, hand-tinted afterimage diagram, as reproduced in volume 1 of Erasmus Darwin, *Zoonomia; or, The Laws of Organic Life* (1796).

Plate 19 J. M. W. Turner, *The Burning of the Houses of Lords and Commons, October 16, 1834*; 1834–1835. Oil on canvas, 36¼ × 48½ inches. John Howard McFadden Collection, Philadelphia Museum of Art (M1928-1-41).

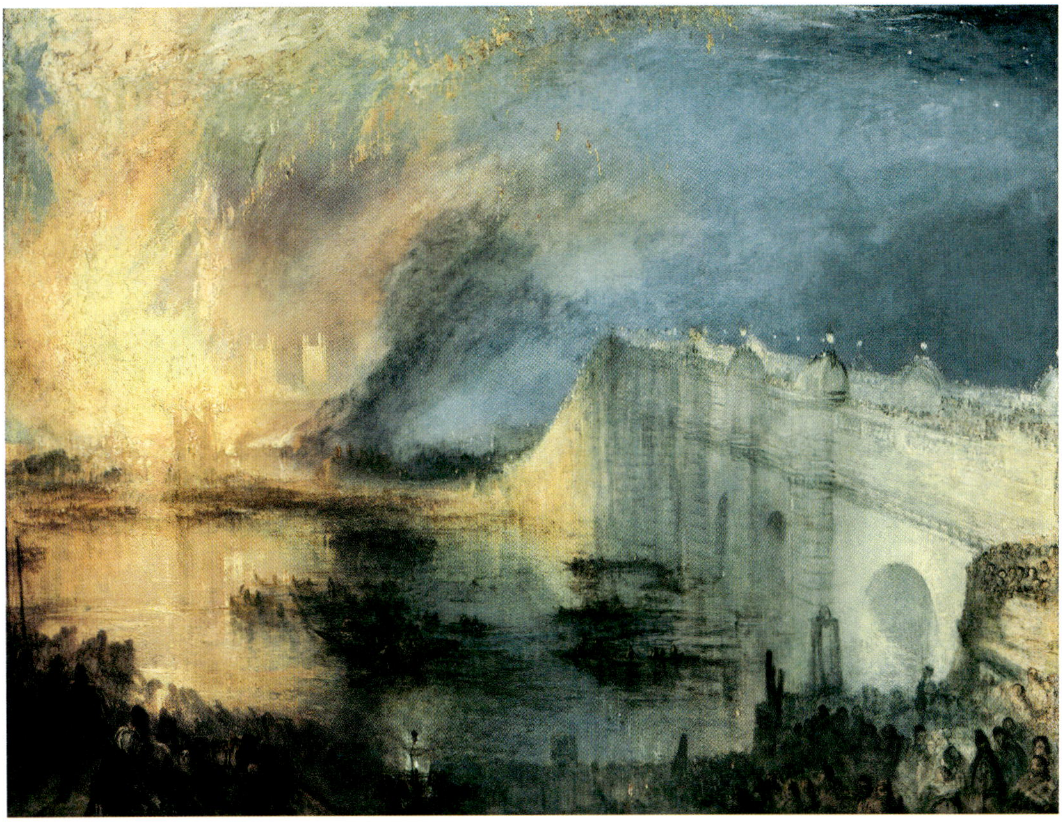

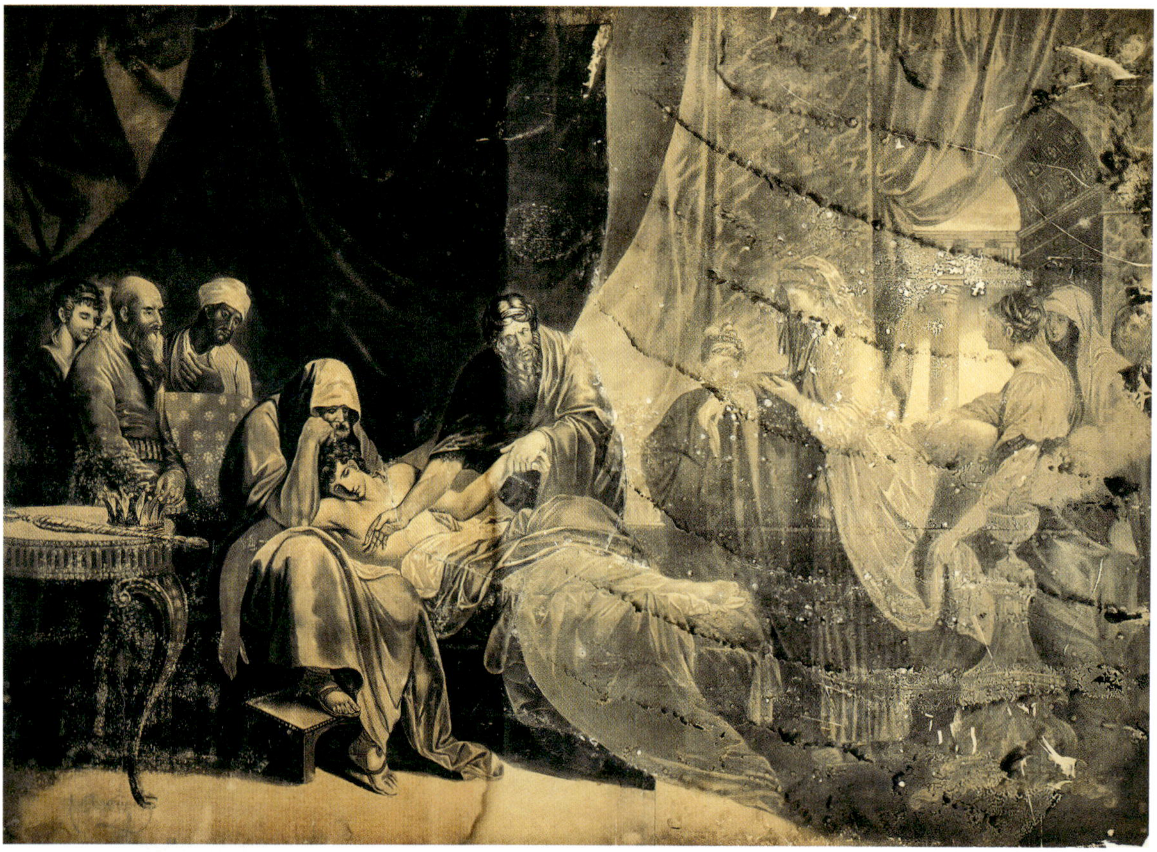

Plate 20 Francis Eginton, *The Physician Erasistratus Discovering the Love of Antiochus for Stratonice*, ca. 1778. Mechanical painting on two sheets laid paper, 77.5 × 110.5 cm. (Formerly inventoried as a sun picture by James Watt.) Science Museum, London (1884–16 pt7).

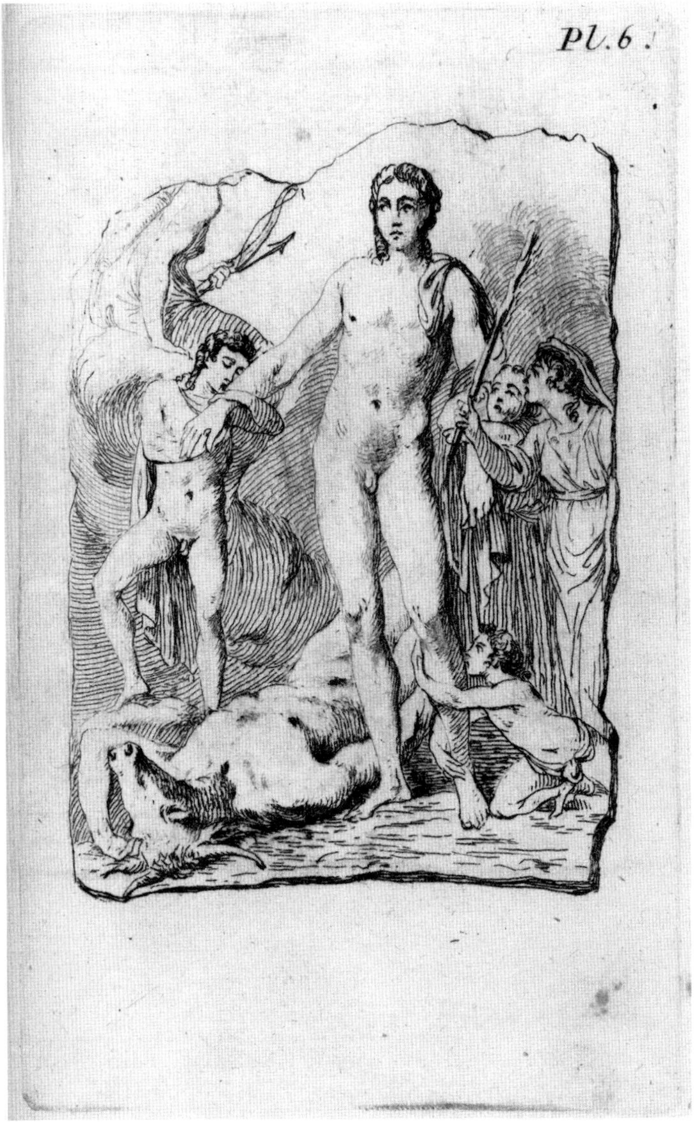

Figure 3.1 Jérôme Charles Bellicard, etching after mural of Theseus slaying the Minotaur. Plate 6 in M. Bellicard, *Observations upon the Antiquities of the Town of Herculaneum* (1753 [1754]).

Superior integrity was a key benefit of ancient wax techniques. Encaustic, Caylus proposes, "will never peel. Neither the heat of the sun nor that of interiors can alter it. The passing of years makes no change to it, which is to say that its works are even more sheltered from all dangers than frescoes and generally they must surpass them in durability when painted on wood."[13] That security against change figured frequently in period conversation. To English audiences (and to Reynolds specifically, as noted in chapter 2), Johann Heinrich Müntz declared in 1760 that pigments bound in encaustic vehicle "will not be liable to fade and change; no damp can affect it, no corrosive will hurt it; nor can the colours crack and fall in shivers from the canvas."[14] Although critical of Caylus and his research, *philosophe* Denis Diderot (1713–1784) too saw improvements to ancient wax painting as a remedy to faults of modern art. Made by combining metallic and earthy pigments—some in vehicles that dried quickly, others that could never fully dry—the colors of modern artists, "which produce such bizarre effects in time in oil painting," Diderot contended, "will not change at all in encaustics."[15] If it could withstand centuries of burial beneath Vesuvius's volcanic ash, then resurrected encaustic painting would offer a material durability sorely needed among modern media.

Physically tough, encaustic would equally restore a clarity muddled by oil paint's louche, unduly malleable handling. Both Caylus and Diderot, Danielle Rice has argued, "aspired to endow painting with an esthetic permanence which they felt could only be achieved through a greater solidity of the forms, and a greater simplicity of line and purity of color."[16] Crucially, chemical means were needed for those aesthetic, physical, and even moral reclamations. Caylus had emphasized that "natural philosophy [*la Physique*] and Chemistry must have equal footing in this discovery," as he detailed his encaustic research with Michel Joseph Majault (ca. 1714–ca. 1790), a doctor in Paris's medical faculty.[17] By their account, encaustic turned upon principles of "analogy" whereby essential oils and other solvents could "penetrate and render wax capable of being spread with a brush."[18] Plausibly informed by the work of Georg Ernst Stahl (1659–1734), personal physician to Friedrich Wilhelm I in Berlin and leading architect of the theory of phlogiston, that analogical operation was a *sine qua non* of chemistry for Caylus and Majault: "All Chymists, even the least enlightened, know that the heavy fluids we call essential oils will penetrate and dissolve analogous bodies, sometimes with the help of fire, often without the help of this agent."[19]

Chemistry also figured prominently in Diderot's promotion of the rival encaustic preparations made by contemporary painter Jean-Jacques Bachelier.[20] True, Bachelier's chemistry of unctuous wax dissolved with alkali to form a pictorial vehicle initially failed to impress Diderot. The results Bachelier obtained with salt of tartar (potassium carbonate) were convincing.[21] "If our eyes can overcome their prejudices and accustom themselves to matte pictures," the *philosophe* concluded, "M. Bachelier will have found the way to preserve the masterpieces of the past and to conserve eternally those of our time in their initial beauty, without allowing any of their colors to change."[22] For leading French, anti-Rococo voices at midcentury, pictorial chemistry stood to resurrect ancient media and to stabilize modern art.[23]

From his time in 1750s Rome, Joshua Reynolds had made meticulous guard of his studio output through the medium of print.[24] He contracted with generations of engravers including James Macardell (1727/28–1765), Valentine Green (1739–1813), and John Raphael Smith (1751–1815) to translate his paintings especially into mezzotint's tonal vocabulary. In his studio, the painter famously kept what Northcote called "a port-folio . . . containing every print that had been taken from his portraits; so that those who came to sit, had this collection to look over, and if they fixed on any particular attitude . . . he would repeat it precisely."[25] An extensive modern literature has documented how Reynolds harvested compositional ideas from the prints he collected; how he used printed reproductions to target expanding audiences among the middling ranks who could never afford his elite oils; and how he continued to be refashioned in print by generations of graphic artists long after his death.[26] By these means, so J. T. Smith (1766–1833) could recall Reynolds declaring, printmakers "would perpetuate his pictures when their colours should be faded and forgotten."[27]

Yet hope burned eternal that color—that great glory of Reynolds's practice, according to many commentators—could *also* be saved. A year after the painter's death in 1792, London's *Morning Chronicle* reported the exhibition in Great Newport Street of James Pearson's full-size replica of Reynolds's portrait of George III.[28] An ingenious species of stained glass made with metal supports "concealed in the folds of the drapery so that it appears one entire plate," Pearson's vitreous king enthroned was as much a feat of preservation as a work of art. "We rejoice to see so fine a specimen of our late admired President of the Academy's colouring," the *Morning Chronicle* put it, "*fixed by such a process that it can never fade*."[29] Pearson was not alone. This chapter argues that, in the revolutionary decade of the 1790s, several leading entrepreneurs and industrialists brought forward new, explicitly chemical means by which to "fix" or otherwise stabilize a body of British grand-manner painting made chemically volatile through Reynolds's influential model. The first section introduces key 1790s chemical replicators and their strategies in reverse-chronological order, an approach taken to subtly resist the impulse to read those techniques as so many progressive steps toward replicative perfection. Instead, highlighting the centripetal role of the second Royal Academy president, Benjamin West (1738–1820), I argue back from the invention of lithography (1796–1798) to the "Venetian system" of painting (disclosed in 1795–1797) and "pollaplasiasmos" (invented ca. 1784) to methods of enamel painting traceable to the ancient world as competing instruments for preserving Reynoldsian painting.[30]

The second section moves from those aspiring echelons of London's art world to practical thinking about replication via chemical change in time among the Lunar Society of Birmingham and its affiliates. Against the backdrop of a Continental triumph for Antoine-Laurent Lavoisier's innovations, known among historians of science as the Chemical Revolution, I highlight chemical practices retaining an older conceptual language and fabric—both at Matthew Boulton's Soho Manufactory in greater Birmingham and at Josiah Wedgwood's Etruria near Stoke-on-Trent. Mingling the fine and the industrial arts, the chapter's collision of chemical painting, printing, and temporal metaphysics takes sanction, ultimately, from the work of Thomas Wedgwood (1771–1805), Josiah Wedgwood's youngest son and "first inventor" of photography in a judgment famously voiced by William Henry Fox Talbot. Tom Wedgwood's drive to visualize a fundamentally antivisual conception of time via chemical supports is, I argue, a supreme expression of the oppositions systemic to the 1790s' ways of making and thinking with temporally evolving chemical objects.

Maybe more resolutely Marxian terms of contradiction would do better than any talk of "oppositions." For, notoriously described by Edmund Burke as "the wild *gas*, the fixed air . . . broke loose," radical politics would become closely linked to chemistry in the era of the French Revolution.[31] Among those considered in this chapter, natural philosopher/theologian Joseph Priestley and copy-printing industrialist James Watt Jr. would be active partisans; enamelist William Russell Birch, Tom Wedgwood, and chemist Humphry Davy had all been connected to radical circles in the 1790s. Contradiction abounds too in the speeds at which the 1790s' chemical images have been taken to operate. Beyond the ongoing, patinated change of oil pictures that Northcote cast as increasing value rather than compromising it, Walter Benjamin's classic writings on technological reproduction plot a new era of politicized acceleration to the 1790s. In its rapid, chemical action on limestone, lithography inaugurated an image-regime of unprecedented speed, driven by the desire of a consolidating proletariat to bring the work of art to its own terms.[32] And yet, anticipated by that stolid, conservative function credited to encaustic's chemistry by the politically free-thinking Diderot, chemical processes of the 1790s have also been read as means by which to slow, arrest, or otherwise still time's fugitive course. Drawing upon Michel Foucault's account of the rupture of a dynamic, modern *episteme* from the taxonomic orders of classical knowledge, Geoffrey Batchen has traced to the years around 1800 a consolidating "desire for an impossible conjunction of transience and fixity." Photography would be that desire's ultimate expression.[33]

Shifting attention from the catalyzing figure of Reynolds to the supports and the supporters of his project, after the manner of those Steven Shapin has called "invisible technicians," this chapter begins mapping a different approach.[34] Whether or not Reynolds could have conceived academic painting in the image of the chemical homunculus, as suggested in chapter 2, the ambition to carry fragile, chemical life through time—and to do so by chemical means that could afford insight into the nature *of* time—explicitly drove practical visual projects just as it haunted imaginative nightmares of leading British figures immersed in 1790s progressive politics. How those endeavors would later come to be rediscovered as photography, what happens when they are now released from it: those are the targets of the next chapter.

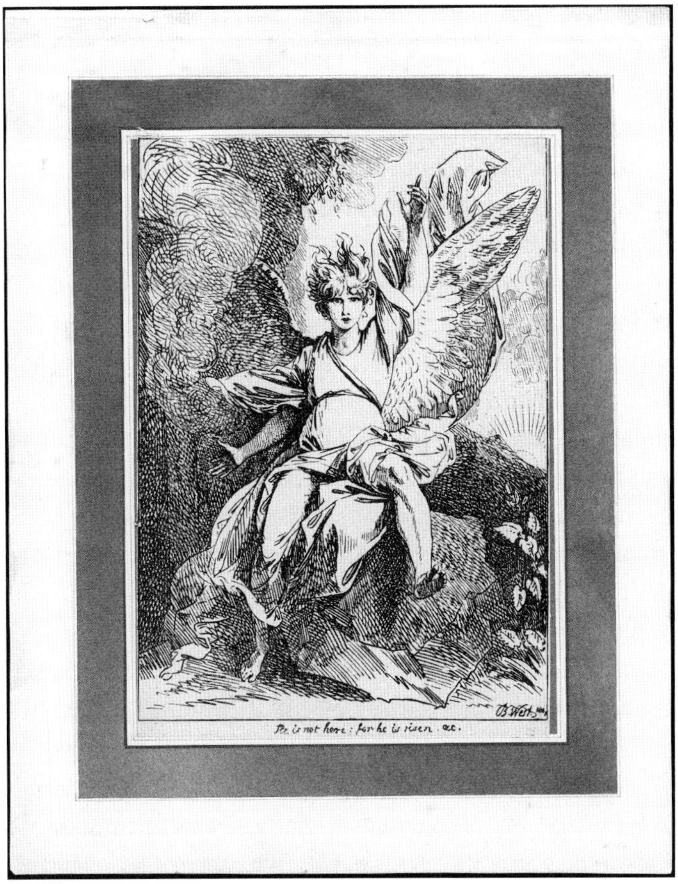

Figure 3.2 Benjamin West, *Angel of the Resurrection*, 1801. Lithograph, as published in Philipp André, *Specimens of Polyautography . . .* (1803–1807).

Chemical Replication in the 1790s

With darkened palm and the shaded sole of an outstretched foot held aloft, Benjamin West's *Angel of the Resurrection* stares directly out from the picture plane (fig. 3.2). Smoke curls in frothy billows from densely hatched darkness at left. That networked intensity of tone answers in counterpoint to the dotted arc and starburst rays of sunlight cresting the sloping hillocks at right. Split down the spine of the image through the licking flames of the angel's unevenly forked locks, West's intent figure appears strangely unsettled. Said differently, the angel figures a struggle for balance, a pursuit of equilibrium between the resplendent plumage of his proper left wing and its obscured twin—between the plant straining upward to find sunlight at right and the inky void at left. Light versus dark, sprouting life versus curling smoke: the image's various efforts at left-right reconciliation call attention to the massive boulder on which the angel perches so awkwardly.

Although its rough hatchings recall the graphic vocabulary of woodcut, West's is a recalcitrantly chemical image. Dated to 1801 and published in Philipp André's *Specimens of Polyautography* (1803–1807), the print is an early example of “l’imprimerie chimique,” or lithography, invented by printer Alois Senefelder between 1796 and 1798. Explaining the oil/water resist at the base of his technique, Senefelder stressed how his preferred Solenhofen limestone possessed “affinity for fat, which is absorbed so deeply that in many cases even extensive grinding will not remove it.”[35] Sinking a greasy, inked design into wet stone and allowing it to “be set aside for a day, that it may take good hold of the stone,” Senefelder’s lithographic artist was then to coat the stone with successive washes of acid and gum arabic, rendering the design invisible.[36] Left immured in stone, the design becomes positively Christological, or so West’s print would have it. Alternately inked and watered—the positive, unctuous forms already immersed in stone attracting successive applications of fatty printers’ ink, which resists the untreated areas that will read as negative—the stone is rolled through the press to reveal the design resurrected on the paper lifted off from it. “He is not here: for he is risen” reads West’s quotation from Matthew 28:6, concisely allegorizing lithography’s action.[37]

With lithography, we could say, the image speeds into the modern age. “The fact that the drawing is traced on a stone, rather than incised on a block or wood or etched on a copper plate,” so would claim Walter Benjamin, “. . . made it possible for graphic art to market its products not only in large numbers . . . but in daily changing forms.”[38] For Senefelder, however, lithography was not simply *technicshen Reproduzierbarkeit*; it counted as an innovation precisely because it was “based not on mechanical but on purely chemical properties.”[39] Senefelder’s chemical principles and poetics were sounded extensively by contemporaries. “The ink is a chemical preparation,” explained one writer in an 1808 contribution to *The Gentleman’s Magazine*, “of which soda, lac, and lamp-black are component parts.”[40] In controversial lectures delivered to London’s Royal Institution in 1806, campaigning engraver John Landseer declared lithography’s “art of producing prints from drawings hatched with lines on calcareous substances” to be a delusion.[41] Fine art as such stood in danger from a technique able to keep chemical time with artists’ preparatory sketches. “It is *not* the painter’s *sketches*, that is most desirable to multiply,” Landseer claimed, “but his *finished performances*. We wish most to see the mercury of his active imagination, amalgamated with the sterling gold of his cultivated understanding.”[42] Privileging the quick, impulsive sketch, lithography’s new chemistry threatened to liquidate that alchemical chrysopoeia essential to finished academic work.

President of the Royal Academy since Reynolds’s 1792 death, Benjamin West appeared in *Specimens of Polyautography* (published by Senefelder’s franchisee in England) alongside some twenty London-based artists. West was also the fulcrum for two other endeavors active in the 1790s, which too would claim to turn on chemical principles: the reclamation of Titian’s

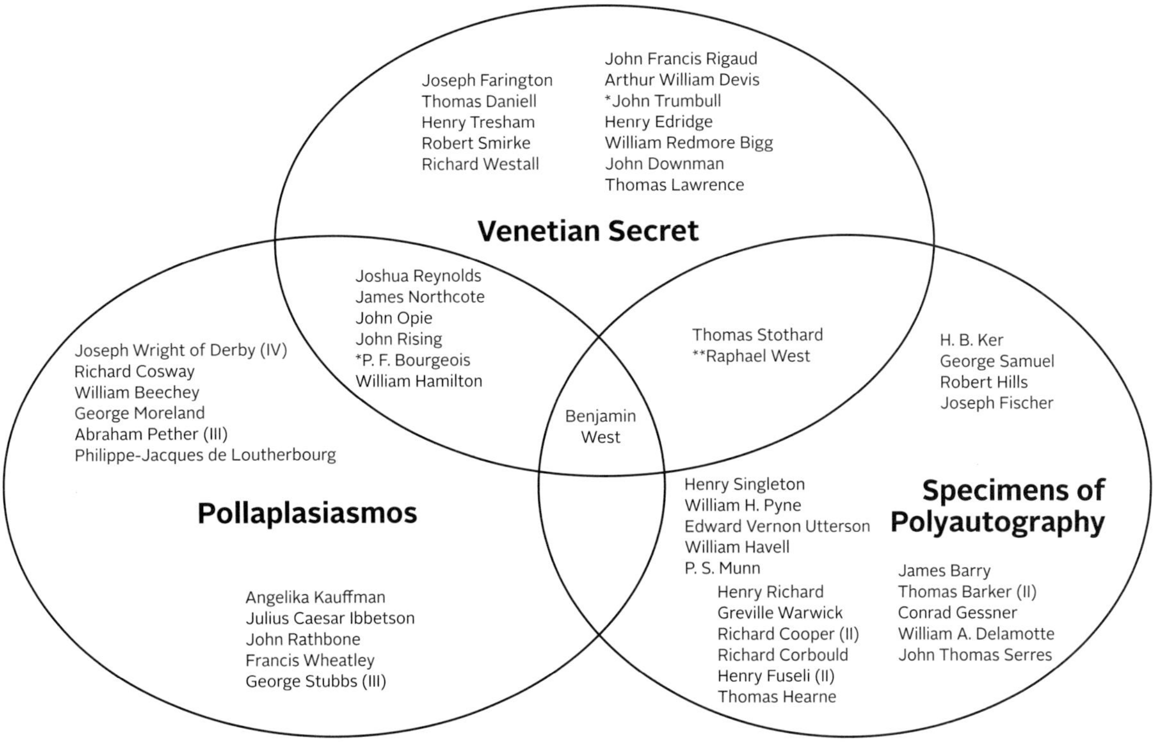

Figure 3.3 Chemical replication in the 1790s. "Venetian Secret": all figures are named or identified in James Gillray's *TITIANUS REDIVIVUS; or, The-Seven-Wise-Men consulting the new Venetian Oracle,—a Scene in ye Academic Grove, No. I.* (1797; plate 16), except * as cited in Mark N. Aronson and Angus Trumble, *Benjamin West and the Venetian Secret* (2008), 12, and ** as noted in entry for January 11, 1797, in *The Diary of Joseph Farington*, ed. Kenneth Garlick and Angus Macintyre (1979), 3:744. "Specimens of Polyautography": all artists listed appear in Philipp André, *Specimens of Polyautography . . .* (1803–1807). "Pollaplasiasmos": all names are listed in Polygraphic Society, *A Catalogue of Pictures Copied or Multiplied (for Sale) by A Chymical and Mechanical Process . . .* (1792). Where present, parenthetical figures after names indicate the number of an individual's contribution to a given project.

painting techniques, known as the "Venetian secret," and the proprietary process of oil-painting replication known variously as "polygraphic" copying or "pollaplasiasmos." Figure 3.3 arrays these enterprises as a Venn diagram, bringing together Henry Fuseli, Angelica Kauffman, Philippe-Jacques de Loutherbourg, and other artists of pan-European repute with marginal figures of the British school. Drawing upon what we will see to be fragmentary, even tendentious evidence, this diagram might be taken to pledge a kind of transitive property between chemical ventures. It could be read to imply a "globally" shared favor for the use of recherché chemical preparations in all artistic pursuits, even though "local" participants such as James Barry (whose work appeared in *Specimens of Polyautography*) were long-standing opponents of chemical tricks in oil painting.[43] Treated provisionally, however, this model allows us sketch a general overlap between chemo-artistic ventures in the 1790s and the position of the Royal Academy's second president among them.

None of these episodes was more salacious than the "Venetian secret."[44] Late in 1795, the father-daughter team of Thomas and Ann Jemima Provis approached West at the suggestion of Richard Cosway, an academician, favorite of the Prince of Wales, and painter with a storied relation to the Society of Arts.[45] The Provises claimed to possess a copy of a manuscript detailing techniques used by Titian, Tintoretto (see plate 3), and their rivals in sixteenth-century Venice to produce brilliant chromatic effects. Negotiating a fee with the Provises but delaying payment, West gained exclusive access to information from the manuscript and instruction in its use by Ann Jemima Provis herself. He then prepared pictures using the Provises' procedures for the Royal Academy's 1797 exhibition at London's Somerset House, where nearly two thousand works would be seen by some fifty thousand viewers.[46] As with his lithographic *Angel*, West thematized his means of production, using the newfound "system" to visualize recovery of knowledge lost to the ages. West's *Cicero Discovering the Tomb of Archimedes* (1796–1797) opens a view into a darkened landscape with Cicero standing at center right (plate 15). Pivoting on a vertical axis of shadow that breaks his russet toga, the orator's elongated arms gather the tentative gestures of his male entourage into emphatic, triangular demonstration. Cicero's reaching fingers point to the stony tomb of Archimedes, hacked back from a canopy of vegetation by bare-chested laborers. Implied by the ominous plume of volcanic smoke rising from the sugarloaf profile of a distant Mount Aetna, nature's might registers in the urn toppled diagonally out of the left-hand corner, the cracked fragment of architectural entablature on which the workman at right plants his splayed foot, and the dark rill in the immediate foreground, separating the modern beholder from the scene's ancient actors. Yet human endeavor is not powerless before nature. It has tethered the horses ready at left; it teems in the sweaty brawn of the laborers. Surpassing the hundred and forty years between Archimedes's burial and Cicero's uncovering of his tomb, the picture implies, secrets lost at the death of Titian in 1576 could be rediscovered some two hundred twenty years later.[47]

Such uncovering was practiced by West himself in a profligate way. By sharing the technical secret with his son Raphael, the painter broke the terms of his agreement with the Provises. Consequently, in early 1797, the aggrieved Provises approached a larger group of academicians, including Joseph Farington, John Hoppner, John Opie, John Francis Rigaud, Thomas Stothard, Henry Tresham, and Richard Westall. They offered to sell by incremental subscription access to the information already passed to the Academy's president. Those seven painters signed a formal contract with the Provises in late January 1797, paying an initial ten guineas each and up to six hundred pounds sterling for disclosure of the process.

What the painters learned was a material lesson in absorption. "This secret requires a new discipline," wrote Farington, quoting West. "Those who are accurate will practice it easily, it will require *practice* also as colours sink in as they do in distemper.—The ground prepared, (a *point* of the discovery) would if seen lead to exposure of part of the System."[48] A first principle of the Provises's system, as recorded in Rigaud's transcription, demands "that the ground on which the picture is painted be of an absorbent nature, that shall imbibe the oil with which the colours are mixed and caused to adhere to the ground; and shall leave them on the surface in all their purity and beauty."[49] To make anew what had been lost to the ages, modern artists would need to remap as much their materials as their timing.

Contemporary response to the Academy's 1797 exhibition made note of these unusual methods: the specially prepared grounds of burnt umber primer mixed in thick linseed oil, spread over repeated layers of five parts of animal glue size per four parts of "Spanish brown." Several paintings in the exhibition, so reported London's *General Evening Post* in late April 1797, were rendered through "a new discovery . . . of the manner in which Titian and other Venetian Masters prepared their canvas; in consequence of which preparation, that force and brilliancy of colouring, which has so long and eminently distinguished the Venetian School of Painting, was produced."[50] But the critics were not kind. Scandalmonger Anthony Pasquin (a.k.a. John Williams) lamented the British artists' knowledge "as to the chemical effect that one body of colour, will precisely and unalterably have upon another, to answer a given purpose," as he traced the origins of their system not to Titian's Venice but to Caylus's encaustic research.[51]

Worse, the Provises were quickly revealed as operators of a confidence game, one dramatized in merciless, satirical detail by James Gillray's *TITIANUS REDIVIVUS; or, The-Seven-Wise-Men consulting the new Venetian Oracle,—a Scene in ye Academic Grove, No. I*, which appeared late in 1797 (plate 16). A crowing phoenix spreads aquiline wings at the apex of Gillray's page, grasping the mythical Venetian manuscript in its talons. The flames from which that legendary firebird ascends are eclipsed by a black obelisk of stretched canvas. Tilted away from the implied beholder to reveal its material heft, the shadowy support faces a diminutive artist at work upon it. Astride her rainbow, Ann Jemima Provis cuts white contours of an exaggerated moustache onto the canvas's darkened ground, figuring a monstrous visage of Titian as three naked Graces heft her peacock-spotted train. An angel trumpets, stars fall, and flatulent jets spray the puffery of critical praise from the buttocks of headless putti. Yet the pleiad of painters seated below on a crescent-shaped bench is oblivious to the celestial spectacle. Turning the backs of their specially prepared canvases to the beholder, these credulous, named painters (Farington, Opie, and other signatories of the ill-starred compact) stare in pool-hall concentration toward the gloomy spectacle of a decapitated Apollo Belvedere befouled by a urinating ape in a liberty cap. As Gillray's print turns exposing light away from that shady target and onto the painters' apish folly—all while implicating by name James Northcote, John Rising, William Hamilton, and other aspirants who scramble up the illusory rainbow—Reynolds himself casts away a ground of flagstones to rise from his grave at lower left.

Dead some five years before the furor unfolded, Reynolds too had to be counted as a sectarian in that chemical cult. Noting how Reynolds had been "so anxious to discover the method used by the Venetian Painters, that he destroyed some valuable ancient pictures by rubbing out the various layers of colour," critic Edmond Malone had first made the Provises' process public in the edition of Reynolds's literary works he published in early spring 1797.[52] Although the late president had never seen the mythical manuscript, Malone allowed, he "undoubtedly attained a part of this process . . . and by various methods of his own invention produced a similar, though perhaps not quite so brilliant an effect of colour."[53] Those comments were silently retracted in subsequent editions of Malone's work, but the Venetian secret lingered on in Reynolds's historiography. According to West's student Charles Robert Leslie (1794–1859), whose biography of Reynolds was completed and published by Tom Taylor in 1865, the first president "believed in the *Venetian secret* as ever an alchymist did in the *philosopher's stone*, and so intense was his love of colour, that he would always hazard the durability of his works rather than give up any chance of attaining its truth and beauty."[54] By Gillray's rendering in 1797, that chemical spirit had brought Reynolds literally up from the dark ground. Unlike the absent Christ risen from the rolled rock of West's subsequent lithograph, Reynolds appears with glasses, ear trumpet, and other frailties of his physical being clearly intact.

Disclosed in the triumphant wake of Lavoisier's Chemical Revolution, was Titian's Venetian secret meaningfully a chemical matter? Or did its vainglorious, specious claims place it instead in more instructive relation to the alchemical pursuits beloved by satirical artists? A month after Gillray depicted him slinking away from the Venetian secret's sordid spectacle, a loaded palette in his hand and trailed by cascading banknotes, Benjamin West went all in with the former interpretation. Because of Titian's elevated social status, West asserted, the Venetian painter enjoyed "easier access to the elegant studies of philosophic science; and he had prosecuted, with great ardour, the science of chemist[r]y, the better to understand the properties of colour, their homogeneous blendings, purity, and duration; as well as the properties of oils, gums, and other fluids, which might form the fittest vehicles to convey his colours upon canvass."[55] The lithographic angel

West produced some five years after the fraud of the Venetian secret was exposed might thus be seen as a resurrection of chemical hopes—the pictorial embodiment of Boyle's "Redintegration," noted in chapter 1. Rather than the failed recapture of Titian's chemical techniques, the print instead imagines a rapid replicative technology enacting rebirth from base, stony substances worthy of the alchemical theology of materials West purportedly shared with leading contemporaries in Britain.[56]

What is crucial to note, though, is that Reynolds, West, and other chemically inclined painters had already been pursuing the immortality of their pictures in material terms through collaboration with enterprising chemical replicators. One of these strategies was an invention known variously as pollaplasiasmos or polygraphic painting. In the spring of 1784, portrait painter and entrepreneur Joseph Booth began a series of bombastic publications trumpeting his replication of grand-manner oil paintings by "a mechanical and chymical process, without any touch or finishing by the hand."[57] Stressing how each pigment, varnish, and vehicle used in his process had been refined through "chymical investigations . . . [that] cost him the study of many years," Booth proclaimed that the display rooms of his Polygraphic Society (first at London's Golden Square, then on Pall Mall) could offer reproductions of Rubens, Claude Lorrain, and other old masters for pennies on the pound.[58]

Leading painters of the British school were cornerstones of the enterprise (see fig. 3.3).[59] Booth made heavy work of his debt to Reynolds, who "with a protecting hand, generously assisted him in his invention in a manner truly great and noble."[60] By the spring of 1792, Booth's venture was offering reproductions of Reynolds's *The Laughing Girl* and a portrait of Whig secretary of state Charles James Fox painted "expressly" for the Polygraphic Society.[61] West was also central to Booth's plans; a picture on the theme of Jupiter's rape of Europa by West was the first painting Booth offered to subscribers in pollaplasiasmic replication.[62] Booth's favor for other painters closely connected to Reynolds is notable as well. In addition to listing four pictures by James Northcote, Booth's sole surviving catalogue offers two historical pictures after Angelica Kauffman (1741–1807), Reynolds's close friend and a founding academician.[63] Although direct collaboration between the Polygraphic Society and the artists it reproduced cannot necessarily be inferred, the overlap between Booth's roster and the painters subsequently enmeshed in the Venetian secret is intriguing: Opie, Bourgeois, Rising, Northcote, Hamilton, West, and Reynolds figure in both.[64]

In fact, connections between polygraphic replications and the Venetian method of painting were visible to contemporaries. In a scathing survey of the 1797 exhibition, critic Pasquin noted how the Renaissance studio administered by Jacopo Bassano had prepared its canvas supports with an underlayer of white chalk and size. Bassano and his sons "kept a sort of *poly-*

graphic manufactory at Venice," Pasquin quips, "by laying a thin coat of black, in water colours, upon the other ground, that they greatly facilitated the finishing of their pictures, which was to them a circumstance of last importance, as they multiplied one subject repeatedly for foreign markets."[65] While Booth and his collaborators were highly secretive about polygraphic painting's proprietary process, a commensurate attention to grounds seems to have been significant to the enterprise. Replicating works such as Philippe de Loutherbourg's *Winter*, they appear to have fabricated a complex printing matrix for stenciling or stamping oil-painted form onto specially prepared canvases (plate 17). Recent technical analysis has identified a surprisingly thick layer of pumice mixed into Booth's primers. "If the paint was applied in blocks, using a stencil, block printing or screen printing" technique, so David Saunders and Antony Griffiths have proposed, "the role of this porous pumice-containing layer might have been to absorb some of the oil, so that the paint would be 'touch dry' more quickly, allowing an adjacent block of color to be applied without smudging."[66] Wallpaper-like in its texture, this facture is visible in Booth's rendering of *Winter*'s central figure: the seated gent in an azure coat who gazes cockily out from the picture plane as a kneeling lackey affixes his ice skates (fig. 3.4). Guided by his thickly outlined, gloved index fingers, a yellowy ground flanks two sides of his chair leg. Contrary to the logic of the picture's fiction, this underlayer reads up through the umber hue of the servant's cast shadow. It is as if the doubled fingers call attention to the invisible ground, which gives support to Booth's claim for pollaplasiasmos's chemical *and* mechanical agency.

Not all period commentators assented to polygraphic painting's chemical credentials, however. Writing to *The Monthly Magazine or British Register* a decade after Booth's 1797 death, a correspondent who claimed to have learned Booth's technique from his father declared the practice "far more mechanical than chemical." He freely grants the necessity of the following materials and procedures to polygraphic picture-making:

> Expressed and volatile oils, alcohol, acids, alkalies, acetites, earths, fossils, minerals, metals, oxyds; mineral, vegetable, and other salts; carbon, hydrogen, oxygen, and nitrogen, sulphates, &c. with the processes of trituration, levigation, cribation, distillation, sublimation, fixation, filtration, concentration, condensation, calcination, crystallization, evaporation, combination, fumigation, agitation, &c. with composition, ignition, decomposition, and more *ations* and *itions* than I can now well call to mind.[67]

Yet pollaplasiasmos can't fairly be distinguished as "chemical," by his estimate, since the above-named materials and techniques were *equally* required for what had become the standard of fine-art painting. Since the only

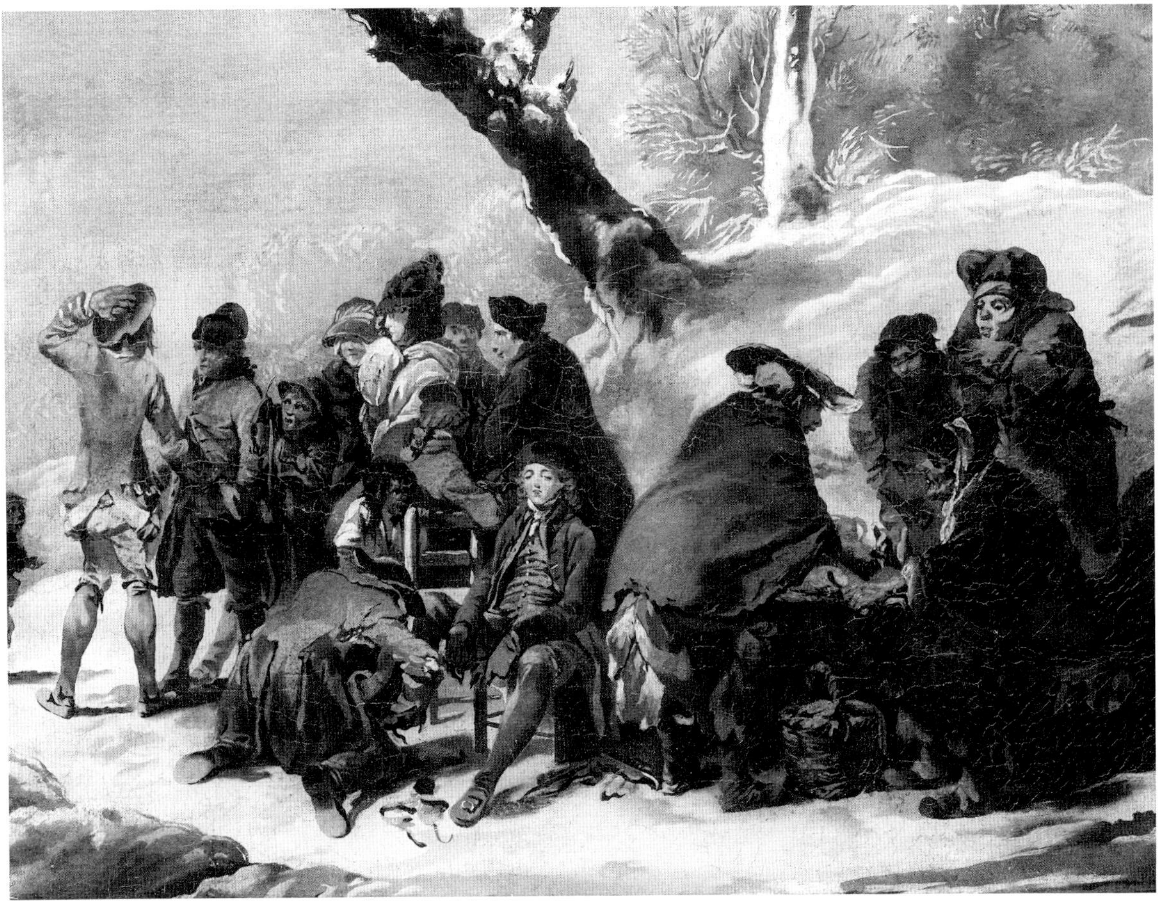

Figure 3.4 Detail of Joseph Booth, "polygraphic" print after Philippe de Loutherbourg's *Winter*, 1780–1790. Mechanical oil painting on canvas. British Museum, London. © The Trustees of the British Museum/Art Resource, NY.

salient difference at the material level between the fine artist and the polygraphic painter is that the former purchases paint media readymade from the colorman, whereas the latter "grinds his own colours," pollaplasiasmos has to be seen as a manual—thus, mechanical—art.[68]

*

It is worth pausing over the vexed speeds of Booth's chemo-mechanical images at their contested join between the fine and industrial arts. On the one hand, the entrepreneurial vision of the Polygraphic Society outlined important benefits ready to flow once paintings by leading British artists were placed in the hands of an imperial public through cut-rate, high-quality reproductions.[69] Hopefully likening London's role in the production of modern art to that of Athens in classical antiquity, Booth positioned his replicative technique as the necessary means to generate art-market elasticity.[70] "By reducing the price of commodities and manufactures," pollaplasiasmos "multiplies customers, and extends their sale."[71] Yet, cognizant of the deep continuity between their quickly made products and the changing chemical substrate of fine-art painting, Booth's publications stressed the freedom of polygraphic pictures from those high-status artworks' vulnerabilities. Pollaplasiasmos would be impervious "to changing, cracking, peeling, or any other of those inconveniencies, which frequently attend first rate pictures painted in the usual way: so that it will multiply pictures in such a manner as to perpetuate the genius, style, and effect of the most celebrated painters, to the most distant ages."[72] Historian Carl L. Becker once claimed that it was Diderot and his fellow *philosophes* who gave newfound centrality to such notions of posterity. For divine love, Enlightenment philosophers "substituted the love of humanity; for the vicarious atonement [they introduced] the perfectibility of man through his own efforts; and for the hope of immortality in another world the hope of living in the memory of future generations."[73] As much as it could open markets, the polygraphic enterprise would secure such futures in material terms. "The original pictures of the most esteemed masters may by this art be preserved for ever," Booth asserts. "And as the inventor has discovered the art of fixing his colours, he is able to imitate and preserve the appearance of age in an old picture, or give the fresh and lively tints of new ones, without the possibility of variation in the copy."[74] Impervious to change in time, Booth's replicas could arrive with time's aestheticizing effects already built in.

Immortalized *and* patinated: competition was fierce for command of that twin-speed chemical hedge. Enamel painting was a key frontier. "Immutable mirror of all that is most beautiful in the world," so asserted accomplished enamelist Jean-Etienne Liotard (1702–1789) in 1781, "painting vanquishes time with its immortal and invariable works that chain the present as they reproduce the past. Enamel painting never changes."[75] Reynolds had had little truck with Liotard's manner when the painter passed through London in the mid-1750s, and those who vied for the president's enamel perpetuation had to do so by different means.[76] Trained at a porcelain factory among Reynolds's Plymouth networks, Henry Bone (1755–1834) began exhibiting at the Royal Academy in 1781 where he showed reproductions of oil paintings in his laborious enameling technique (fig. 3.5). As with his remarkable preparatory study of Reynolds's self-portrait in academic robes, Bone prepared crisp graphic renderings of the portrait to be immortalized on a gridded, numbered sheet. Once plotted to the dimensions of a commissioned miniature (in this case, for the Prince of Wales, a steady patron), Bone's drawings could be readily scaled in size. Prepared for use and kept for reuse, a drawing would be chalked on the verso, then traced and fired onto his supports.[77] Next, Bone applied pigmented metal oxide powders suspended in vegetal oils onto metal supports, firing them at upwards of eight hundred degrees Fahrenheit.[78] Unlike oil painting where pigments

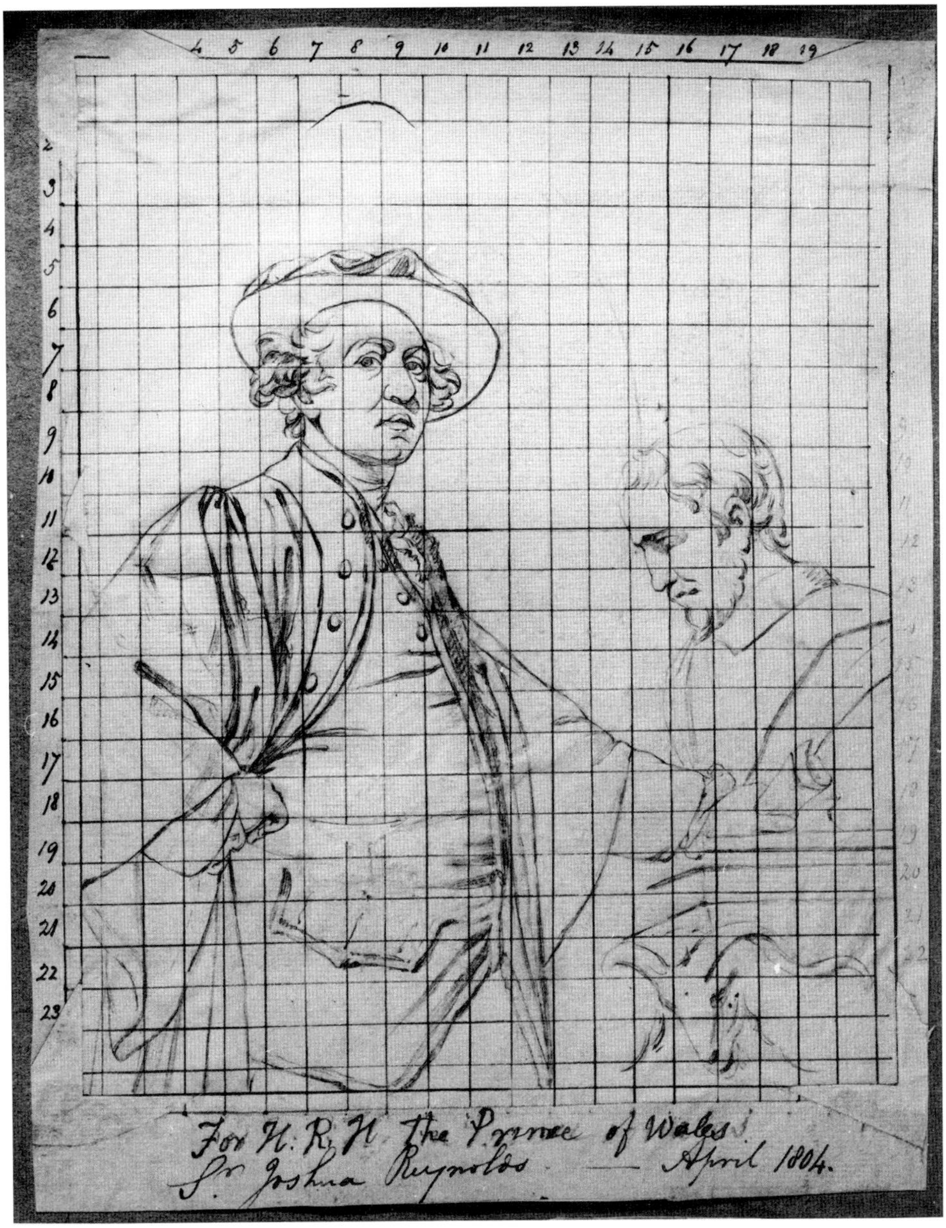

Figure 3.5 Henry Bone, *Study of Sir Joshua Reynolds*, April 1804. Pencil drawing squared in ink for transfer. National Portrait Gallery, London (NPG D17345). © National Portrait Gallery.

could easily be mixed, revised, and otherwise improvised, each chromatic layer of an enamel painting had to be fired independently and in sequence, such that the color with the highest melting point was fired first. Bone's reproductive transpositions were further enhanced by the innovative, glossy flux he added as a prophylactic fixing layer.[79] Made brilliant and durable by these repeated processes—campaigns of red heat that spectacularly "changed the colors, from blue to green, from red to brown, or otherwise," as one visitor to his studio in Soho's Berners Street put it—Bone's enamels were recognized by contemporaries as the persisting standard through which subsequent generations could know Reynolds's excellence as a colorist.[80] "As much as of the interest of Sir Joshua's pictures is annually lessened by the fading of his colours," J. T. Smith put it, ". . . the surest means of handing down to posterity that great Artist's fascinating style of colouring, [are] the correct copies which Mr. Bone has made of them in enamel."[81]

A challenger to Bone's supremacy was William Russell Birch (1755–1834). Born in Warwick, shipped to Birmingham to train in the metal works, Birch was educated by his first cousin William Russell (1740–1818), who placed him in an apprenticeship with London jeweler Thomas Jefferys around 1771. Attaching himself to London-based miniature painter Henry Spicer, Birch mined the joint vein between painting and metalwork that, under fire, became enamel.[82] Birch achieved novel effects through experiments with what he called "calcined Terrovert compounded with the soft well-melted Fluxes from crucibles of the imitative Jewellers' manufactory."[83] Late in 1784, the Society of Arts judged his new enamel colors "in Imitation of Burnt Umber in Oil Painting" worthy of the prize of a silver palette.[84] The palette's form would linger in Birch's endeavors, its ovular contours organizing the "Complete Sett" of enamelist's colors he would sign and date to 1815. Now housed in a private collection, Birch's enameled pigments stand ever-ready for use, fixed forever by fire.[85]

Enamel was a "chymical practice" ideally suited for artistic conservation: "The Unique art of hightening and preserving the beauty of tints to futurity as given in the Works of the most celebrated Masters of Painting," as Birch defines it in his autobiographical *Life and Anecdotes of William Russell Birch, Enamel Painter* (composed between 1815 and 1834). Enamel, he writes, operates "without a possibility of . . . changing." Birch made much of his service to Reynolds in that eternalizing capacity. Because Reynolds's colors suffered "much from the use of Meterials he knew himself would soon rob his pictures of their beauty," Birch contends, "there was no other Pencel in Enamel that reserved his tints before they faded then that of my own."[86] Yet, as with Booth's polygraphic painting, such chemical ingenuity compels Birch both to embrace the Reynoldsian mythology of what Alois Riegl would call "age-value" and to subtly unmoor it.[87] A key innovation of his technique was, by his account, "a thin layer of yellow Enamel under the last coat of white, before the flux for the reception of the Colours is laid on the plate; it gives warmth of colouring not otherwise to be obtained for the picture, a practice hitherto peculiar to myself and effects the beauty of Age seen in many of the old paintings in oil by the former masters."[88] Impervious to change but manufactured to render the optical effects of aging oil paintings through a distinctive yellow flux—Birch thus endorses the aura of the authoritative object passed through time even as he explains how it can be speedily simulated by chemical techniques.

What drove Birch, Booth, Senefelder, West, and many others into these complex chemical preparations and their strange games with time? Highlighting what he calls the "paradox of eighteenth-century taste," Adrian Forty has interpreted neoclassical form as a steady disavowal of social and technological worlds in accelerating motion. The British leisure classes desired "to see classical principles, and classical designs, applied to contemporary life," he argues, ". . . partly out of a wish to suppress their consciousness of its disturbing tendency to change."[89] Forty's reading finds resonance when we consider Birch's politics. For Birch penned his paean to Reynolds's enameled afterlife not in London but in greater Philadelphia, where he had arrived in 1794, bearing a letter of introduction from native-son Benjamin West. In that capital of the early United States, he was soon joined by his cousin William Russell and by Joseph Priestley. Both sympathizers of the American and French Revolutions, Russell and Priestley were, more pointedly, refugees from reactionary mob violence that had destroyed their Birmingham homes in July 1791.[90] "Dr. Phlogiston": so decried a loyalist print from that combustible summer (fig. 3.6). There, Priestley figures as a Bible-trampling agitator who braces for combat against church and crown with incendiary pamphlets alight, more chemical weapons ready in his bulging pockets. Writing decades after the fact, Birch would downplay his own politics of the 1790s—although he does confess to trafficking some five hundred copies of *The Rights of Man* (1791), Tom Paine's rejoinder to Burke's *Reflections* and a book ruled treasonous in 1792.[91] As with the immortalizing supports to chemically decaying images of Britain's ruling classes that he promoted, Birch's contradictory case compels us to see how chemical replicas and their makers could also be agents of speedy, rupturing break, per the 1790s epochal redefinition of "revolution" itself.[92]

"Such multiplication of copies, as shall place them beyond the reach of accident"

When teaching in the 1760s at the catalyst of dissenting thought that was the Warrington Academy, Joseph Priestley had devised an ingenious tool for rendering time visible to students of history: "The abstract idea of Time, though it be not the object of any of our senses, . . . admits of a natural and easy representation in our minds by the idea of a measurable space, and particularly that of a Line."[93] Unfolded from his octavo volume *A Description of a Chart of Biography* (1765), this linear vision of time was dwarfed by the three-foot-by-two-foot wall chart Priestley simultaneously designed to encompass some two thousand names and nearly three thousand years of

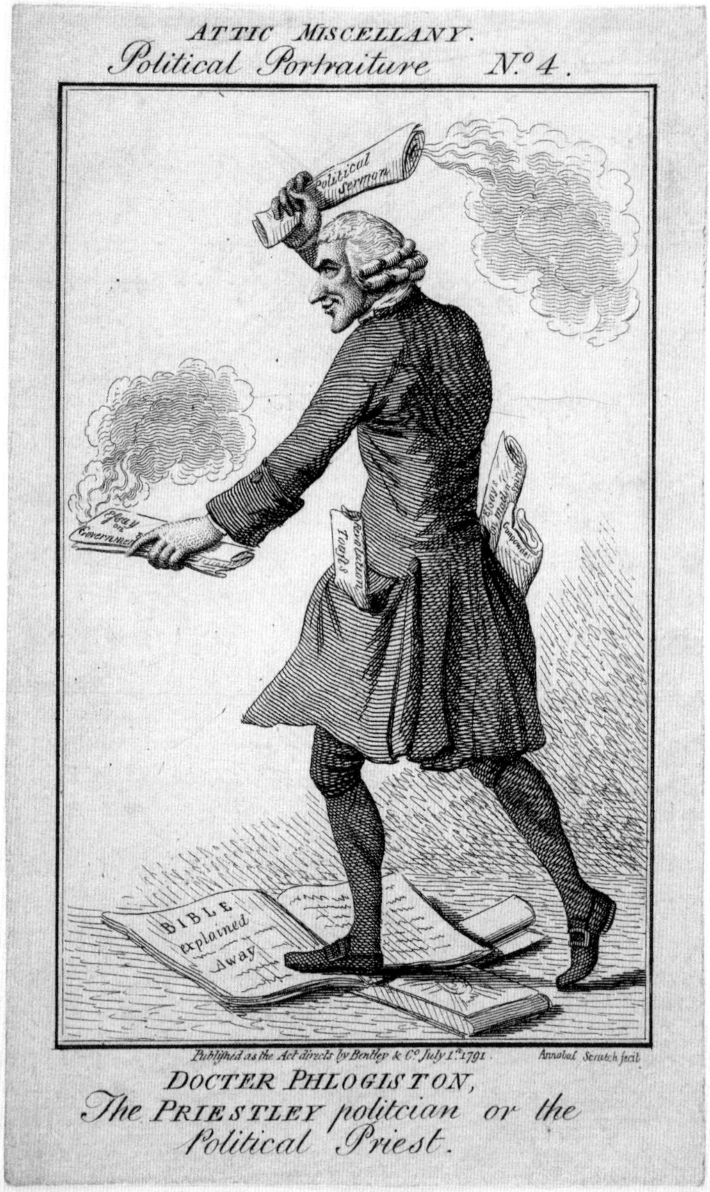

Figure 3.6 Annabel Scratch, "Doctor Phlogiston, the PRIESTLEY politician or the Political Priest," July 1, 1791. From *Attic Miscellany./Political Portraiture N° 4*, "Annabal Scratch fecit," "Published as the Act directs by Bentley & Co." (In British Museum copy of *Attic Miscellany*, "W. Locke" is substituted for "Bentley & Co.") British Museum, London (1873,0712.1150). © The Trustees of the British Museum/Art Resource, NY.

history (fig. 3.7). Nonetheless, prototype specimen and full chart share core principles. Both render time's movement from left to right in even punctuations of decades. Thin, horizontal tracks of ink denote the span of individual lives. "The first chronographer to conceptualize his charts in terms similar to those of scientific illustration," so argue recent interpreters, Priestley "was the first to lay out systematic principles for the translation of historical data into a visual medium."[94] The contemporaneity of great figures active across geographical region reads easily when history's horizontal progress is split into vertical registers of human endeavor. The system also concretizes a primordial intuition about time itself. "*Long* and *short* are so univer-

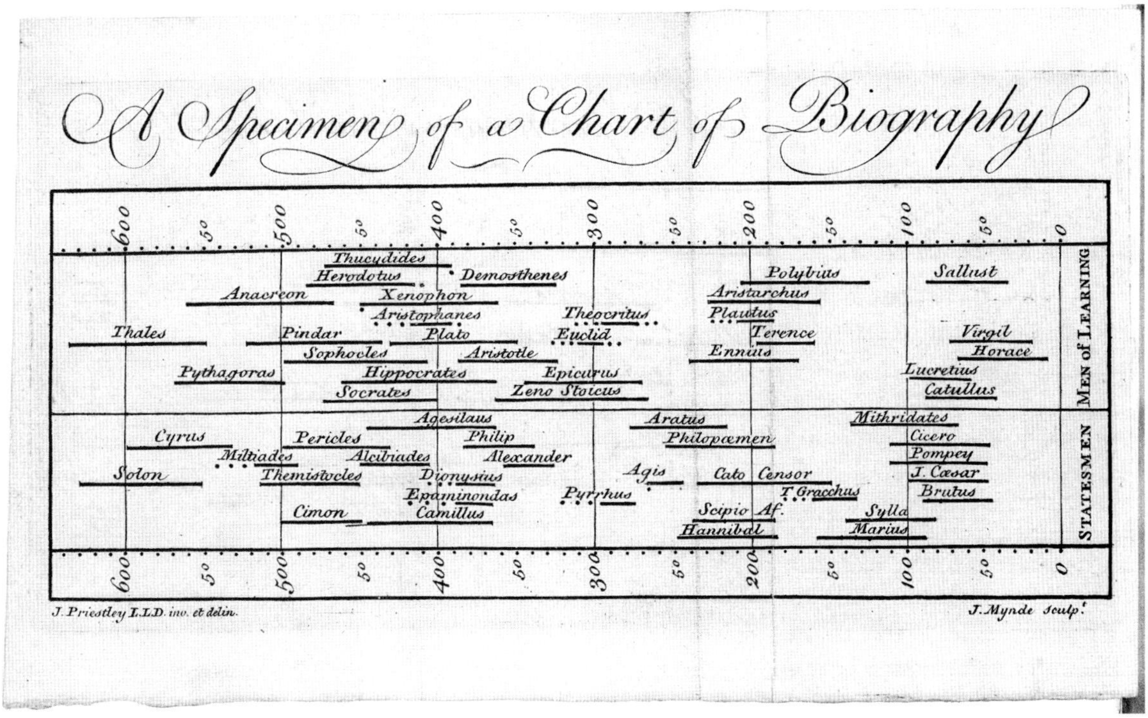

Figure 3.7 J. Mynde after Joseph Priestley, "A Specimen of a Chart of Biography," as published in Joseph Priestley, *A Description of a Chart of Biography . . .* , 2nd ed. (1765).

sally applied to time," Priestley proclaims, "that . . . it never occurs to us that there is any figure in the use of them, and that they are borrowed from any other subject." Imprinted onto the mind as the printmaker's cuts pressed ink into paper, a linear sweep of time would become "fixed in the imagination, not in succession, but at once," Priestley promised. Linear time lodged with such force that the student "can never wholly lose the idea of it."[95]

Polygraphic printer Joseph Booth made his own ventures into history. Touting the social benefits of chemo-mechanical paintings in 1788, Booth likened his newly invented technique to the cotton mills introduced into Lancashire—that crucible of the Midlands's emergent Industrial Revolution—that Alois Senefelder would eagerly eye for the application of his lithographic chemistry.[96] Booth's version of history acknowledges how the mills "at first excited among the laboring manufacturers of cotton, yarn, and stuffs in Lancashire a general alarm." While pollaplasiasmos might too look like a threat to the oil painter's livelihood, Booth claims, industrial history teaches otherwise:

> The extension of trade which those mills have occasioned, has rather increased the number of hands employed in the manufactures of cotton than diminished them. In like manner, may we not fairly suppose, that the multiplication of pictures . . . may cherish and diffuse a general taste for

painting, and thereby give employment and encouragement to the masters and adepts in that imitative art?[97]

In point of fact, Booth may well have taken his technique from another site of Midlands industrial production: Matthew Boulton's Soho Manufactory of greater Birmingham. Founded in 1765 and soon employing a thousand men on the outskirts of a city that doubled its population (to forty-eight thousand) in the eighteenth century's third quarter, Soho was arguably the largest factory in the Western world prior to the industrial expansion of Booth's Lancashire cotton mills at century's end.[98] In 1776 Boulton had begun what was planned as a fourteen-year partnership with Edward Jee, John Eginton, and his brother Francis Eginton (a leading designer and glass-painter at Boulton's Soho site) in the production of mechanical paintings.[99] As would Booth's Polygraphic Society a decade later, the Jee/Eginton/Boulton collaboration foregrounded replicas after leading academicians, including Reynolds, West, and, especially, Angelica Kauffman, vending their wares as decorative furniture for ceilings and coaches (see plate 20, figs. 4.7–4.13).[100] The products of that Soho endeavor will figure centrally in the following chapter, but the partnership itself was a financial disappointment; it had been abandoned by 1781.[101]

Far greater worldly success was achieved by a more prosaic conjunction of temporalized chemical ingenuity, also born of a Soho partnership: James Watt's copying machine. Relocating to Boulton's Soho from his native Scotland in 1774 with the aim of exploiting his patent on the steam engine's independent condenser, Watt (1736–1819) developed an instrument for duplicating writings and drawings that became a standard article of bureaucratic culture well into the twentieth century.[102] Watt characterized his invention—mythically sprung from a desire to best the double-nibbed pen or "bigrapher" developed by fellow Lunar Society member Erasmus Darwin in the late 1770s—in unambiguously chemical terms: "I have fallen on a way of copying writing chemically, which beats your bigrapher hollow."[103] More practicable and speedier than Darwin's mechanical pantograph, Watt's copy press spoke to a consolidating vision in which records would be the public property of posterity. No more powerfully was this sentiment articulated than by American statesman Thomas Jefferson (himself an avid user and improver of Watt's copier) as he observed the careless handling of the early United States' founding documents. "Let us save what remains," Jefferson wrote to former postmaster general Ebenezer Hazard in 1791, "not by vaults and locks which fence them from the public eye and use, in consigning them to the waste of time, but by such multiplication of copies, as shall place them beyond the reach of accident."[104]

Patented by Watt in May 1780, the copying press went into industrial production under the name of James Watt & Co., a partnership held by Watt,

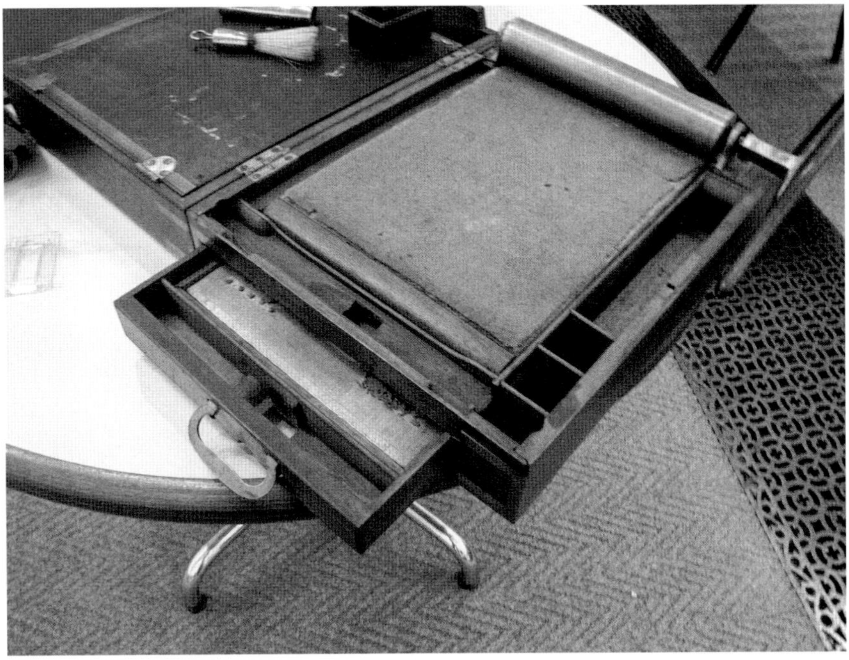

Figure 3.8 Portable copying press, fitted for use, manufactured by James Watt & Co. of Birmingham. McLean Museum, Inverclyde Council, Greenock, Scotland (1917-09-00/1).

Boulton, and chemist James Keir. By the early 1780s, the Watt & Co. copying press was in use by John Smeaton for replicating engineering drawings.[105] Operation of the company shifted in the mid-1790s to sons of the Soho operation, James Watt Jr. and Matthew Robinson Boulton, who developed a portable version of the device just as Watt's original patent expired. A sleek piece of executive gadgetry, the portable press is a tabletop device with a keyhole tooled below the seam encircling its horizontal waist. Brass handles set into the copier's sides collapse to run flush with the polished contours when the device is not in use. Once opened, two opposing grounds lie before the user: a black, oilcloth-covered panel on the underside of the lid and a green felted ground on the near side of the central spine (fig. 3.8). The black ground is itself hinged; releasing two spring-loaded hasps, the user can access a well or "secretary" designed to accommodate a cache of specially prepared materials. Copying papers, dried ink cakes (fig. 3.9), an inkwell manufactured at Josiah Wedgwood's Etruria, "Drying" and "Wetting" books, and writing implements—all were included inside.

Kept aloft upside down when the copier is closed, this black-bound secretary is the dry companion to the moist, green ground below it. At the head of the felt are two rollers (available in various materials, at different price points) to which a crank can be attached to rotate in circular motion. With the rectangular crankshaft removed from the foot of the felt, the user extracts a tray bearing a metal-lined chamber called the damping box

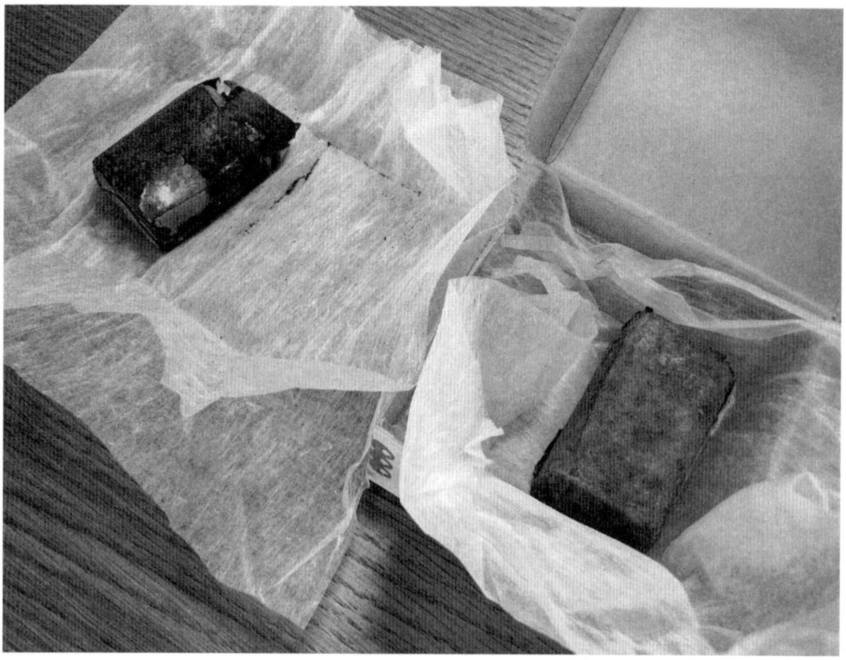

Figure 3.9 Ink manufactured for use with James Watt & Co.'s portable copying press. Archives of Soho, Library of Birmingham (MS 3147/20/11).

where papers can be kept premoistened and copy-ready for upward of two weeks. The device works like this. Preparing the proprietary ink sold with the device in a specified proportion of water, the user writes a letter on well-sized paper.[106] Once the document has dried thoroughly (but no more than twenty-four hours later), a moistened sheet of copying paper is withdrawn from the damping box and placed on the face of the dried letter. Dry original and moistened replicator are sandwiched between sheets of oiled pasteboard, gently fed and pressed between the two rollers. "*The greatest art in taking an impression*," the manufacturer's instruction manual advises, "*is to turn it very slowly*."[107] Removing the pressed papers, teasing the copy away from the original, the replica is placed in the drying book and pressed again under an even weight. Crucial to the success of the technique is its controlled transfer of ink under pressure. While the dried copy will show written characters of the original letter in reverse, the copying paper is so thin that the supplied ink bleeds through it; the text is legible simply by flipping the replica over and reading the verso. The device is thus a multiplier, but not infinitely so. Should the letter writer wish to control the duplication of the original, the user is instructed to make "*a Copy himself* before he lets the Writing go out of his Possession; for, *as only one Impression can be taken*, by this Method, any Person, who uses this Invention, will infallibly prevent others from obtaining a Copy."[108]

The materials supporting the copying press—the paper and proprietary ink—were fundamental to its success. From the early 1780s, Watt worked intensively with London-based stationer James Woodmason to source and perfect copying papers. The challenge was not simply to produce quantities sufficient for the copiers' expanding markets, but to resolve what Watt called the "very ticklish matter" of crafting papers that would receive the moistened ink's transfer without spreading and staining uncontrollably.[109] To resolve these challenges, Watt made contact with at least five papermakers, including the operation of James Whatman II. At the forefront of several important developments in the production of high-quality artists' papers, the Whatman firm engineered against color-change in time, adding blue pigments to its rag-linen papers from 1765 to correct against their eventual yellowing.[110]

Watt developed chemical additives of his own. These he sent to London, with instructions to Woodmason, not on how to make the concoctions, but on how to incorporate them into the paper. As reconstructed by historians Barbara Rhodes and William W. Streeter, "the copying tissue was impregnated with a solution containing tannins. When the faintly colored, partially reacted ink on the original document penetrated into the copying tissue, it would theoretically react with the tannins therein, and eventually oxidize to form the dark insoluble ferric precipitate within the fibers of the paper."[111] Referring to these chemical additives by cryptic or abbreviated titles ("SHG," "C+," "G+"), Watt directed Woodmason in how he was to apply the solutions.[112] "I send by Coach this night 4 8 oz. vials of Strong Preparation," Watt wrote in early September 1780, "each of which must be mixed with 30 Gallons of pure water with which the paper must be prepared. I have made up this preparation myself, and believe you will find it have much less colours than the former and also be less subject to spread."[113] Crucial as collaboration with Woodmason was in producing the needed papers, the specifics of their chemical composition were closely guarded by Watt and Keir at Soho.

Equally vital to the success of the copier was its ink. If perversely absent from recent historiography of the book, as Adrian Johns has noted, ink was fundamental infrastructure to the Enlightenment's Republic of Letters.[114] Recipes for making it abound in commonplace books of the long eighteenth century.[115] Cognizant of that fact, the instruction manual supplied with James Watt & Co.'s portable press urges caution: "As no article is prepared in so many various ways, the greater part of which are unfit for producing good ink, we have found ourselves under the necessity of preparing an ink for the use of our friends."[116] In particular, the manual highlights the unlikely visual appearance that their special preparation initially yields: "This ink will be very pale and brown at first, but will become black by keeping; the writing and copies done with it, will both become black in time."[117] Understood in modern terms, the metallic salts in the ink-preparation mixed with tannins from gall nuts would form ferrous tannate and then blacken into enduring visibility. "Writing would be done with this almost colorless solution of newly prepared ink (*i.e.* the ferrous tannate)," as Rhodes and Streeter put it, "which would penetrate into the fibers of the paper. There the newly written text would oxidize to form deep black, insoluble letters (the ferric form of the compound) within a few days."[118]

In private correspondence, the company acknowledged to Woodmason (who was required to pay for the privilege of vending the ink to the device's metropolitan retailers) the challenge of adjusting customers' expectations to the visual appearance of copier-machine ink precisely because it would change in time. The stationer was asked to send to Soho "a bottle of Westerman's ink by the Coach [so] that we may compare it with ours, . . . we believe ours is better for the reason that Westerman's is apt to be preferred which is as you say that it is blacker at first."[119] Against rival inks that would appeal immediately but then fade, Woodmason's mission was to accustom users to a preparation that appeared pale in use but blackened attractively in time. Many other features of the ink—from its reconstitution by the user to its viscosity under the pen—remained objects of laborious refinement. "The number of faults to which the ink is liable," the company lamented in the autumn of 1780, "is so great that by remedying one fault we are apt to fall into another."[120]

Numerous, perhaps intentionally misleading recipes for making the ink were developed and distributed by James Watt & Co.[121] An undated draft entitled "Directions for making Copying Ink & for Copying Drawings for which this ink is used" reads this way:

> The Ink proper to be used is thus made—Gum Arabic 240 grs or ½ oz.—added to (& together well pounded) 20 grs of Spanish Liquorice—these dissolved in 720 grs. or 1½ oz. of water; thus made, should be boiled by more boiling water. Then take 60 grains, or ⅛ oz. of best lamp black, mix this with one tea spoonful of Lisbon white wine—rum or brandy—& also add two tea spoonfulls of the liquid first mentioned.[122]

Gradually combining these liquids, the maker was to strain the ink through muslin and then keep it sealed for upward of four months. The practical chemistry specified for the ink's production (and literally built into the paper support through which it would move) is then foregrounded at the heart of the copying process. Noting that the copy press could be used to replicate drawings as well as written documents, the text instructs the user to moisten the original images as well as the paper prepared to receive them. Sandwiching the moistened original and copy paper between bolstered pasteboards, the operator passes "the board *very slowly* (in order to give the water on the paper time to dissolve the gum of the ink, for if the board be moved twice, there is a chance of the moving of the papers & a double copy on the same side of the copying paper) under the rollers."[123] The copy machine acts not only by exerting physical pressure by way of rollers that transfer ink from original image to replicator-surface. Rather, it requires those rollers to be operated sufficiently slowly to enable the action of water to dissolve the gum in an ink designed to darken in time.

Water and its delayed actions were matters of no small interest to Watt in the 1780s. Expanding this point allows us both to situate Watt in the larger intellectual milieu of the Midlands' Lunar Society and to foreground his chemical commitments, which will return in the following chapter. Soon after relocating to Birmingham in 1780, Priestley had begun experimenting with earthenware retorts made expressly for him by Lunar Society associate Josiah Wedgwood.[124] Ostensibly, Priestley was replicating experiments of London-based chemist Henry Cavendish, wherein mixtures of dephlogisticated air (oxygen) and inflammable air (hydrogen) were exploded inside receiving vessels. Priestley, Cavendish, and other researchers sought to understand the yield of such explosions: their unexpectedly loud sound, alterations in weight of the materials and apparatus they produced, and the identity of the dewy substance found deposited on the interiors of the experimental receivers after firing. Upon learning of Priestley's research in late 1782, Watt interpreted the experiments into terms of latent heat, a concept central to the Scottish tradition of philosophical chemistry that he shared with Joseph Black, professor in the University of Edinburgh's vaunted medical faculty and a frequent correspondent with Watt on the chemistry of the copying press.[125] Asking why a block of ice does not immediately melt on exposure to warmth, Black had developed his doctrine of latent heat. Changes in chemical state, he argued in the 1750s, require far greater transfer of heat than registers with static thermometer measurement. In contrast to available measures of sensible heat, Black's concept of latent heat foregrounded the dynamic transfer of heat from, as historian of science Henry Guerlac elaborates, "one body to another. It is the *time* involved in the process of bringing bodies to a given temperature that gives a clue to the relative amount of heat lost or taken up by a body."[126]

In Priestley's experiments on the chemistry of airs in Wedgwood retorts, Watt saw latent heat's possibilities radicalized. As he then proposed to Black, "Steam parts with its latent heat as it acquires sensible heat or is more compressed . . . when it arrives at a certain point it will have no latent heat and . . . will again change its state and become something other than steam or water."[127] Watt's thoughts on the possibility that time- and heat-intensive conversions could not only transform water's state but could also decompose it into component parts would figure centrally in subsequent debates advancing his priority in the claim for water as a chemical compound, not an irreducible element.[128] Such high-level debates need not be assigned to the practical use of the copying press, but they serve to underscore the context, circa 1780, in which the device for "copying writing chemically" was explicitly forged and refined. Its ink dissolving through the newly minted compound of water, then darkening handsomely in time by progressive, photosensitive response to its oxidized ferrous particles in a chemically engineered paper support: the copy press, like the inkstand sold with it, bears intimate relation to cultures of chemical experiment at Wedgwood's Etruria.

The Cup of Time

Change, progress, betterment in time: those, argued philosopher William Godwin in *An Enquiry Concerning Political Justice, and Its Influence on General Virtue and Happiness* (1793), were inexorable tendencies of individual human life. So why do vast portions of humanity remain poor, ignorant, exploited by a few idle rich who lord the resources? Godwin cast unjust governments as the retarding stay on human happiness. Such governance "prompts us to seek the public welfare, not in innovation and improvement," Godwin charges, "but in a timid reverence for the decisions of our ancestors, as if it were the nature of mind always to degenerate, and never to advance."[129] In mid-November 1795, James Watt & Co. received a return on its own chemical-copying innovation destined for none other than Godwin.[130] The return came from Thomas Wedgwood, youngest son of pottery industrialist Josiah Wedgwood, who addressed his complaint to James Watt Jr.

Both sons of Lunar Society luminaries, the correspondents knew each other from then-recent travels in revolutionary Paris. There, Watt Jr. had been an ardent supporter of the Jacobin faction; he had been introduced to the club by Robespierre himself—and publicly denounced for it by Edmund Burke on the floor of Parliament.[131] Amid such *Sturm und Drang*, the exchange that unfolds around Wedgwood's efforts to gift a Watt copier to Godwin seems especially absurd.[132] Wedgwood objects to Watt that the copying press sent to him was fitted with rollers of brass, "which you recollect informing me were 'worse & more expensive.'" Instead, Watt is instructed to dispatch a copier with iron rollers to Wedgwood's address in Soho. A week later, Wedgwood writes to Watt again; he has now received the replacement copier, but not at the desired foolscap paper size.[133]A third machine, sent in an additional week's time, has, like the first, the unwanted bronze rollers. Wedgwood returns it too, along with the following note to Watt: "Sorry to give any farther trouble respecting the copying machine; but I find you have now sent one with *bronze rolls*. I will repeat the order. A portable copying machine with iron rolls, foolscap size ½ ream Foolscap paper, ½ ream quarto paper."[134] A final letter complicates this comedy of logistical errors

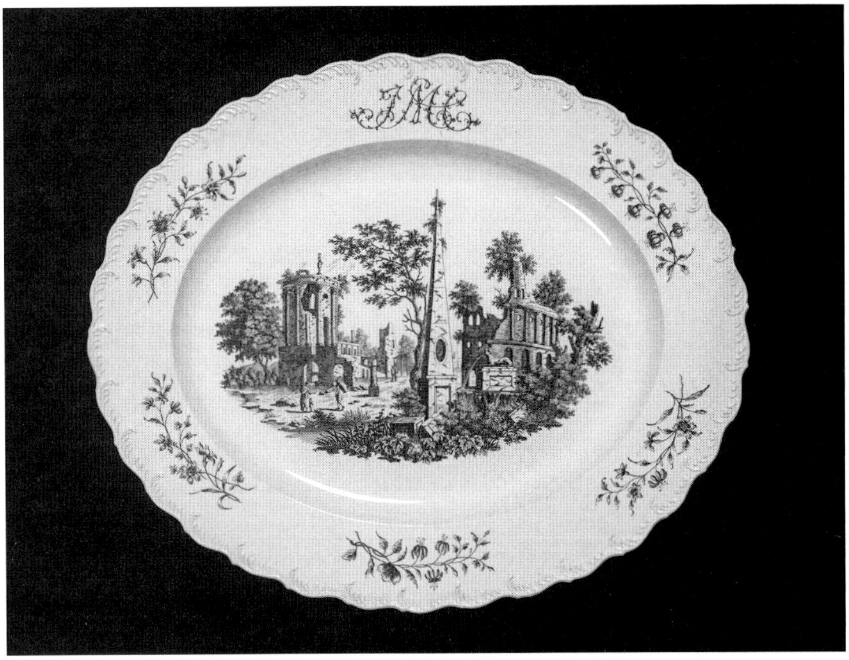

Figure 3.10 Oval earthenware dish transfer-printed in black, ca. 1780. Made by Josiah Wedgwood and Sons Ltd., Etruria; printed in Liverpool by Sadler and Green. Victoria & Albert Museum, London (414:1133-1885). © Victoria and Albert Museum.

still further, as Wedgwood acknowledges his own complicity in the confusion: "The fact was, I only just looked with the box without taking out the machine & therefore did not discover the error in size."[135]

Wedgwood's persistent, if not always attentive, interest in obtaining a copying device speaks to the high regard for technologies of expedient chemo-mechanical transfer actively cultivated at Etruria, the manufacturing site Josiah Wedgwood and partner Thomas Bentley established on the outskirts of Stoke-on-Trent in 1769.[136] An inveterate experimentalist, Josiah Wedgwood embraced numerous innovations, including the new technique of transfer printing for ornamenting fired earthenware (fig. 3.10). Since 1761, he had contracted with the Liverpool-based partnership of John Sadler and Guy Green, who replaced hand painting with a two-stage "bat printing" process.[137] Transfer printing offered an "enlightened" industrialist like Wedgwood significant opportunities to control and quickly change his wares' iconographic features. Correspondence shows him not only updating print sources to keep pace with metropolitan fashion, but employing a partnership capable of unrivaled speed. Sadler and Green could reportedly do more in six hours than could one hundred skilled pot painters.[138] And through this Liverpool partnership, Wedgwood and Bentley were introduced to Peter Perez Burdett, who claimed secretive possession of "a chemical, & not a mechanical operation" for replicating designs.[139]

A skilled mapmaker, Burdett is best remembered for his friendship and collaboration with Joseph Wright of Derby, who was also living in Liverpool in the early 1770s.[140] Likely draftsman for the sketch of Wright's *Alchymist* (see fig. 2.14), Burdett developed a version of the aquatint process known in France since the late 1750s. Therein, acid-resistant rosin was dispersed over an etching plate, which was then suspended into aqua fortis, and variously bitten and inked or masked to fashion richly modulated, chiaroscuro renderings of images like Wright's *Two Boys Blowing a Bladder by Candlelight* (fig. 3.11).[141] "The effect of a stained drawing attempted, by printing from a plate wrought chemically, without the use of any instrument of sculpture": such was the rubric under which Burdett exhibited his aquatints at London's Society of Artists of Great Britain in 1773.[142] Wedgwood was initially impressed with Burdett's technique, which he perceived to offer as much superior speed and delicacy as "a *new effect*" to transfer-printed ware.[143] He was reluctant, however, to pay Burdett for explaining a process susceptible to discovery by practical experiment, should the enquirer "provide himself with every dissolvent of copper, acid or alkaline, combine & mix them in various proportions . . . brush a little of each upon a copper plate & let it rust a longer or shorter time."[144] Wedgwood's commonplace book indeed records multiple accounts of how to perform "Aqua-tint Engraving."[145] Aggrieved, Burdett quarreled with Wedgwood over payment before fleeing his Liverpool creditors for the Continent in late 1774.

Wedgwood was highly secretive about his own technical experiments. He devised a numerical system of encryption to identify the material substances employed in his exhaustive trials of body, glaze, and other features of fired earthenware.[146] Perhaps no secret was more closely guarded than Wedgwood's own version of encaustic painting, the sole process he ever patented. In the partnership's first catalogue from 1773, Wedgwood and Bentley would cite as inspiration the methods of painting on "Etruscan" vases published by Sir William Hamilton and then donated to the British Museum—pictorial techniques supposedly lost since the era of Pliny the Elder's death at the eruption of Mount Vesuvius.[147] Contrary to the wax-based preparations descended from Caylus, the catalogue proclaims, "our Encaustic *Colours* can be applied with great Ease and certainty; they change very little in the Fire; are not liable to run out of Drawing; are perfectly durable, and not glassy; they have all the Advantages of Enamel, without its essential Defects."[148]

Skeletal details on that technique are provided in the patent specification for encaustic that Wedgwood signed on November 16, 1769. Stipulating the proportions of "Ayoree" clay, which Wedgwood had sourced from Cherokee peoples of modern-day North Carolina, to be combined with powdered bronze dissolved with gold in aqua regia (a mixture of nitric and hydrochloric acids), antimony ground with tin ashes and white lead, and

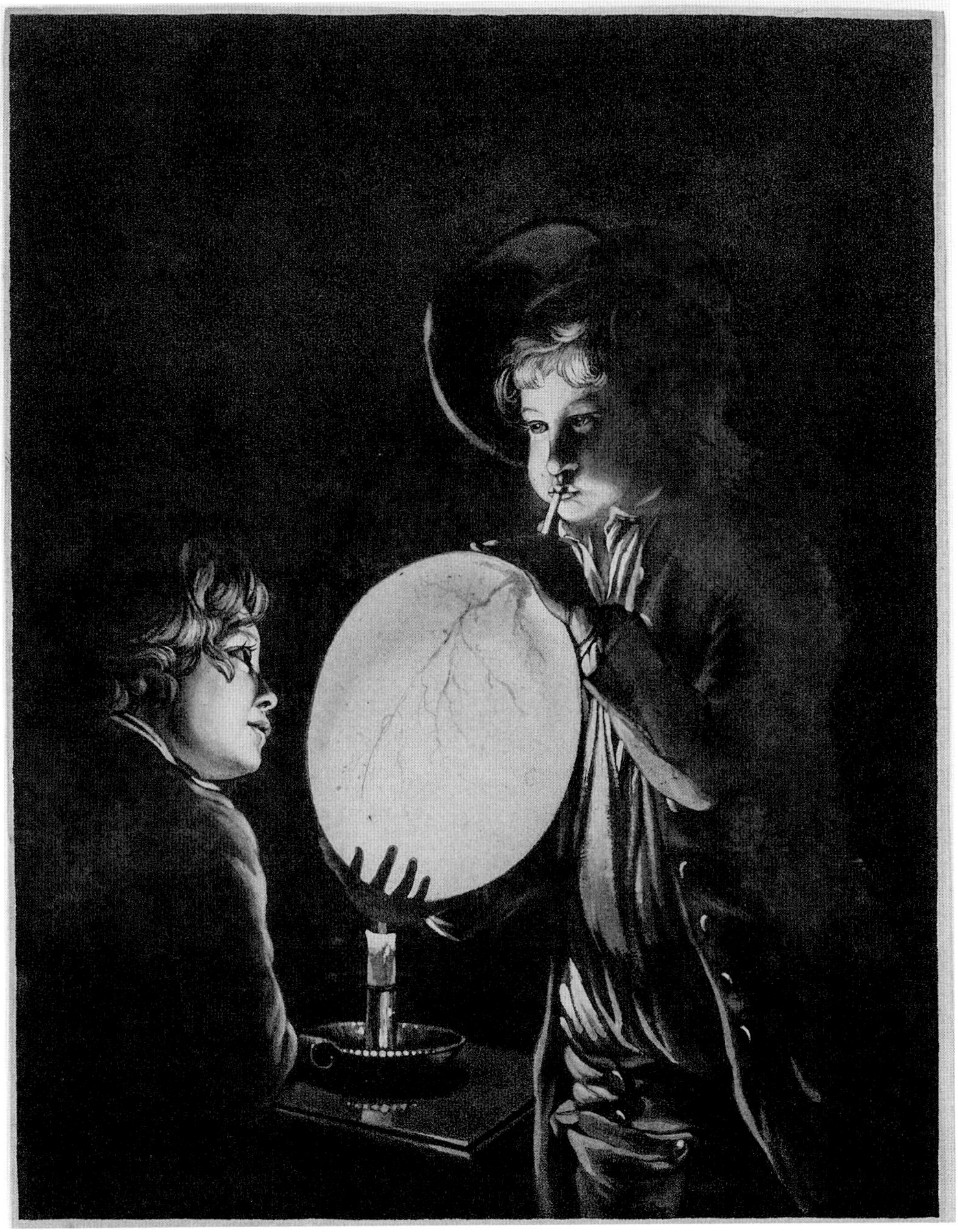

Figure 3.11 Peter Perez Burdett, *Two Boys Blowing a Bladder by Candlelight*, 1773. Aquatint. British Museum, London. © The Trustees of the British Museum/Art Resource, NY.

other pigmenting materials, the patent indicates steps needed for painting, shading, and firing designs in red upon a dead coloring ground of black biscuit (fig. 3.12). It is an open question, though, whether knowing that the pigments were to be "ground in oyl of turpentine, and burnt upon the vessels in a muffle or enamel kiln" could offer anything more than an accent of

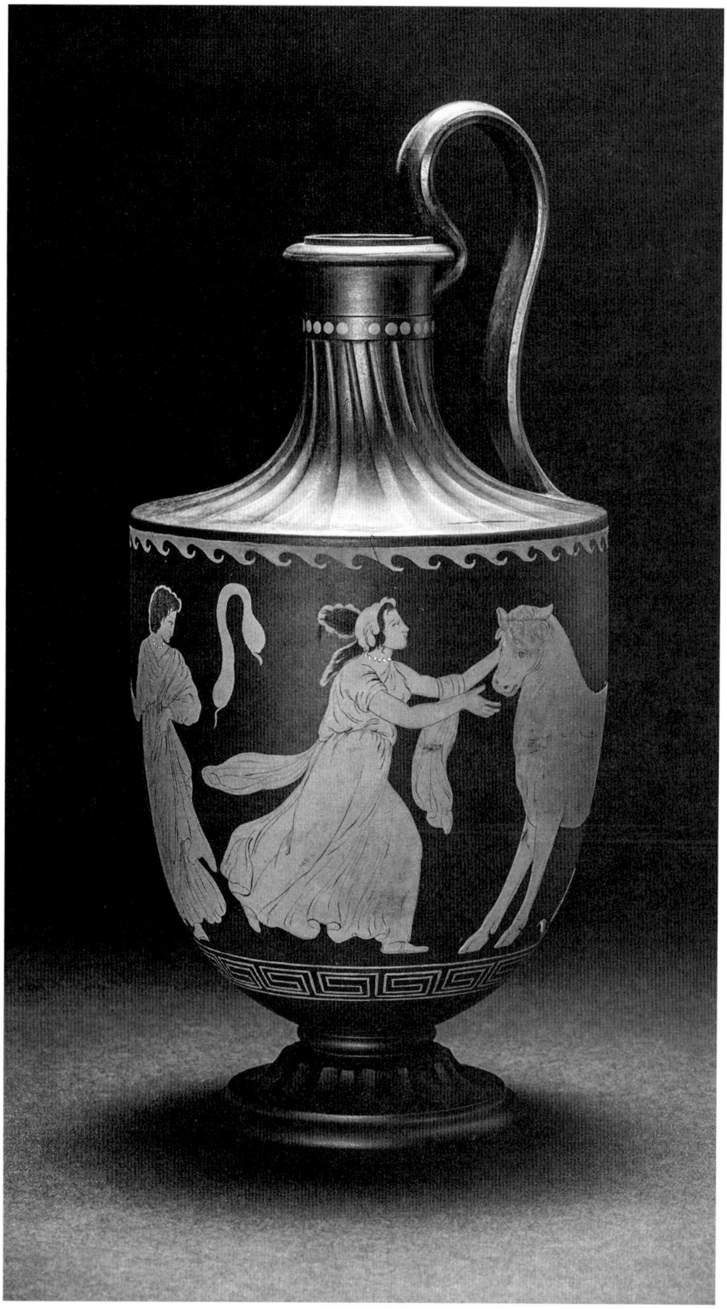

Figure 3.12 Wedgwood and Bentley, black basalt ware ewer with encaustic painting of Europa and the bull, ca. 1770. British Museum, London (1909,1201.121). © The Trustees of the British Museum/Art Resource, NY.

marvel to a beholder of the factory's encaustic-painted works.[149] Certainly, the opportunity to work in this ingeniously reclaimed and seemingly timeless medium was attractive to many artists. Prior to their rift, Wedgwood had extolled how Burdett was "very fond of the idea of having some of his most capital performances rend'rd imortal by our encaustic painting."[150]

Chemical processes that could render timeless effects or immortal subjects were not the only preparations of interest to Wedgwood, however.[151]

Writing in January 1779, he addressed color change as a thoroughly chemical problem. "Colours which take place in bodies in the fire—change with a quick succession, or totally disappear, by a continuance or small increase of the heat which produced them," he noted in his commonplace book, "are called by some of the chemists *Passing colours*." Wedgwood took these fugitive chromatic states as sites for industrial exploitation:

> From some experiments lately made, I have been led to extend the idea of *passing colours*, much farther, perhaps, than the chemists who have hitherto treated them would warrant me; and apply it to the discharging even of metallic colours from our clays & stones in which we wish to have the purest white; and particularly to the Cornish moorstone & clay, which are so much impregnated with iron as to burn in our kilns to a purplish brown color.[152]

Thoughts like these led Wedgwood to consideration of the causes of coloration in stone in the first place. A contribution to his commonplace book likely cowritten with Alexander Chisholm (the Aberdeen-born chemist who, as noted in chapter 2, came to Etruria in the early 1780s after the death of William Lewis) sets out an explanation of lithic pigmentation.[153] Carried in solution and then deposited by water, calcareous sediments "attract one another and unite firmly together, into crusts or masses," slowly forming stones with brilliant chromatic properties. Other pigmented effects occurred absent these causal circumstances. Exposed to "dry air" for several years, mounds of fine white potters' clay would turn brown. Broken open, the clay's discoloration could be found penetrating into the earth not "gradually paler or more dilute, the further it advances inward; but on the contrary it forms rings or concentric circles of different shades, some of the inner ones of which are deeper than the outer." Equally, flint extracted from mines and left exposed to the elements would gradually become useless as it came to exhibit iron's reddish hue, which would be communicated into fired wares. The obvious explanation for such phenomena, Wedgwood proposes, is a decay of the stone through chemical contact with air so as to make "the exterior part porous and permeable to the subtile ferrugineous earth which appears to be universally diffused."[154]

By Wedgwood's own terms, that explanation would hardly do. Concentric, colored rings in stones like flint did not simply propagate from the outside in, but emerged from the center outward. A more plausible explanation, then, would assign color change not to air but to light. "It would still be difficult to account for the regular correspondence of the coloured circles to the surface of the pebble," the document proposes, "unless we could suppose that light, whose action is known to influence remarkably the production of color in certain metallic solutions and precipitates, <may> penetrate these semitransparent stones, perhaps be concentrated in virtue of their form, and contribute to the production of the iron calx and its colour."[155] Light, Wedgwood speculates, must effect chemical actions even through lithic layers that become visible over extended periods of time.[156]

*

No one in Wedgwood's circles was more engaged with that nest of chemoluminous problems than Godwin's would-be friend Tom Wedgwood. Drawn to metaphysical philosophy during his years at the University of Edinburgh in the late 1780s, Wedgwood had been immersed in chemistry from his youth. Through progressive, home-based pedagogy developed by his father and Erasmus Darwin, Wedgwood had been taught primarily by chemist Chisholm, with whom he continued correspondence while at university.[157] Chemical cure, collaboration, and collapse have all shaped Wedgwood's fate among historical interpreters. As therapy for his chronic illness, Wedgwood sought treatment with nitrous oxide and other psychotropic preparations at Thomas Beddoes's controversial Pneumatic Institution at Bristol in the years around 1798.[158] There, he met youthful chemist Humphry Davy (1778–1829), who published "An Account of a Method of Copying Paintings upon Glass, and of Making Profiles, by the Agency of Light upon Nitrate of Silver. Invented by T. Wedgwood, Esq. with Observations by H. Davy" in 1802.[159] That paper has secured Wedgwood an enduring place in the history of photography, while also condemning him to failure. Cementing available opinion, Talbot declared in *The Pencil of Nature* (1844) that Wedgwood and Davy had to be counted as "the first inventors of the Photographic Art."[160] Such status was based on their supposed pursuit of, but inability to enact, aims Talbot's photographic processes had since putatively brought to practice.[161] Where, in 1844, Talbot could place a vegetal specimen on paper soaked in a solution of table salt and brushed with silver nitrate to darken on contact with sunlight, and then desensitize the surface with a sodium hyposulphide fixative, Wedgwood and Davy "found it impossible to fix or preserve those pictures: all their numerous attempts to do so having failed."[162]

As with his father's meditations on industrial exploitation of chemists' passing colors, however, Tom Wedgwood's interests were hardly restricted to fixed chemical images. In an undated note (fig. 3.13), Wedgwood recalls a dream in and about printing: "Dreamt of looking at a print. Wakened half open my eyes, perceived the image of the print. Without being sensible of any change in my state of mind or body, the objects of the picture began to extend their edges which also became indistinct."[163] Wavering pencil lines denote this oneiric episode as successive printed states. At left, Wedgwood frames the contours of a prominent tree and a *repoussoir* screen of vegetation, forms standard to picturesque landscapes of the 1790s. Under dream's erasure, that first state bleeds and evaporates, morphing into a sec-

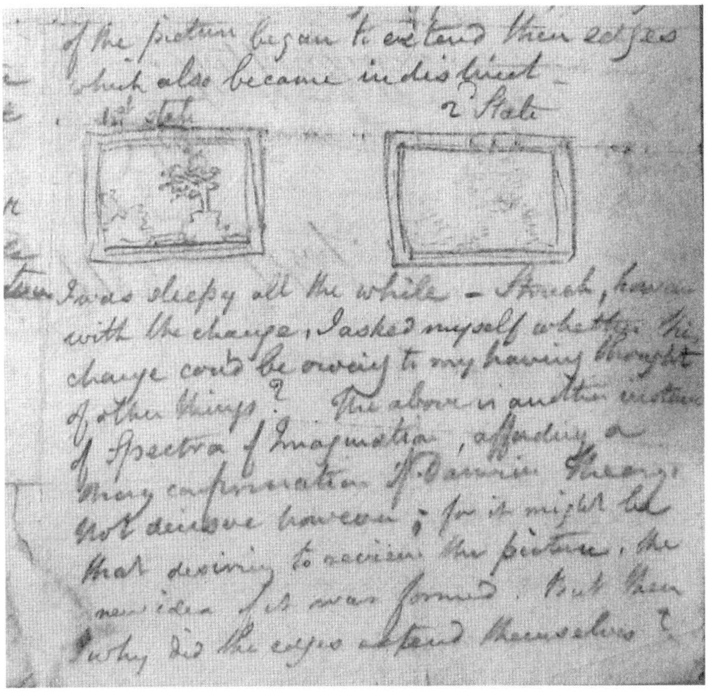

Figure 3.13 Thomas Wedgwood, "A Dream in Two States," ca. 1800. Pencil on paper. Victoria & Albert Museum, Wedgwood Collection (MS E40 28515/23, 2_r).

ond state at right. These dream-states are, he claims, "Spectra of Imagination, affording a strong confirmation of Darwins theory."[164] The theorist invoked here is Robert Waring Darwin (1766–1816), Wedgwood's brother-in-law and future father of Charles Darwin, who had proposed a sequence of afterimage experiments to support his claims for retinal activity (plate 18).[165] Darwin charges the reader to stare at his bull's-eye figures through a tube for half a minute. Once the experimenter's eyes are closed, the target pattern will persist before optical attention, giving what Darwin takes to be proof of the retina's muscular character "excited into action by irritations . . . strengthened or fatigued by exertion."[166] Wedgwood likely encountered that theory and its hand-tinted, experimental orbs as republished by Darwin's father, Erasmus Darwin, in his *Zoonomia; or, The Laws of Organic Life* (1796), a work he was then recommending as essential reading to William Godwin, among others.[167]

Unsurprisingly, Wedgwood gives no indication whether the mythical printing technique on which he gazed in dream had been lithography, pollaplasiamos, aquatint, or some combination of the chemical processes examined in this chapter. Nonetheless, in a 1792 paper that constituted exactly half of his lifetime publishing output, Wedgwood explored a potential candidate. "White paper, when dipped in a solution of sal ammoniac, and slowly dried," he writes, "becomes black upon the heater, and then gives out much less light than common paper."[168] Wedgwood was then writing on phosphorescence, phenomena that he—like Priestley—traced via the accidental discoveries of Vincenzo Cascariolo through Robert Boyle and C. A. Balduin. As Boyle had done, Wedgwood too tested the production of phosphorescent light *in vacuo*; he illuminated powdered earths and metals heated on a red-hot iron plate in an air-pump's evacuated chamber, and he controlled the effects of varying "airs" upon those experiments.[169] Marveling at the brilliant colors produced as agitated blue fluorspar's bright green luminescence "quickly changes into a beautiful lilac, which gradually fades away," Wedgwood was also interested in the joins and ruptures between production of light and combustion proper: "The light of boiling oils proceeds, probably, from some kind of inflammation, as it is scarcely visible unless the vessel be agitated; and, if a little oil be thinly spread on the heater, a subtle lambent flame, of a bluish hue, instantly arises."[170] Dreaming of prints that change, testing chemicals that alter paper's visual appearance under experimental trial: Wedgwood's reveries of flaming, oily motion will recall conjunctions sounded by the seventeenth-century experimentalists. They will also recur later in this story.

All that said, a conspicuous air of belatedness hangs over Wedgwood's surviving writings. The Molyneux problem, visual apprehension of distance, childhood perception—questions key to John Locke and George Berkeley nearly a century earlier—loom in his papers, with time as their *Schwerpunkt*.[171] An undated "Essay on Time" explores how language ascribes names to entities that are in fact compounds of sundry sensorial impressions registered sequentially. English speakers use the word "honey" to identify that sweet, yellow, sticky granulated syrup detected by discrete acts of perception; the percipient gains time consciousness by comparing different intensities of a given entity. Noting how "the honey in the spoon is diminished . . . or the tide makes 3 times as much progress on the Sands during the 3rd occasion as in the first & ⅓ more than during the second," Wedgwood explains, the perceiver is "led to call this perception tide (*Zeit*)—being time."[172]

When defending his biographical charts in the 1760s, Priestley had imagined the apprehension of time via length as a transfer "universally applied." Wedgwood too takes the coding of perceived time into quantitative measure as an innocuous concession to linguistic convenience.[173] In a letter to Godwinian-apostate political philosopher James Mackintosh from early 1801, he gives powerful expression to Priestley's intuition.[174] A line of time begins at left, taking the writer's journey across the page as its form (fig. 3.14). It propagates from the tiny humanoid figure (possibly an infant, more likely a mother bearing that child) crouching beneath the word "Christ" at the gutter of the letter's fold. Punctuated in recent time to five evenly spaced days, that straight line stretches underfoot to conclude in an upright male traveler who strides away from his diminutive origins, gazing into the blank page of future before him. Temporal linearity is ontology. "Give me

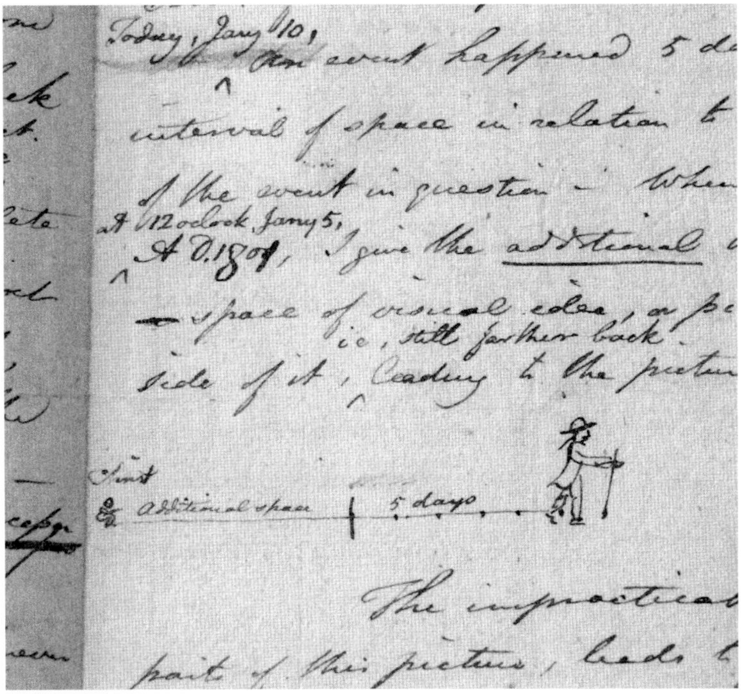
Today, Jany 10,
the event happened 5 da
interval of space in relation t
of the event in question — When
At 12 o'clock Jany 5,
A.D. 1801, I give the additional
space of visual ideas, or pi
i.e., still farther back
side of it, leading to the pictu
& additional space | 5 days
The impractical
part of this picture, leads t

Figure 3.14 Diagrammatic rendering of time as reducible to visual metrics of distance, from letter of Tom Wedgwood to James Mackintosh, January 1801. Victoria & Albert Museum, Wedgwood Collection (MS 28489 a $f2_r$).

therefore <in description> a regular line of all your <past> visual perception," Wedgwood charges Mackintosh, "& you give me your life. All is *picture*—consequently, relation is another word for Position."[175]

Such elliptical thoughts significantly simplify the quantitative conception of measure advanced around Robert Hooke's chemo-mechanical model of phosphorescent temporality, examined in chapter 1. And that is to say nothing of their relation to the conception of time as an *a priori* category of human cognition, as articulated in the philosophy of Immanuel Kant, which Wedgwood might have known through poet Samuel Taylor Coleridge, one of his several literary clients.[176] In a striking letter to Mackintosh from April 1801, Wedgwood seems to have become intensely cognizant of those limitations. Time undulates in its intensity, he now proposes. But it is profoundly misapprehended when parsed into familiar divisions: "In *Time* there is no *Past*, *present*, or *future*. These terms express qualities, & relations of visual perception. To speak of *past time*, is just as correct as of coloured time or Sound."[177] Where he had previously allowed the transfer of distance's quantitative values onto time to be a harmless linguistic convention, Wedgwood here casts such rendering as a pernicious capitulation to vision. Phrases like past or future time "are figurative terms drawn from the affections of sight. Time never *has* or *will be* but always *is* & all its variations are analogous to those of a continual sound which is sometimes swelling, sometimes becoming fainter & sometimes unvarying . . . at one time *more*, at another *less* intense & sometimes *uniform*."[178]

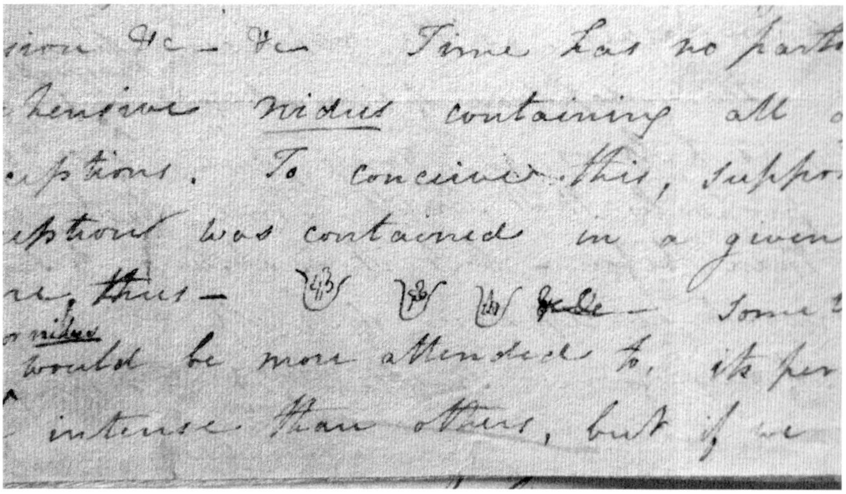
sion &c – &c Time has no parts
hensive nidus containing all
eptions. To conceive this, suppo
eption was contained in a given
re, thus – &c – Some t
or nidus would be more attended to, its per
intense than others, but if we

Figure 3.15 Thomas Wedgwood, the "nidus of time," April 1801. Pen and ink on paper. Victoria & Albert Museum, Wedgwood Collection (MS E40 28492, $f1_r$).

Yet, in immediate opposition to—or perhaps, contradiction of—those antivisualizing principles, Wedgwood again appeals to a graphic figure. "Time has no parts," he instructs Mackintosh, "it is a comprehensive *nidus* containing all our other perceptions. To conceive this, suppose by way of illustration every *visual* perception was contained in a given & constant figure like this."[179] The letter gives three, successive nests depicted in cross-section, each containing an individual figure (a tree, a leaf, perhaps a church) before ceding back to linguistic symbolization, also repeated: &c., &c. (fig. 3.15). "Nidus," according to Nicolas Salmon's 1796 Latin etymological dictionary, is a hive for bees, a cradle for a child, a nest, or a drinking cup.[180] "Sometimes the *Cup* or nidus would be more attended," Wedgwood explains to Mackintosh, "its perception would be more intense than others, but if we were always seeing, [it] would, like Time, always be."[181]

Nest, hive, or bed: it is hard not to see Wedgwood's organizing temporal motif outside the shadow of the pots and cups made so ruthlessly at Josiah Wedgwood's Etruria. Its industrial wares, rendered in figure 3.16 by poet/engraver William Blake, made "Time . . . the new idol," as historian Neil McKendrick once put it."[182] Indeed, it is against the factory's incessant division of time into hours and its emphasis on future production that Tom Wedgwood variously rails in his papers. He vows "to avoid all wavering and superficial examination of future urgencies & conduct. Give myself to the present moment. Oh grand, inestimable secret of human happiness!!!"[183] More compelling still is his obsession with visualizing his temporal ideas, even when they are framed as explicitly antivisual. In draft notes apparently for revision to his "Essay on Time," Wedgwood writes, "What is fixed

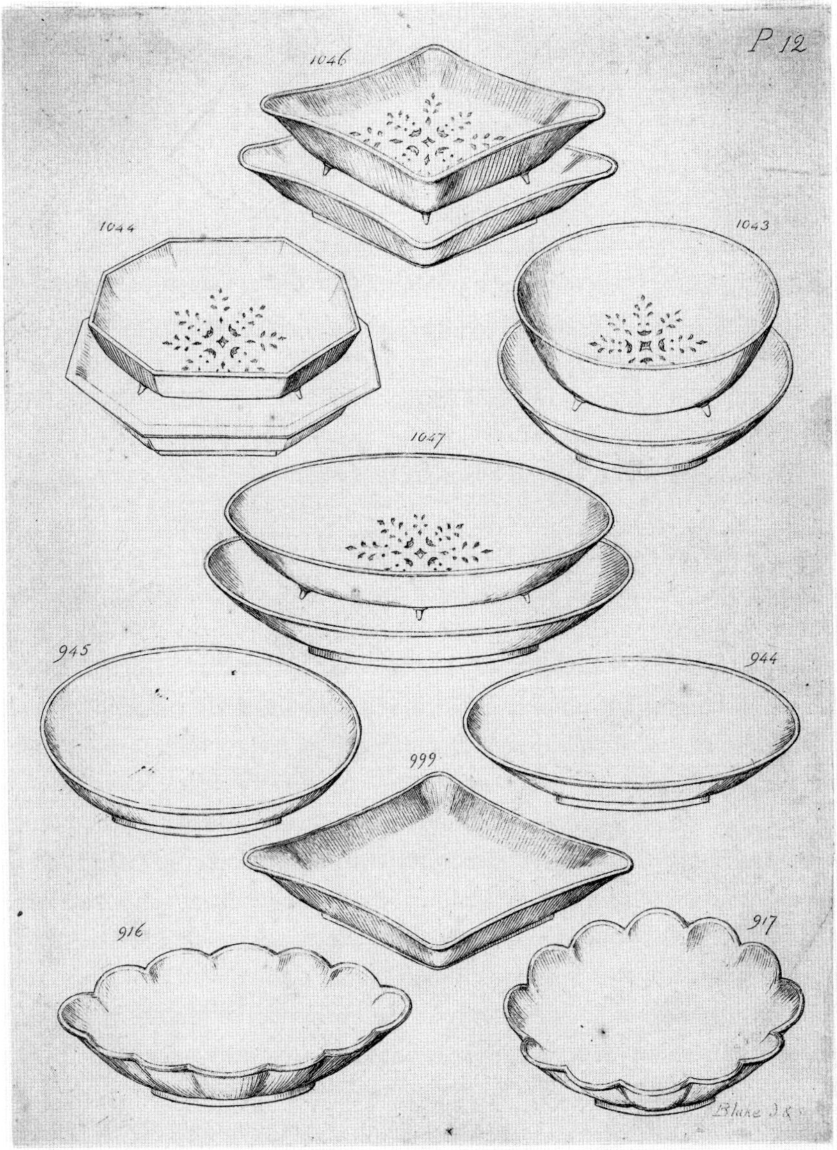

Figure 3.16 William Blake, plate 12 in J. Wedgwood & Sons, *Catalogue of Earthenware and Porcelain* (ca. 1816). Engraving. Huntington Library.

cannot represent [the] variable. [We say that] time moves, flows, creeps, passes, [etc. These expressions] represent moving objects concomitant of time, which are the language of *time*, but to give no idea of *time*."[184] What Wedgwood seems to want is a nonlinguistic means capable of possessing and highlighting the undulating intensity of time. He wants a way to exemplify what he calls "the perpetually evolving series of visual perception," without reducing temporal conception to quantitative terms.[185] As with Watt's chemically engineered copy paper or the special canvas supports required for pollaplasiasmos and painting in the "Venetian system," it is as if Wedgwood needs a peculiar nest, cup, or bed receptively mobilized to absorb time.

In this mood, consider the opening lines of Davy's famous article of 1802 presenting Wedgwood's investigations:

> White paper, or white leather, moistened with a solution of nitrate of silver, undergoes no change when kept in a dark place; but, on being exposed to the day light, it speedily changes colour, and, after passing through different shades of grey and brown, becomes at length black. The alterations of colour take place more speedily as the light is more intense. In the direct beams of the sun, two or three minutes are sufficient to produce the full effect. In the shade, several hours are required.[186]

Now, it is irresponsible not to acknowledge that the 1802 article does go on to specify how, from these principles, it would be possible to copy "the images formed by means of a camera obscura . . . [which] was the first object of Mr. Wedgwood"; that aim was hindered for want of a viable fixative.[187] Yet, published through the new, metropolitan institution in which Davy was shedding his 1790s radical skin to reemerge as fashionable proponent of establishment order, that *ex post facto* account hardly exhausts the matter.[188]

What, then, if we take an interpretative cue from the contradictory drives of the 1790s' broader work with temporally evolving chemical objects, expressed supremely in Wedgwood's aim to visualize antivisual temporal ideas, rather than presuming he was marching toward some "protophotography"? By that approach, the 1802 article could well be informed by Davy's contemporaneous research on the chemistry of leather-tanning.[189] Given its stress on how "outlines and shades of paintings on glass" might be transferred in minutes to chemically prepared supports and become "in a high degree permanent," the article should surely be placed in dialogue with bat-printing, aquatint, and other speedy, chemical innovations Josiah Wedgwood had ushered into industrial pottery.[190] Yet, just as he had sought release from the quantification of his father's factory time, so Tom Wedgwood might be seen to imagine a prepared bed or cup not fitted to bind time in illusory, quantitative extension, but equipped to receive it as a swelling, qualitative undulation. Perceivable images overflow the cup's lip; they spill out of attentive range as the support blackens into invisibility. Rather than failing to achieve fixity per a self-refashioning Davy—let alone a photographic patriarch like Talbot more than four decades later—a figural depiction that fades from an unfixed chemical support to visualize time's undulating ontology could, in fact, be a successful achievement of Wedgwood's project.

With these patriarchal stories in play, it is instructive to return to a crucial term at the center of Wedgwood's cup of time: "nidus." According to the venerable Aristotelian theory of generation, the male partner in sexual reproduction provides the form (ειδος) and principle of movement (αρχην της κινησεως) while the female supplies body (σωμα) and matter (της υλην).[191] Rendered through the Latin *nidus* by way of many early modern accounts, Aristotelian notions of feminine, incubative passivity found significant medical endorsement in the mid-1790s from the very text that Wedgwood recommended to William Godwin and that would guide his own dream in print (see plate 18, figure 3.13).[192] An early British partisan for Lavoisier's Chemical Revolution, Erasmus Darwin explained in *Zoonomia* (1796) how the mother acts as chemo-nutritive provider for the embryo during sexually initiated gestation.[193] Bearing a tiny speck of matter already living, male seminal ejaculate "is received into an appropriate nidus, in which it must acquire two circumstances necessary to its life and growth."[194] This maternal nidus gives food to the growing seed. It provides "that part of the atmospherical air . . . which by the new chemistry is termed oxygene."[195] A false inference is drawn, Darwin contends, when we thus conclude that the nidus—swelling, intensifying, pregnant with time, as Wedgwood seems to have it—is anything more than an inert support of the kind James Northcote had dismissed. Beyond supplying the embryo "with a proper nidus, with sustenance and oxygenation," female contribution to the generation of life is modest in Darwin's assessment: "The idea of the semen of the male constituting only a stimulus to the egg of the female, exciting it into life, (as held by some philosophers) has no support from experiment or analogy."[196]

What, though, of life *not* made by sexual means? Said differently, if we see Tom Wedgwood here mapping his metaphysical forays on chemical supports back into a sexualized ontology of time richly sounded by the Shandean clock-winders noted in chapter 2, what of the Reynoldsian alternative in the abiding dream of the chemical homunculus? What would a shape of time born from a *chemical* nidus look like? No more powerfully would those vicissitudes be explored than in the imaginative fiction of the author born from the sexual union of the radical philosophers to whom Tom Wedgwood had gifted the Watt chemical copier in 1795. After all, the daughter of William Godwin and Mary Wollstonecraft, Mary Shelley has her doomed protagonist Victor Frankenstein reading deeply in the alchemical writings of Agrippa, Paracelsus, and Albertus Magnus.[197] Those childhood encounters incline him not to "the more rational theory of chemistry which has resulted from modern discoveries," but to ambitions of an older chymical tradition: "the search of the philosopher's stone and the elixir of life."[198] Devoting himself to "natural philosophy, and particularly chemistry," Frankenstein sets out aims continuous with the chemical immortality trafficked by Reynoldsian replicators of the 1790s. "'I thought, that if I could bestow animation upon lifeless matter,'" he tells Shelley's narrator, Walton,

"'I might in process of time (although I now find it impossible) renew life where death had apparently devoted the body to corruption.'"[199]

Frankenstein's chemical research succeeds in bringing inanimate matter to life, of course. To his horror, he finds that the creature he has made without feminine nidus is freakishly strong, nearly impossible to destroy. Where it had been a selling point for Booth, Birch, Bone, and other supporters ready to carry the fragile body of Reynoldsian painting into the future, chemically induced "immortality" becomes the bane of Frankenstein's existence, as the creature systematically destroys member after member of the creator's own biological family. And where backers of encaustic or enamel had looked to fire to render their chemical preparations indestructible, Shelley's chemically created life looks to fire for release. "'Soon these burning miseries will be extinct," so Walton quotes the creature in the novel's penultimate paragraph: "'I shall ascend my funeral pile triumphantly, and exult in the agony of the torturing flames. The light of that conflagration will fade away; my ashes will be swept into the sea by the winds. My spirit will sleep in peace.'"[200] An elemental chemistry is here still in play. Fire releases spirit; air scatters earthy matter over water. A lifetime born from a chemical nidus is once again unfixed. Where Mary Shelley's fiction looks back from 1818 to see the 1790s' aspirations to chemical immortality as the stuff of nightmare, let us now look ahead to the earth and this same chemical flame unearthed in the 1860s *as photography*.

4

Space, Time, and Chemistry

Making Enlightenment "Photography" in the 1860s

By boat and by the banks of the River Thames, Joseph Mallord William Turner watched the Palace of Westminster burn through the night of October 16, 1834. Turner (1775–1851) made numerous sketches and, later, two oil paintings of the destructive spectacle. Flames explode in scintillating fireworks from the combusting palace at left in Turner's first exhibited oil version, a picture both shown and partially made in the gallery of the British Institution in February 1835 (plate 19). Positioned at avian height above the south bank, the beholder looks north across the Thames flanked below on either side by swarms of shadowy observers. Masses of tiny human forms crowd the luminous span of Westminster Bridge, whose ghostly figure juts in at exaggerated height and elongated recession from the picture's right edge.

If the river's reddened, murky waters had doubled the palace's incandescence in muted mirror on the night of combustion, then so the picture had appeared on Turner's canvas as if conjured by "a magician, performing his incantations in public."[1] That was the claim of contemporary painter E. V. Rippingille. Published posthumously in 1860, Rippingille's recollection framed Turner's semipublic act of painting *The Burning of the Houses of Lords and Commons, October 16, 1834* through contrast to the "processes, nostrums, and pigments" of none other than Joshua Reynolds. Forgoing his predecessor's secretive approach, Turner had entered the gallery of the British Institution with a canvas virtually bare. Using "a small box of colours, a few very small brushes, and a vial or two," so Rippingille put it, the painter "worked almost entirely with his palette knife . . . rolling and spreading a lump of half-transparent stuff over his picture."[2] The spectacle of experimental making was nearly as dramatic as Turner's picture itself.

Turner was as unconventional a technician as the British school of painting had produced; yet the architectural destruction he rendered effectively prompted a pictorial *retour à l'ordre* in material terms.[3] For it was writing as secretary of the commission charged with ornamenting the reconstructed Houses of Parliament that painter and polymath Charles Lock Eastlake produced his landmark *Materials for a History of Oil Painting* (1847). Future president of the Photographic Society and director of the National Gallery, Eastlake (1793–1865) traced a history of artists' vehicles, pigments, and grounds forward from the ancient world. Little of the evidence for Eastlake's narrative comes from conversations heard or practices witnessed in the accounts of technique given by James Northcote, Joseph Farington, J. T. Smith, or indeed Rippingille himself. Instead, Eastlake turns to documents. In step with the epistemological privileging of records among the period's printing clubs and emergent archives, he quotes works from numerous languages, ancient and modern; he cites learned contributions of his contemporaries as well as materials extracted from manuscripts and modern translations.[4] Such erudite inquiries lead Eastlake to a more measured account of Rippingille's secretive empiric. "The notes which Sir Joshua Reynolds kept of the materials employed, and the order of the processes adopted by him," Eastlake asserts, "are important links in the technical history of painting."[5] Transcribing several pages from Reynolds's ledger books and quoting recipes inscribed on one of his experimental canvases (see fig. 2.12), Eastlake's Reynolds is an inheritor of Continental traditions, not their liquidator.[6] Contrary to the familiar view of Reynolds's experiments as "entirely novel," so claims Eastlake (himself a native of Plymouth, and an alumnus of the Plympton Grammar School once administered by the Rev. Samuel Reynolds), they represent "a judicious and generally successful union of the Italian and Flemish practice."[7]

Eastlake's volume was heralded by contemporaries as a means to technical retrenchment for a British school of painting grown "of too great importance to waste any of its powers . . . in the perishing and weak materials of our various meguilps."[8] But Eastlake's tempered view of Reynolds also stood on the shoulders of more frankly partisan interpreters. In 1844 William Laxton's *Civil Engineer and Architect's Journal* had published an account declaring that Reynolds "kept *no* journal"—that his painting practice was little more than "an ever vacillating, muddled dash at crude experiment."[9] To this, Benjamin Robert Haydon (1786–1846) took exception. A true believer

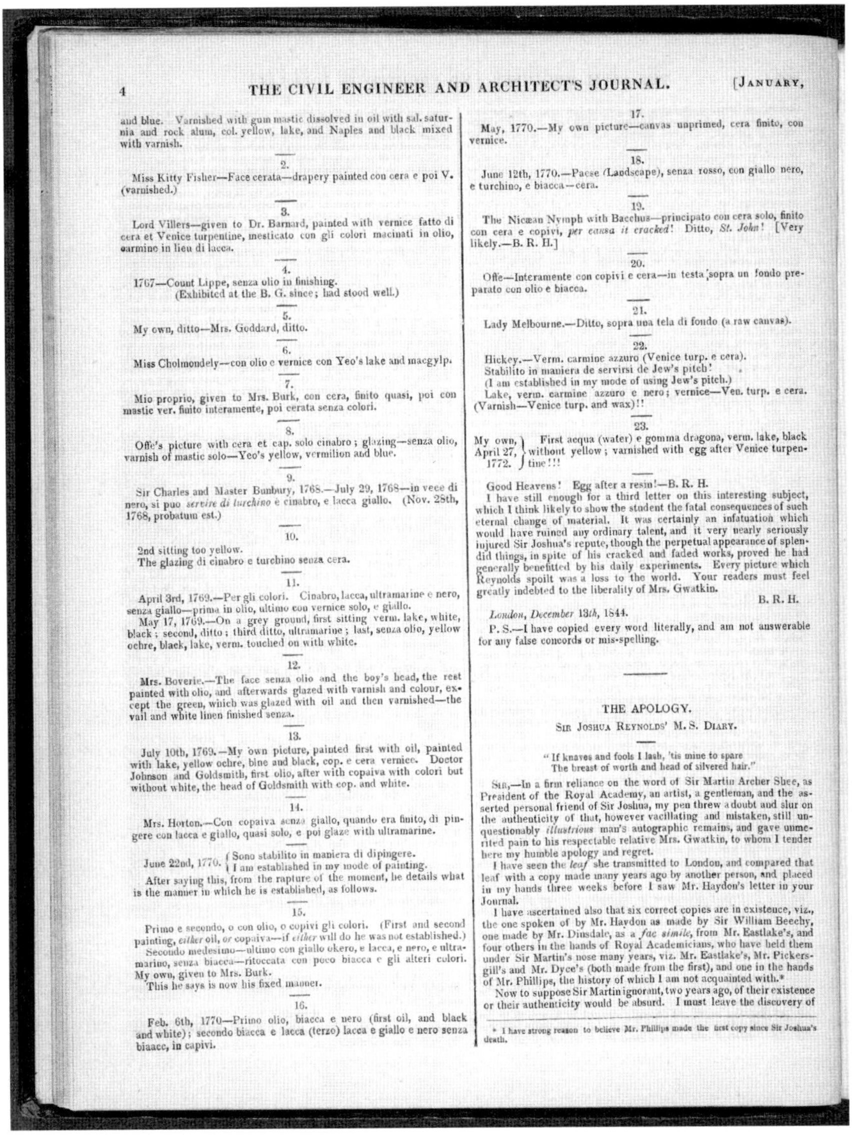

4 THE CIVIL ENGINEER AND ARCHITECT'S JOURNAL. [JANUARY,

and blue. Varnished with gum mastic dissolved in oil with sal. saturnia and rock alum, col. yellow, lake, and Naples and black mixed with varnish.

2.

Miss Kitty Fisher—Face cerata—drapery painted con cera e poi V. (varnished.)

3.

Lord Villers—given to Dr. Barnard, painted with vernice fatto di cera et Venice turpentine, mesticato con gli colori macinati in olio, carmine in lieu di lacca.

4.

1767—Count Lippe, senza olio in finishing.
(Exhibited at the B. G. since; had stood well.)

5.

My own, ditto—Mrs. Goddard, ditto.

6.

Miss Cholmondely—con olio e vernice con Yeo's lake and macgylp.

7.

Mio proprio, given to Mrs. Burk, con cera, finito quasi, poi con mastic ver. finito interamente, poi cerata senza colori.

8.

Offe's picture with cera et cap. solo cinabro; glazing—senza olio, varnish of mastic solo—Yeo's yellow, vermilion and blue.

9.

Sir Charles and Master Bunbury, 1768.—July 29, 1768—in vece di nero, si puo *servire di turchino* è cinabro, e lacca giallo. (Nov. 28th, 1768, probatum est.)

10.

2nd sitting too yellow.
The glazing di cinabro e turchino senza cera.

11.

April 3rd, 1769.—Per gli colori. Cinabro, lacca, ultramarine e nero, senza giallo—prima in olio, ultimo con vernice solo, e giallo.
May 17, 1769.—On a grey ground, first sitting verm. lake, white, black; second, ditto; third ditto, ultramarine; last, senza olio, yellow ochre, black, lake, verm. touched on with white.

12.

Mrs. Boverie.—The face senza olio and the boy's head, the rest painted with olio, and afterwards glazed with varnish and colour, except the green, which was glazed with oil and then varnished—the vail and white linen finished senza.

13.

July 10th, 1769.—My own picture, painted first with oil, painted with lake, yellow ochre, blue and black, cop. e cera vernice. Doctor Johnson and Goldsmith, first olio, after with copaiva with colori but without white, the head of Goldsmith with cop. and white.

14.

Mrs. Horton.—Con copaiva senza giallo, quando era finito, di pingere con lacca e giallo, quasi solo, e poi glaze with ultramarine.

June 22nd, 1770. { Sono stabilito in maniera di dipingere.
I am established in my mode of painting.

After saying this, from the rapture of the moment, he details what is the manner in which he is established, as follows.

15.

Primo e secondo, o con olio, o copivi gli colori. (First and second painting, *either* oil, *or* copaiva—if *either* will do he was not established.)
Secondo medesimo—ultimo con giallo okero, e lacca, e nero, e ultramarino, senza biacca—ritoccata con poco biacca e gli alteri colori. My own, given to Mrs. Burk.
This he says is now his fixed manner.

16.

Feb. 6th, 1770—Primo olio, biacca e nero (first oil, and black and white); secondo biacca e lacca (terzo) lacca e giallo e nero senza biaacc, in capivi.

17.

May, 1770.—My own picture—canvas unprimed, cera finito, con vernice.

18.

June 12th, 1770.—Paese (Landscape), senza rosso, con giallo nero, e turchino, e biacca—cera.

19.

The Nicæan Nymph with Bacchus—principato con cera solo, finito con cera e copivi, *per causa it cracked*! Ditto, *St. John*! [Very likely.—B. R. H.]

20.

Offe—Interamente con copivi e cera—in testa sopra un fondo preparato con olio e biacca.

21.

Lady Melbourne.—Ditto, sopra una tela di fondo (a raw canvas).

22.

Hickey.—Verm. carmine azzuro (Venice turp. e cera).
Stabilito in maniera de servirsi de Jew's pitch!
(I am established in my mode of using Jew's pitch.)
Lake, verm. carmine azzuro e nero; vernice—Ven. turp. e cera. (Varnish—Venice turp. and wax)!!

23.

My own, April 27, 1772. } First acqua (water) e gomma dragona, verm. lake, black without yellow; varnished with egg after Venice turpentine!!!

Good Heavens! Egg after a resin!—B. R. H.
I have still enough for a third letter on this interesting subject, which I think likely to show the student the fatal consequences of such eternal change of material. It was certainly an infatuation which would have ruined any ordinary talent, and it very nearly seriously injured Sir Joshua's repute, though the perpetual appearance of splendid things, in spite of his cracked and faded works, proved he had generally benefitted by his daily experiments. Every picture which Reynolds spoilt was a loss to the world. Your readers must feel greatly indebted to the liberality of Mrs. Gwatkin.

B. R. H.

London, December 13*th*, 1844.
P. S.—I have copied every word literally, and am not answerable for any false concords or mis-spelling.

THE APOLOGY.

SIR JOSHUA REYNOLDS' M. S. DIARY.

"If knaves and fools I lash, 'tis mine to spare
The breast of worth and head of silvered hair."

SIR,—In a firm reliance on the word of Sir Martin Archer Shee, as President of the Royal Academy, an artist, a gentleman, and the asserted personal friend of Sir Joshua, my pen threw a doubt and slur on the authenticity of that, however vacillating and mistaken, still unquestionably *illustrious* man's autographic remains, and gave unmerited pain to his respectable relative Mrs. Gwatkin, to whom I tender here my humble apology and regret.
I have seen the *leaf* she transmitted to London, and compared that leaf with a copy made many years ago by another person, and placed in my hands three weeks before I saw Mr. Haydon's letter in your Journal.
I have ascertained also that six correct copies are in existence, viz., the one spoken of by Mr. Haydon as made by Sir William Beechy, one made by Mr. Dinsdale, as a *fac simile*, from Mr. Eastlake's, and four others in the hands of Royal Academicians, who have held them under Sir Martin's nose many years, viz. Mr. Eastlake's, Mr. Pickersgill's and Mr. Dyce's (both made from the first), and one in the hands of Mr. Phillips, the history of which I am not acquainted with.*
Now to suppose Sir Martin ignorant, two years ago, of their existence or their authenticity would be absurd. I must leave the discovery of

* I have strong reason to believe Mr. Phillips made the first copy since Sir Joshua's death.

Figure 4.1 Extracts from Reynolds's ledger books, as printed in Benjamin Robert Haydon, "Sir Joshua Reynolds," *Civil Engineer and Architect's Journal* 8 (1845): 4.

in Reynolds's vision of grand-manner history painting and self-declared mentor to Eastlake, Haydon promptly contacted Reynolds's surviving niece Theophilia "Offy" Palmer Gwatkin (1757–1848), then also living in Haydon's native Plymouth. Gwatkin was keen to support the case; she cut several pages from the painter's ledgers (see figs. 2.10–2.11) and mailed them to Haydon in London.[10] Haydon transcribed the polylingual accounts of practice and printed them through several installments of Laxton's journal in 1844–1845 (fig. 4.1).[11] Accompanied by his own lively marginal comments, Haydon's transcriptions also retain and relay annotations made by a still-earlier painter, Sir William Beechey (1753–1839).[12] While mention of Haydon's publications is conspicuously absent from Eastlake's 1847 volume (the older painter had committed suicide in the summer of 1846), that collective attention to documentary evidence of Reynolds's practices would only expand in the following decade.[13]

What should we make of this meticulous documentation of material practices pursued by a painter then dead for half a century? According to familiar histories of art, to express sympathy for Reynolds in the later 1840s was to fly in the face of vanguard opinion. Among the Pre-Raphaelite painters who would unite in purpose after the massive Chartist demonstration on south London's Kennington Common in the revolutionary year of 1848, Reynolds was "Sir Sloshua": supreme exponent of the louche, unstable paint-handling to be overcome by recourse to older methods and surer techniques.[14] More surprising still is the prominence afforded Reynolds among theorists and practitioners of the new art of photography that had arrived with such *éclat*, with the rival inventions of Louis Joseph Mandé Daguerre and William Henry Fox Talbot both announced publicly in 1839. Less had the popularity of the new art grown than exploded. From a single professional practitioner in 1842, claimed Lady Elizabeth Rigby Eastlake (Charles Lock Eastlake's wife) in an influential essay, London's army of photographers had multiplied to an estimated one hundred fifty-seven by 1857.[15] And Reynolds soon headed that battalion. In a serialized essay for the *Photographic News* beginning in 1860, painter/photographer William Frederick Lake Price (1810–1896) proposed "a concise sketch of the rise and progress of pictorial arts" for photographers' instruction, with Reynolds's principles and portraits figuring prominently.[16] The painter's Dr. John Hunter peers quizzically out from the darkened, wood-engraved depths of his study at the inception of Lake Price's eighth installment (fig. 4.2). Surrounded by fragmentary attributes of his anatomical learning—canisters of wet samples immersed in chemical preservative, interrupted tibias of a hanging skeleton installed in the fictive niche behind him—Reynolds's Hunter is, Lake Price explains, an exemplary model for the photographer, who could emulate the portrait's general composition and characterizing accessories with equal facility.

Victorian photographers' gravitation toward Reynolds seems weird given the new art's embrace by the Pre-Raphaelites and their insurgent ilk.[17] "What can it mean," historian of photography Steve Edwards has asked of Lake Price's essay, "when the major text published on photography in 1860—one that played a highly significant role in defining terms, establishing oppositions, and setting out categories—opens with an appeal to the authority of Reynolds?"[18] This chapter poses a different question: how should we understand the relations between chemically inclined, Reynoldsian painting; its chemical prolongation through the reproductive likes of Joseph Booth, William Russell Birch, James Watt, and others; and that as-

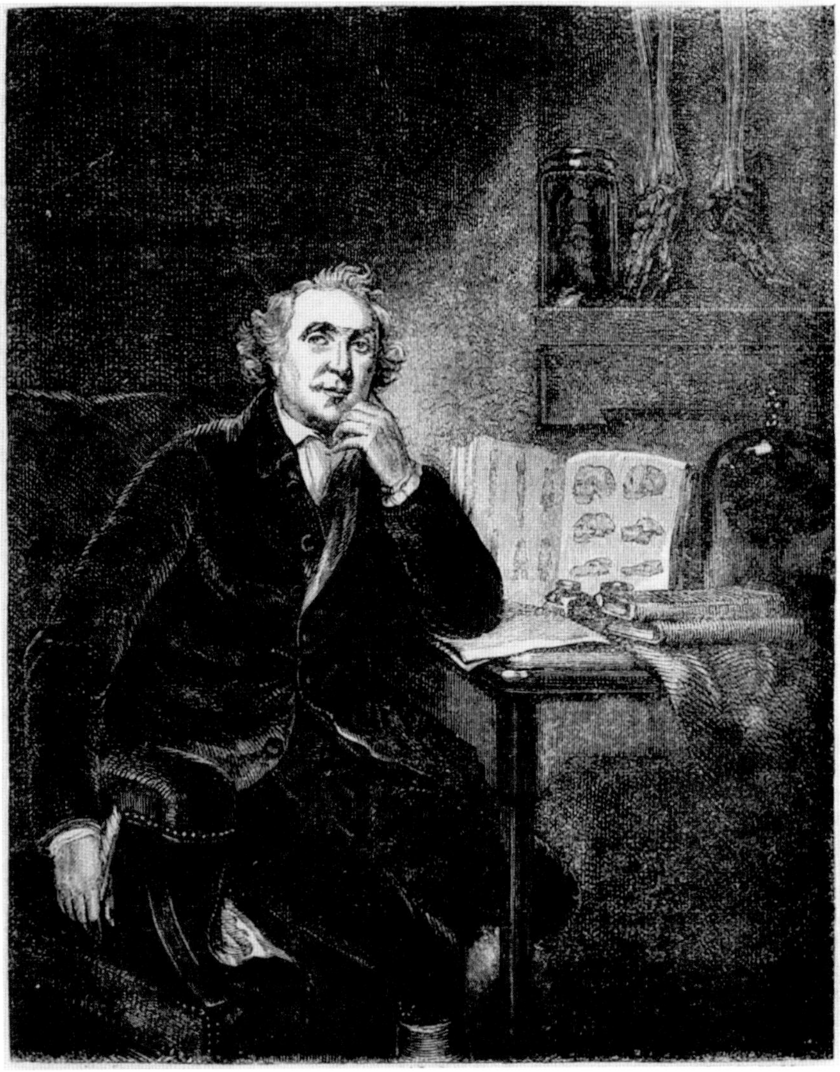

Figure 4.2 Unknown printmaker, after Joshua Reynolds's *John Hunter*, as published in William Lake Price, "On Composition and Chiar'oscuro. VIII," *Photographic News* 3, no. 83 (1860): 367.

semblage of competing techniques inescapably bound to chemistry, which would come to be called "photography"? Said more simply, where does photography sit in the British Enlightenment's ways of making and thinking with temporally evolving chemical objects? As will be clear, this line of inquiry is different from the pathways historians have carved through photography's vexed origins. I make no recourse to any claims for a *Kunstwollen* directed toward "a new and fundamentally modern pictorial syntax of immediate, synoptic perceptions and discontinuous, unexpected forms" that yielded photography up from the depths of artistic drives.[19] Nor do I appeal to the consolidation of "the hitherto strange and unfamiliar desire to have images formed by light automatically fix themselves" that Geoffrey Batchen has ascribed to several of my protagonists in his account of photography's advent.[20] I do not pursue these options for the simple reason that my story aims at a different target. It is not an explanation of photography's coming, but a framing of chemical photography within a longer history of making and thinking with chemical materials that change in time, a trajectory British painting itself had joined in the era of Reynolds.

Any such effort must confront a formidable obstacle: a storied intuition whereby painting and photography are fundamentally—ontologically—different.[21] Painting, such narratives often claim, is a species of representation made constitutively by the human hand; photography is otherwise. Roland Barthes put it this way: "Painting can feign reality without having seen it. . . . In Photography, I can never deny that *the thing has been there*."[22] Stripped of cultural encoding, so argued Rosalind Krauss as she built on Barthes, "it is the order of the natural world that imprints itself in the photographic emulsion. . . . This quality of transfer or trace gives to the photograph its documentary status, its undeniable veracity."[23] Akin to a fossil, Walter Benn Michaels has claimed more recently, "the thing the photograph is of is causally indispensable to the photograph in a way that the thing a painting is of need not be."[24] In ways that even the most faithful drawing could never do, so André Bazin once proposed, photography shares "by virtue of the very process of its becoming, the being of the model of which it is the reproduction; it *is* the model."[25] Kaja Silverman expands these sentiments, declaring photography "the world's primary way of revealing itself to us—of demonstrating that it exists, and that it will forever exceed us."[26] The late anthropologist Alfred Gell summarized these convictions in a striking way. Where the painter transforms pasty pigments to make images, "the alchemy involved in photography . . . [is] regarded as uncanny, but as uncanny processes of a natural rather than a human order, like the metamorphosis of caterpillars into butterflies."[27] We commit category violations—we break taboos—if those profound differences between painting's chemistry and the alchemy of photography are not respected.

When thinking with a hammer, the best way to approach such idols is less to break them than to make them ring. I aim to do that in the present chapter by considering a moment in the 1860s when examples of Enlightenment Britain's chemical tradition would be massed as material evidence for a surprising intervention into photography's consolidating historiography, a claim that placed the new art's invention not in 1820–1830s France or England, but back in the English Midlands of the 1780s–1790s. My first section follows the construction of evidence used to support the "sun pictures" made by Francis Pettit Smith, curator at London's newly formed Patent Museum, and his chief supplier at the derelict Soho Manufactory, Edward Price.[28] Situating the case built by Smith and Price as much in photography's gestating histories as in Victorian Britain's imaginings of its industrial past, the chapter follows these sun pictures up to their largely successful presentation before a leading photographic body in November 1863. Thereafter, this revisionist history would meet the stern opposition of Matthew Piers Watt Boulton, the Soho industrialist's grandson and namesake.

Why was M. P. W. Boulton so opposed to the sun pictures? What satisfactory explanation can be given of the reasons that drove him not only to deny awarding photography's inventive priority to his grandfather, but to press his disavowals through threatened litigation and a sequence of ad hominem, ever-expanding pamphlets issued against the sun pictures' backers in 1863, 1864, and 1865? To answer such blunt questions, I turn not to the periodic table of elements invented by Russian chemist Dmitri Mendeleev in 1869, but back to Empedoclean tradition. With the classical elements, I deal the deck, so to speak, in two different directions. If, as philosopher of science Gaston Bachelard has put it, "earth . . . is first and foremost characterized by *resistance*," a preliminary, down-to-earth dealing canvasses a story told in the period: Boulton's opposition was the class-conscious move of a landed man of leisure intent on masking his family's industrial wealth.[29] Sounding this story offers an occasion to examine competing conceptions of photography's ontology operative in the 1860s—its fomenting negotiations between industrial mechanism and natural productivity—amid attempts to position the new medium within the elevated domain of the fine arts. But a second, frankly unearthly dealing reveals a very different account. "With air," Bachelard contends, "movement takes precedence over matter."[30] Through this aerial cut, the sun pictures emerge at a crucial juncture of chemistry, mechanism, and motion, one equally implicating Boulton, Smith, and James Watt himself. Shifting the patrimonial stakes from Boulton primus to Watt—seeing M. P. W. Boulton's opposition to the sun pictures not as an attempt to shake off the disfiguring, industrial hand of his long-dead grandfather but as a stab against a rival technologist—this aerial approach affords three larger lessons for rethinking the sun pictures, Reynoldsian painting, *and* photography. Making good on Tom Wedgwood's project, the air allows us to release temporally evolving chemical objects from photography and, perhaps, photography from itself. And all this release, as the conclusion argues, has made the air of what some now call the Anthropocene.

Unearthing the Sun Pictures

In the fall of 1862, a curator from London's Patent Museum named Francis Pettit Smith traveled to Birmingham on a collecting mission.[31] Smith (1808–1874) was seeking to acquire models and early prototypes of James Watt's steam engines. A sheep farmer turned engineer, Smith knew the value of

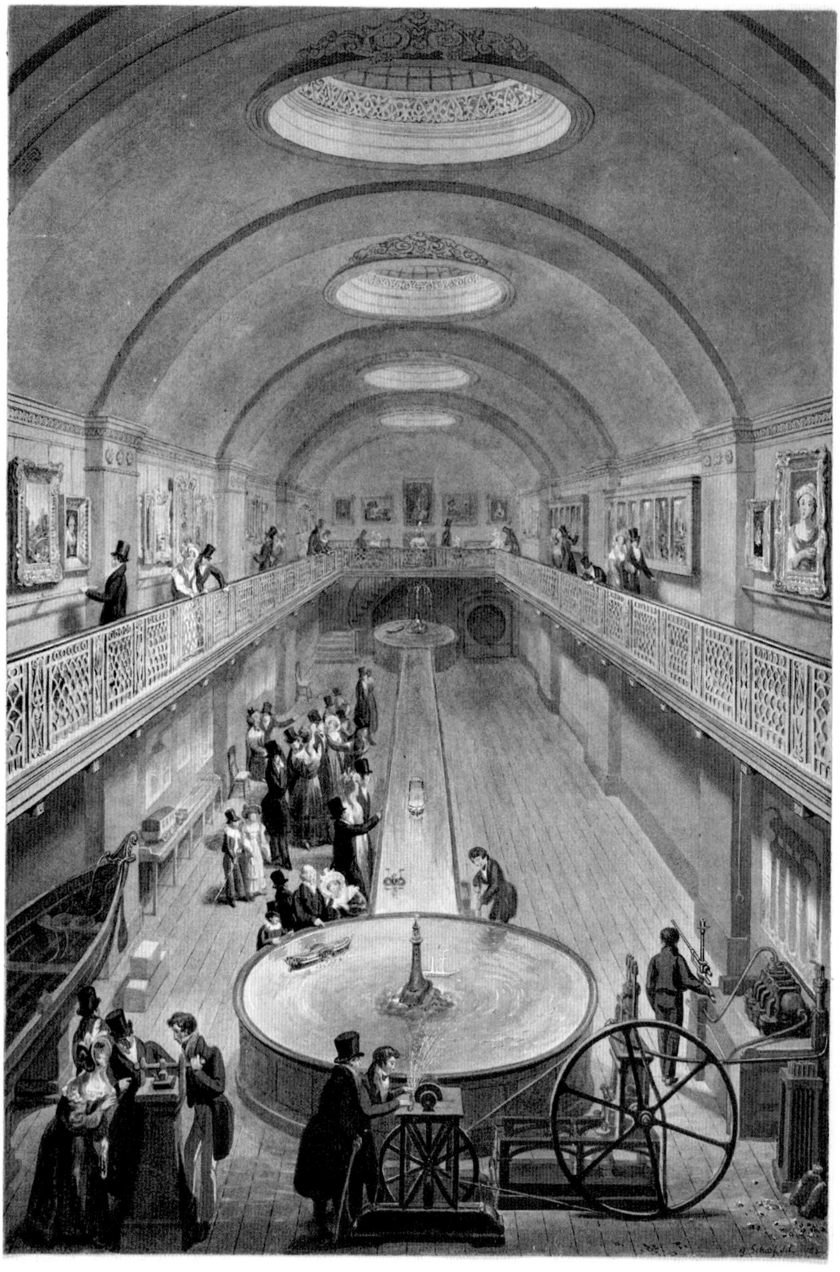

Figure 4.3 George Scharf, *Interior of the National Gallery of Practical Science, Adelaide Street*, 1832. Watercolor. British Museum, London (1862,0614.689). © The Trustees of the British Museum/Art Resource, NY.

models.[32] By the mid-1830s, he had experimented with and then patented designs for the screw propulsion of steamships, designs he exhibited at Jacob Perkins's National Gallery of Practical Science on Adelaide Street near London's Charing Cross.[33] As depicted in George Scharf's watercolor of 1832, the Adelaide Gallery featured a watercourse where Perkins and other inventors could demonstrate prototypes of their ships (fig. 4.3).[34] In 1836 a

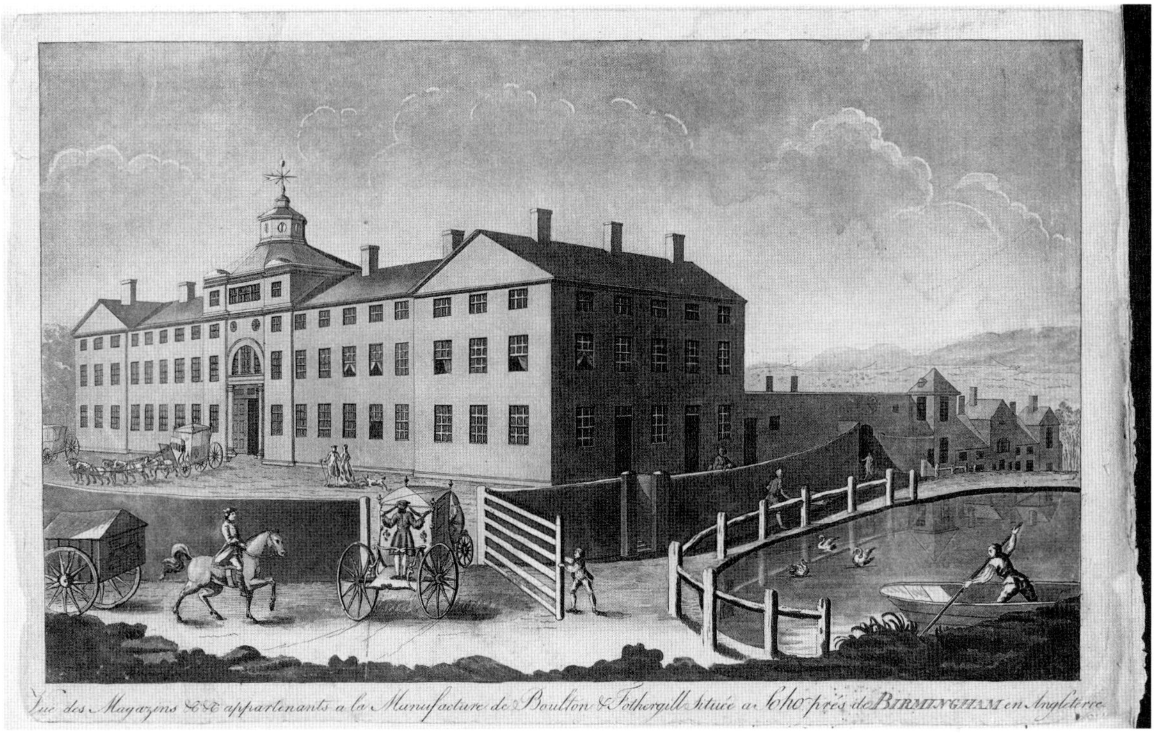

Figure 4.4 Francis Eginton, *Vue des Magazins & appartenants a la Manufacture de Boulton & Fothergill Située a Soho prés de Birmingham en Angleterre*, 1773. Etching and aquatint. British Museum, London. © The Trustees of the British Museum/Art Resource, NY.

screw-propelled model made to Smith's specifications caught the eye of Sir John Barrow, second secretary of the Admiralty. A ship aptly named *Archimedes* was eventually commissioned by the Royal Navy as a 237-ton trial of the concept.[35] Disappointment and litigation followed such that, in 1862, Smith found himself seeking to locate what one local correspondent would call "the old 'rocket'" from the Soho Manufactory established in 1765 by Matthew Boulton in Handsworth parish, on the outskirts of Birmingham.[36]

An aquatint from 1773 by Francis Eginton suggests the Soho complex that James Watt himself would have encountered when he arrived from Glasgow just one year later (fig. 4.4). Coaches parade through the gates flanking Soho's two-story edifice, a smart, Palladian front to the Hephaestian forge of metalworking industry hidden down a row of smoking shops to the right. While a punter and swans gliding on the pool in the foreground give a hint of leisure to the crucial source of water power that had attracted Boulton to the Soho site, an arching Diocletian window offers the slimmest of trim to the edifice's double-height vehicular exit. Below it, coaches groaning under the weight of buckles, sword-hilts, and other manufactured metallic "toys" make their way to metropolitan markets.

That was not the Soho that Francis Pettit Smith saw in 1862 (fig. 4.5).

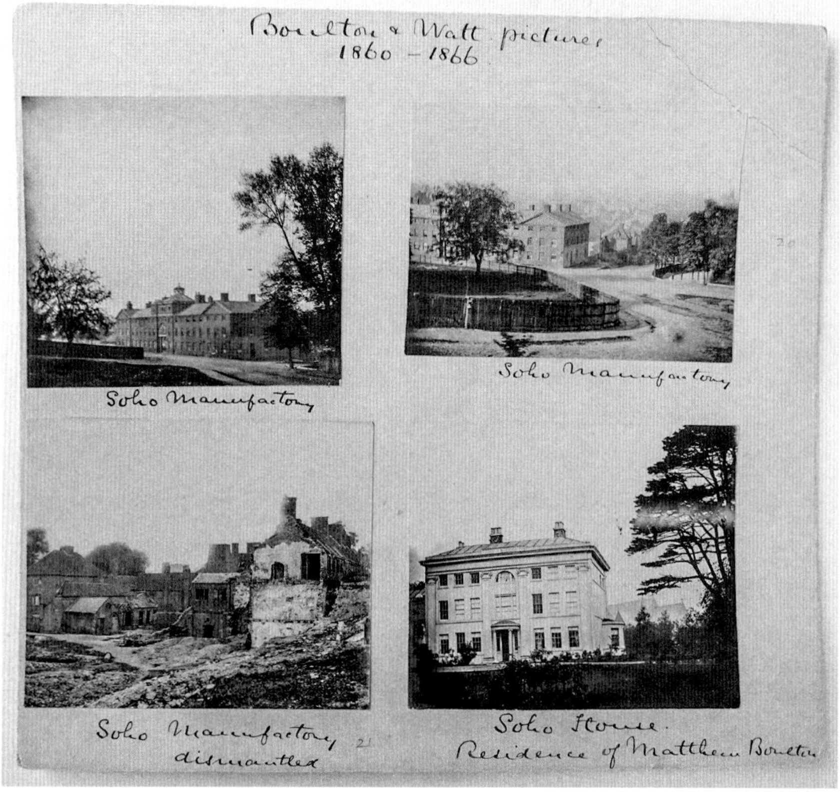

Figure 4.5 Unknown photographer, Soho House and Manufactory in Handsworth, ca. 1860–1866. Stone Collection, box 14, print 33 (1/113). Birmingham City Council/LSH.

Shuttered since 1848, demolished in 1863, Soho's glory days had long passed when Smith arrived to meet an agent of the Boulton family named Edward Price (b. 1822). Surviving evidence about Price is sketchy. Apprenticed to Soho site manager Joseph Westley in 1836, he appears to have assumed his mentor's functions as rent collector and general factotum after Westley's death in 1844. Certainly, by 1849, Boulton's grandson Matthew Piers Watt Boulton was leaving directions for "Price respecting the moving of some rhododendrons &c also of the picture of my grandfather & the Drawing Office Clock."[37] Claiming to F. P. Smith that he had worked at Soho since the 1830s, the third generation of his family to do so (all assertions disputed by his employer), Price cast himself as "a person who seemed delighted at the idea of anything belonging to Soho being taken care of."[38] Among the "relics" dredged up from some "five tons of old Books and papers," Price offered a folio in which Smith "discovered these sun pictures—and begged them of me."[39]

What were "these sun pictures"? An example is instructive. A cortege of male attendants has entered from the left, clustering behind a starry chaise (plate 20). Crown and scepter—instruments of terrestrial rule—are set aside as two trains converge to meet in a dark truth. His hooded head lit against an umbrous drape, king Seleucus of Hellenistic Syria supports the bared, slumping torso of his son Antiochus in melancholy resignation as anatomist Erasistratus palpates the youth's chest and wrist. In this subject set for prize competitions at both the British and French art academies in the mid-1770s, Erasistratus will soon diagnose the malaise clouding the male line at left as Antiochus's forbidden love for the woman standing at the foot of his bed: his stepmother, Stratonice.[40] Modest, even ablative, when paired against the contemporaneous vocative of Jacques-Louis David's version (which finally won him the Prix de Rome in 1774), the drama of the composition lies in its anticipated remaking of binary orders: dark/light, male/female, horizontal/vertical, child/parent.[41] But those binary orders of pictorial fiction are also cloven by the image's own recalcitrant surface. Seven sloping trails of resinous material sweep across the illuminated female entourage at right, as if washed by so many dark waves. Where dark laps light, so light lacerates light; it breaks the luminous, framing entablature at center right into a white frost of crystalline encrustation and perforates the fiction entirely as bright, jagged scratches carve out the tonal ground beneath Stratonice's feet.

Composed of two sheets of Whatman paper glued together to span some thirty-one by forty-four inches, this largest of the sun pictures acquired by Smith in the early 1860s is also broadly indicative of the set. First, it is a copy. Exhibiting the familiar left-right reversal found in so many early modern intaglio prints, the image takes its composition directly from a picture exhibited at the Royal Academy in 1775 by Benjamin West (fig. 4.6).[42] Second, the artists featured are a familiar cast of characters (see fig. 3.3). In addition to the *Antiochus*, West is the source for two surviving sun-picture versions of *Venus and Adonis*, one in monochrome, a second with the protagonists in flowing robes of azure and scarlet, respectively (figs. 4.7–4.8). Academician Angelica Kauffman (who figured significantly in Joseph Booth's chemo-mechanical schemes for picture replication at the Polygraphic Society) inspired the set's *Graces Awakening Cupid* and *Nymphs Adorning the Statue of Pan* (figs. 4.9–4.12). Third, the collection of objects amassed by Smith was larger than what now survives; it included a depiction of Soho House on silver plate possibly made before its architectural alterations in 1791, from which a hand-retouched print was then made (fig. 4.13).[43]

In a certain sense, these sun pictures are not mysterious at all. The *Antiochus*, along with the *Venus and Adonis* variants, are likely species of polygraphic replication strategies produced at Soho through partnership with Francis Eginton and others, noted in chapter 3. But were they "really photographic"? That question inflects much of the scholarship available on these artifacts.[44] I resist it. The questions I want to ask are historical, not ontological. Said differently, establishing the ontology of these artifacts *as* photo-

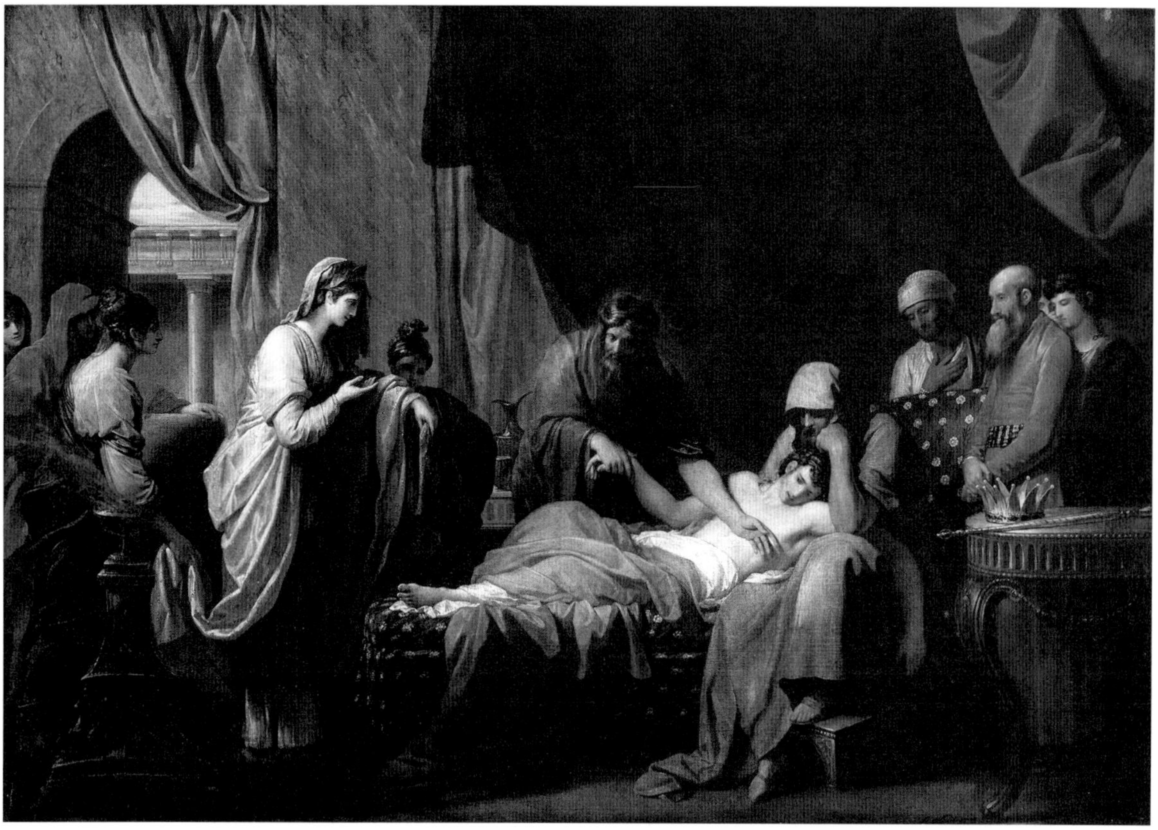

Figure 4.6 Benjamin West, *Erasistratus the Physician Discovers the Love of Antiochus for Stratonice*, 1772. Oil on canvas. Collection of the Birmingham Museum of Art (museum purchase with funds provided by Mr. and Mrs. Houston Blount, Mr. and Mrs. Michael Bodnar, John Bohorfoush, Mr. and Mrs. Percy W. Brower Jr., Mr. and Mrs. Thomas N. Carruthers Jr., Catherine Collins, Mr. and Mrs. Henry C. Goodrich, Mr. and Mrs. Hugh Kaul, Harold and Regina Simon Fund, Mr. and Mrs. William M. Spencer III, Mr. and Mrs. Lee Styslinger, and other donors).

graphic was the expressed concern of Smith, Price, and other historical agents; we obscure their arguments and short-circuit the stakes of their claims when we project the surety of our categories backward, presuming that we know definitively what photography "really" was then or should be now. Certainly, as parsed by the logic of the institution that now owns them, these works on paper and metal—reproductions of academic paintings and a depiction of domestic architecture—make a heterogeneous lot. London's Science Museum currently classifies five of these objects as "mechanical paintings" by Francis Eginton. The two other works on paper after Kauffman appear as "stipple engravings" attributed to William Wynne Ryland. Yet, from their discovery in the 1860s through their transfer to the newly opened Science Museum in 1886, all were discussed as "sun pictures."[45] That phrase was first introduced into Smith's correspondence in November 1862

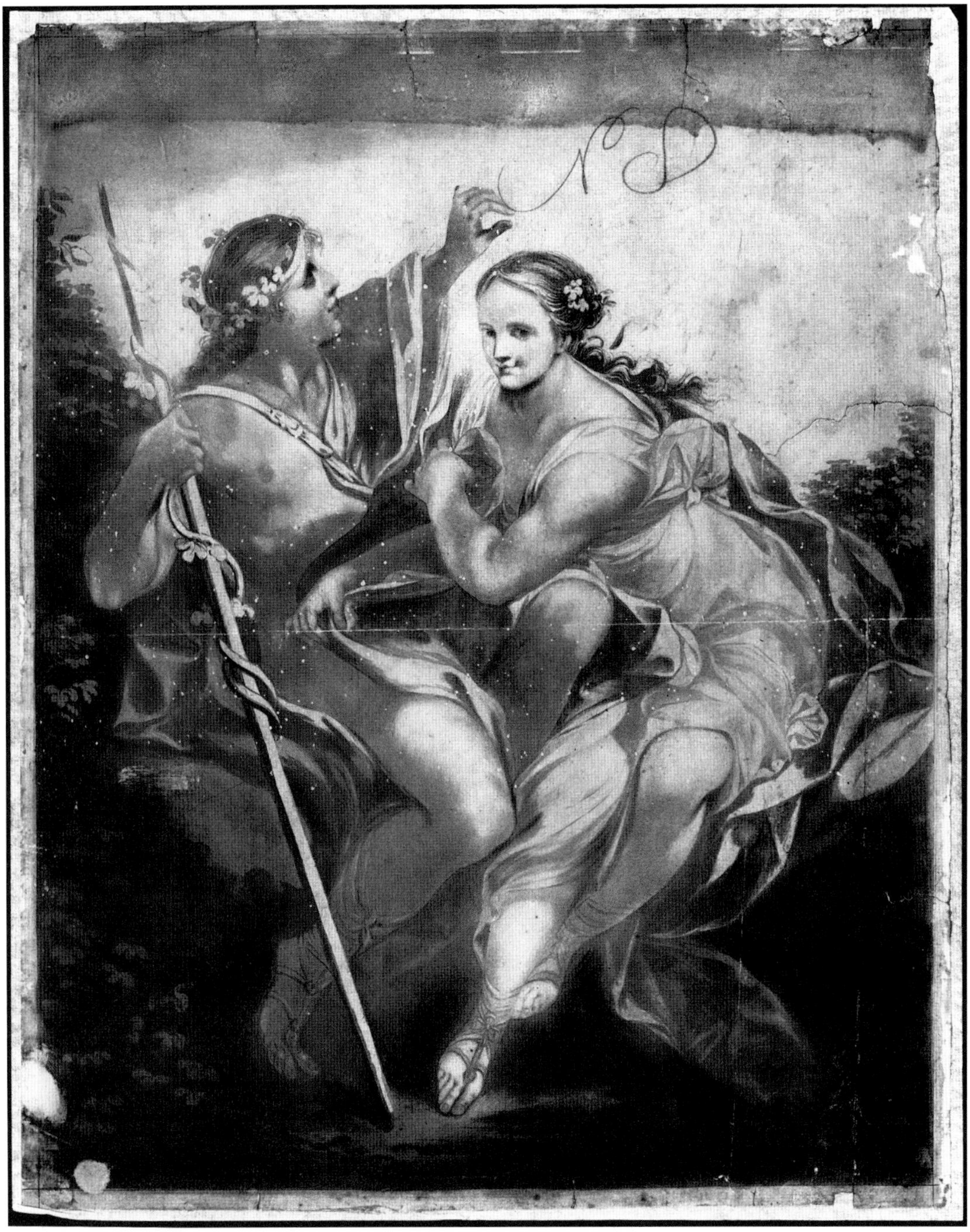

Figure 4.7 Francis Eginton, *Venus and Adonis*, ca. 1778. Mechanical painting. Science Museum, London (1884–16 Pt2).

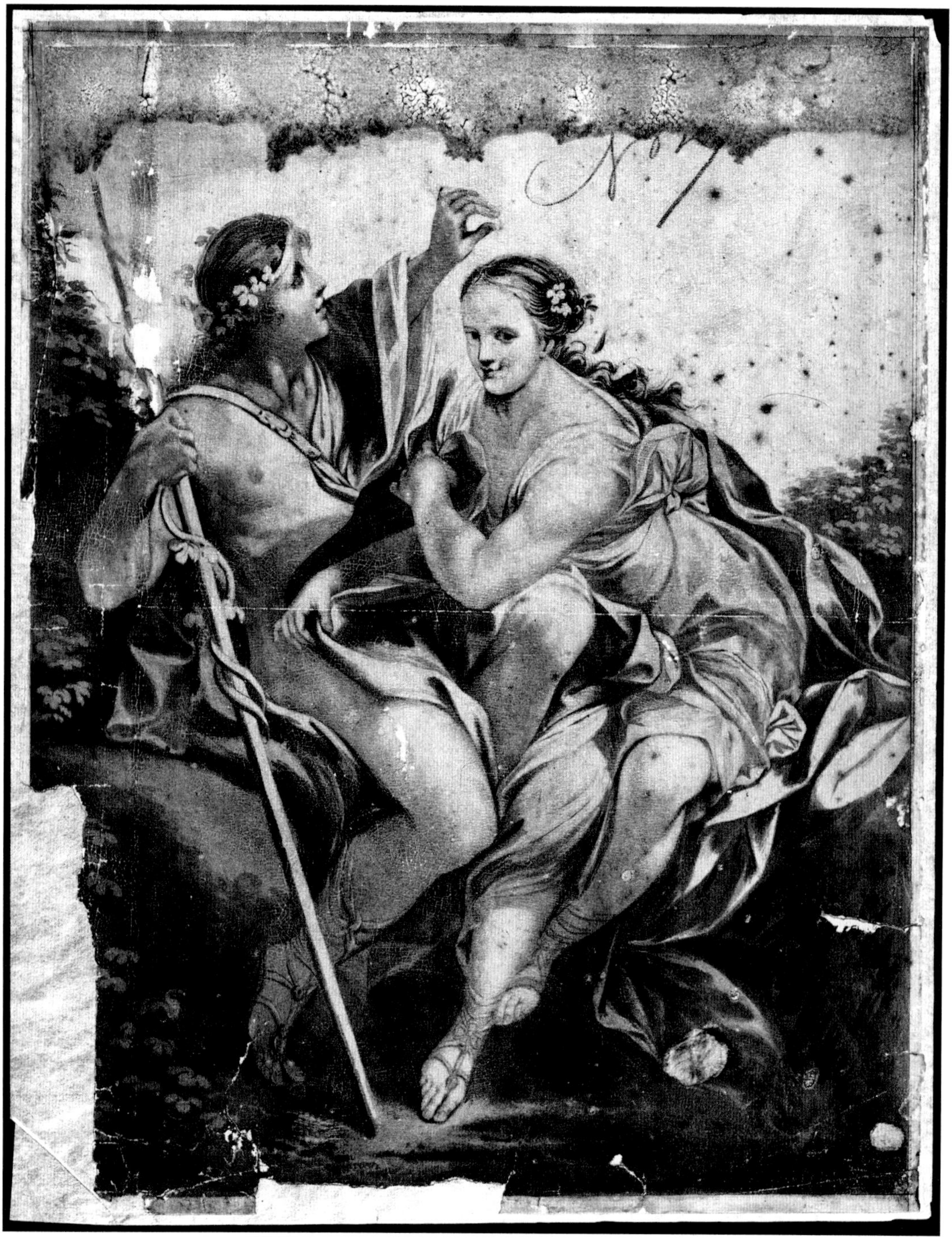

Figure 4.8 Francis Eginton, *Venus and Adonis*, ca. 1778. Mechanical painting. Science Museum, London (1884–16 Pt1).

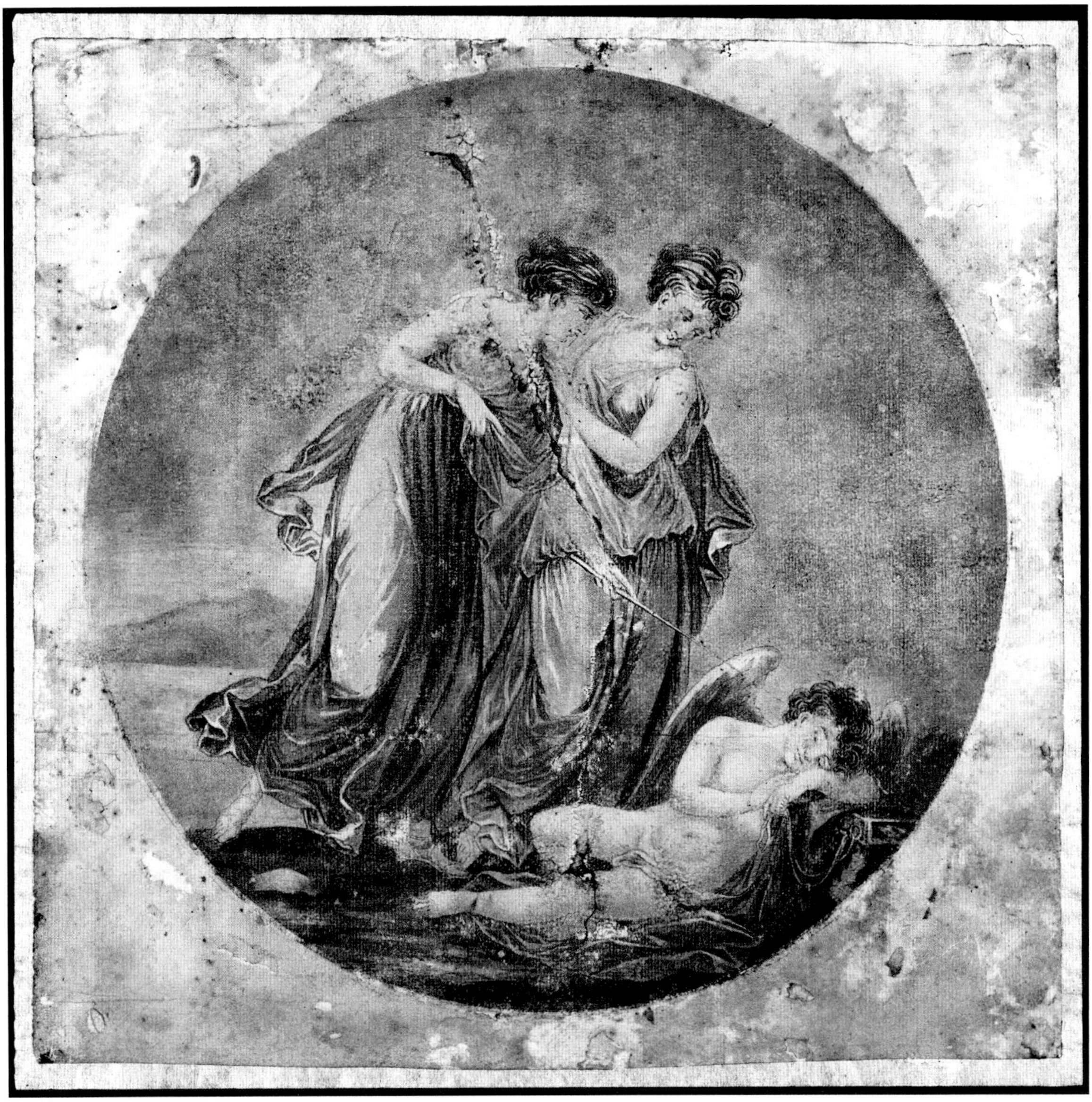

Figure 4.9 Francis Eginton, *Graces Awakening Cupid*, ca. 1778. Mechanical painting. Science Museum, London (1884–16 Pt3).

by Edward Price, who implies it had been employed by Boulton and Watt. Price claims to have seen a draft entry "in Mr. Watts own hand-writing" of a chapter on Staffordshire's Handsworth parish entered into the fifth edition of Samuel Lewis's *Topographical Dictionary of England* (1844) that "refer'd to these Sun pictures."[46] The only reference to image-making at Soho there, however, is to "the art of copying pictures in oil colours, called Polygraphic, [which] was also invented and pursued here under the direction of Mr. Francis Eginton, to whom it was subsequently resigned, and who be-

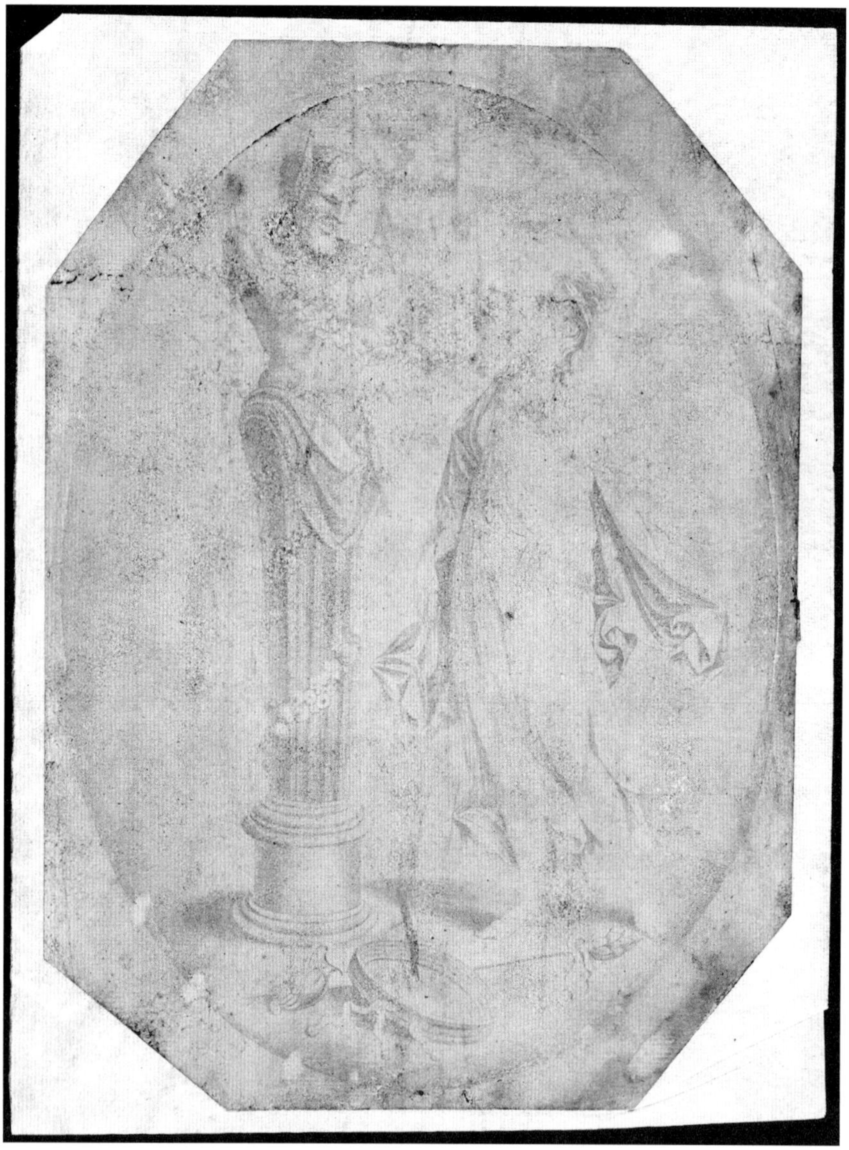

Figure 4.10 Francis Eginton, *Nymph Adorning the Statue of Pan*, ca. 1778. Mechanical painting. Science Museum, London (1884–16 Pt5).

came celebrated for his paintings on glass."[47] Lewis's entry makes no mention of "sun pictures" as such.

That was not Price's only recourse to texts. Subsequently, he instructs Smith that more about the sun pictures' provenance can be found in "the Record of the International Exhibition, part 12 page 565 [where] reference is made to experiments by Davy and Wedgwood for copying paintings on glass."[48] In that catalogue's concluding chapter, entitled "Photography and Chemical Manufactures," a concise history of photography by physician Hugh Diamond indeed begins with a suggestively decorated initial (fig.

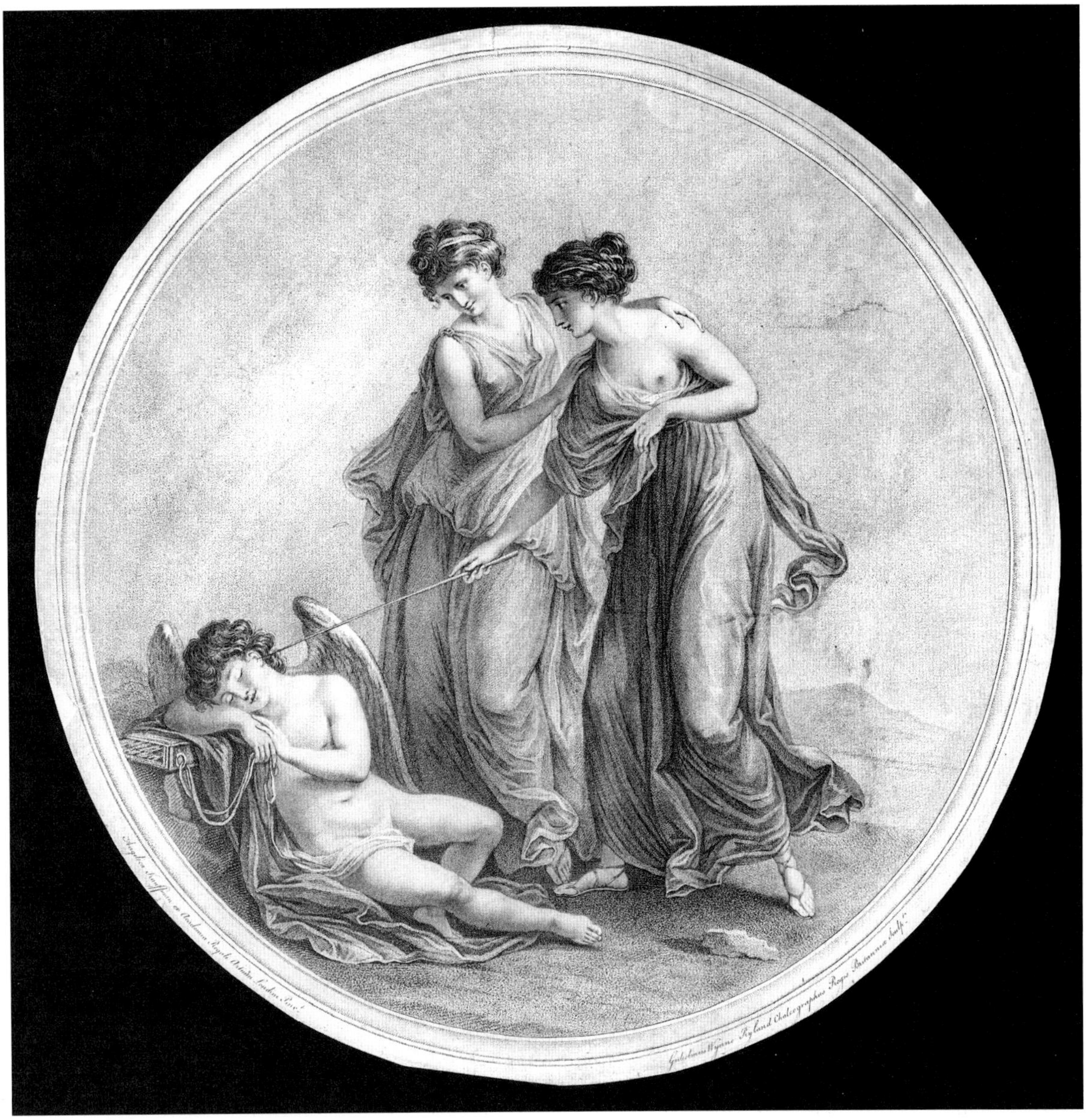

Figure 4.11 William Wynne Ryland, *Graces Awakening Cupid (Dormio Innocuous)*, issued [May 21] 1776. Stipple engraving after a painting by Angelica Kauffman. Printed in red, on laid paper; cut to roundel. Science Museum, London (1884–16 Pt4).

4.14). From the solar starburst at the lower left of this tiny wood engraving, the god Mercury sails through luminous heavens toward the stylized letter "I" with which Diamond's text begins. This pictorial transit nicely encapsulates the compression of light (*photo*) and writing (*grahos*) that composes photography linguistically, but Mercury is also depicted performing an act of photographic image-making. Turning from the solar illumination required for photosensitive exposure, the god breathes a white cloud onto

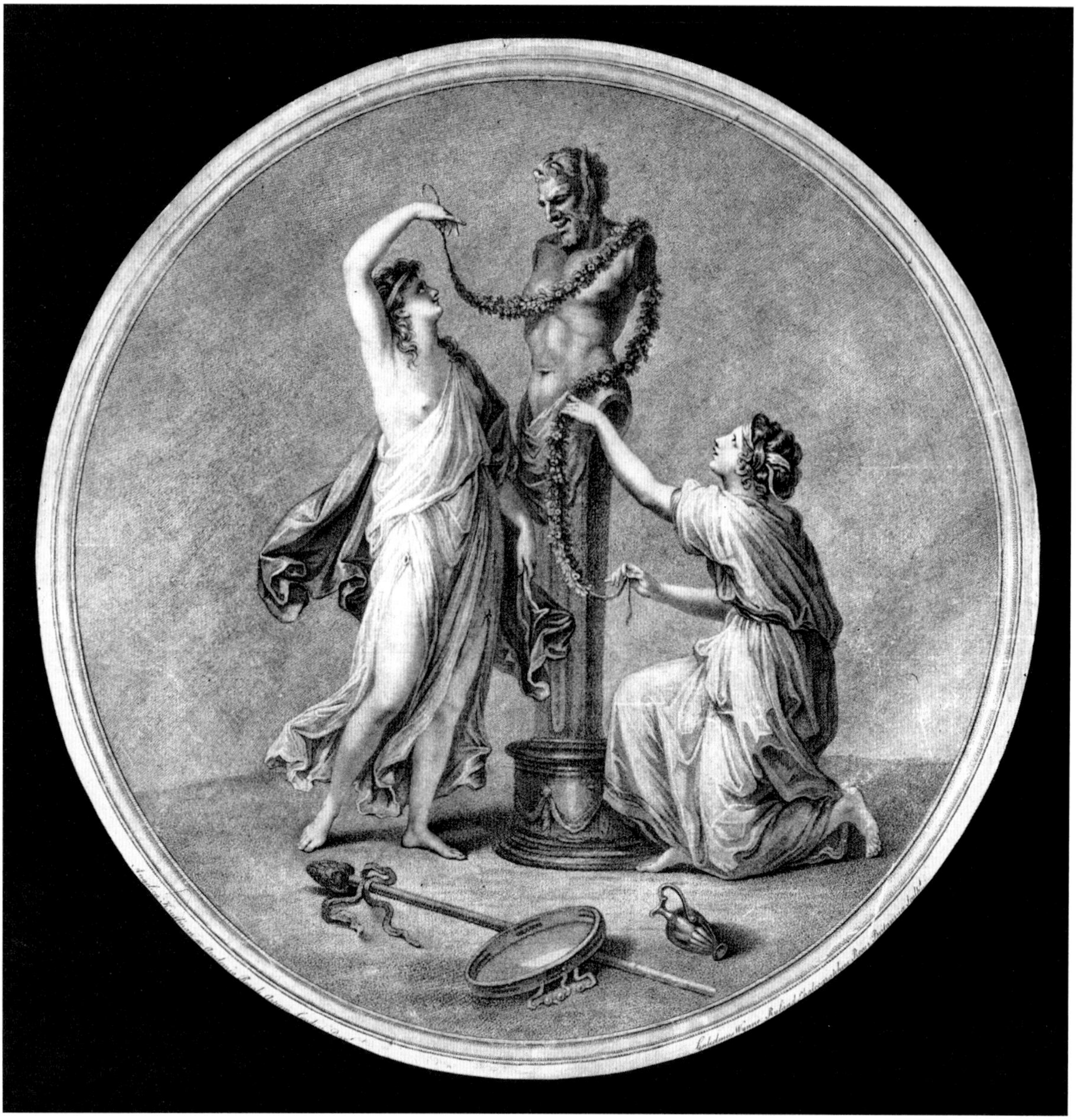

Figure 4.12 William Wynne Ryland, *Nymphs Adorning the Statue of Pan*, issued 1776. Stipple engraving after a painting by Angelica Kauffmann. Printed in red, on laid paper; cut to roundel. Science Museum, London (1884–16 Pt6).

the plate clutched in his left hand, thus delivering the mercury vapors required to make the latent image appear per Daguerre's process.[49] Secretary of the Photographic Society of London and pioneering photographer of the insane, Diamond then pits photography's putative origins among "alchemists of old . . . familiar with the properties of chloride of silver, then known as horn silver, which blackened when exposed to the rays of the sun." He finds more secure footing amid the trials of Wedgwood and Davy, despite

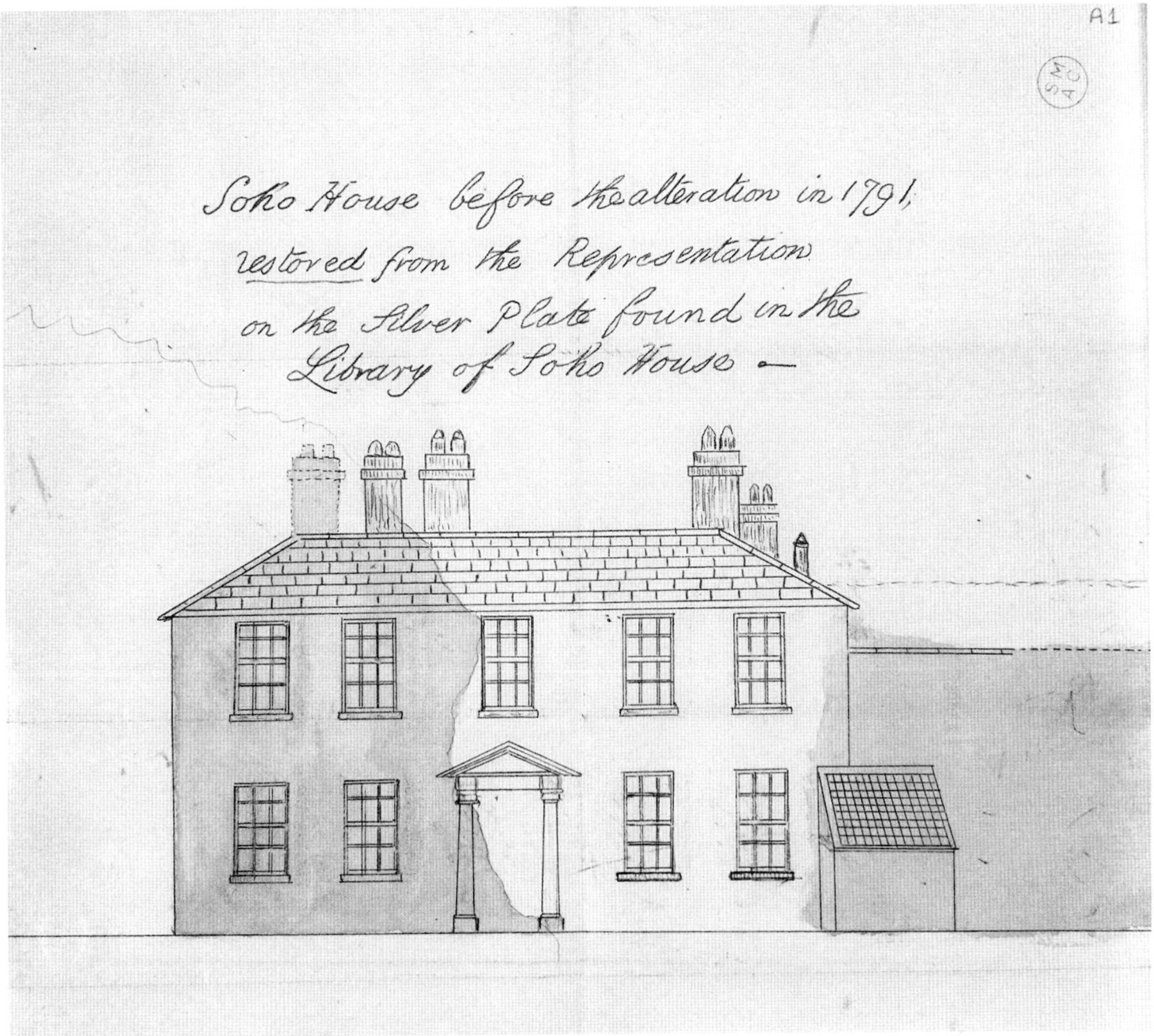

Figure 4.13 Unknown maker, "Soho House before the alteration in 1791, restored from the Representation on the Silver Plat found in the Library of Soho House," ca. 1865. Line drawing with green wash. Science Museum Library & Archives (MS 2117 A 1).

the fact that their figures were defaced by the same solar "power which produced the images."[50] Canvassing Joseph Nicéphore Niépce's art of "heliography" and the broader progress of photography through Daguerre and others, Diamond uses "sun pictures" as a generic term for images yielded by a consolidating lineage of photochemical techniques.[51]

I belabor the terminology and documentary sourcing to ask this question: with his talk of "sun pictures," had Edward Price innocently confused a sequence of practices—the "polygraphic" picture-making employed variously by Eginton at Soho and Joseph Booth in London; Thomas Wedgwood's unfixed experiments with photosensitive chemical preparations; Niépce's *héliographs*—that could foreseeably be accommodated to the marches of

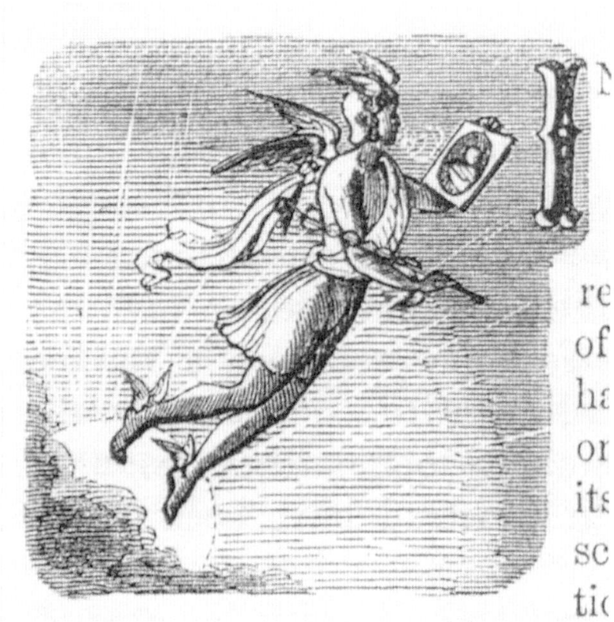

Figure 4.14 Unknown printmaker, decorated initial showing Mercury making a Daguerreotype, as printed in Dr. [Hugh] Diamond, "Photography and Photographic Apparatus," in *The Record of the International Exhibition* (Glasgow: William Mackenzie, 1862), 564.

progress then marshaled as photography's early history? After all, from Sir David Brewster's talk of "drawing by the agency of light" to the phrase "sun pictures" proper, made familiar through Talbot's *The Pencil of Nature* (1844–1846) and *Sun Pictures in Scotland* (1845), a language of solar productivity was central to emergent photographic discourse.[52] Or was Price being knowingly ambiguous, even intentionally duplicitous?

Either way, Price proved supremely skillful in sourcing an expanding array of sun picture–related artifacts over the winter of 1862–1863. A chance encounter with a Birmingham auctioneer reminds him, he tells Smith, that "some years ago when I turned out all the rubbish and waste paper from the Library at Soho House, he bought the old scrap paper and amongst it was a very *curious picture*! which he could not make out! He says it is neither chalk, Crayon, Indian Ink print or painting."[53] At the same time, Price was tempting Smith in a different direction. "I don't want to tease you too much," he avows in December 1862, "but suppose I could give you a clue to the Camera which made these pictures! I had it once!! and did not know what it was for."[54] As the correspondence unfolds, Price delivers dimensions for that camera ("twelve inches cube, made of oak, roughly and . . . the lens was not more than 2½ or 3 inches diameter") and an account of how the device had been standard at the Lunar Society's meetings.[55] The members stood "in a dark tent, and [there was] nothing to be seen except a *picture on the table*. By some process *they* secured this shadow."[56] This sets Price, a local auctioneer who had trafficked the artifact, and agents mustered by Smith on an ultimately quixotic quest for the missing camera. As Price ruefully laments, "it

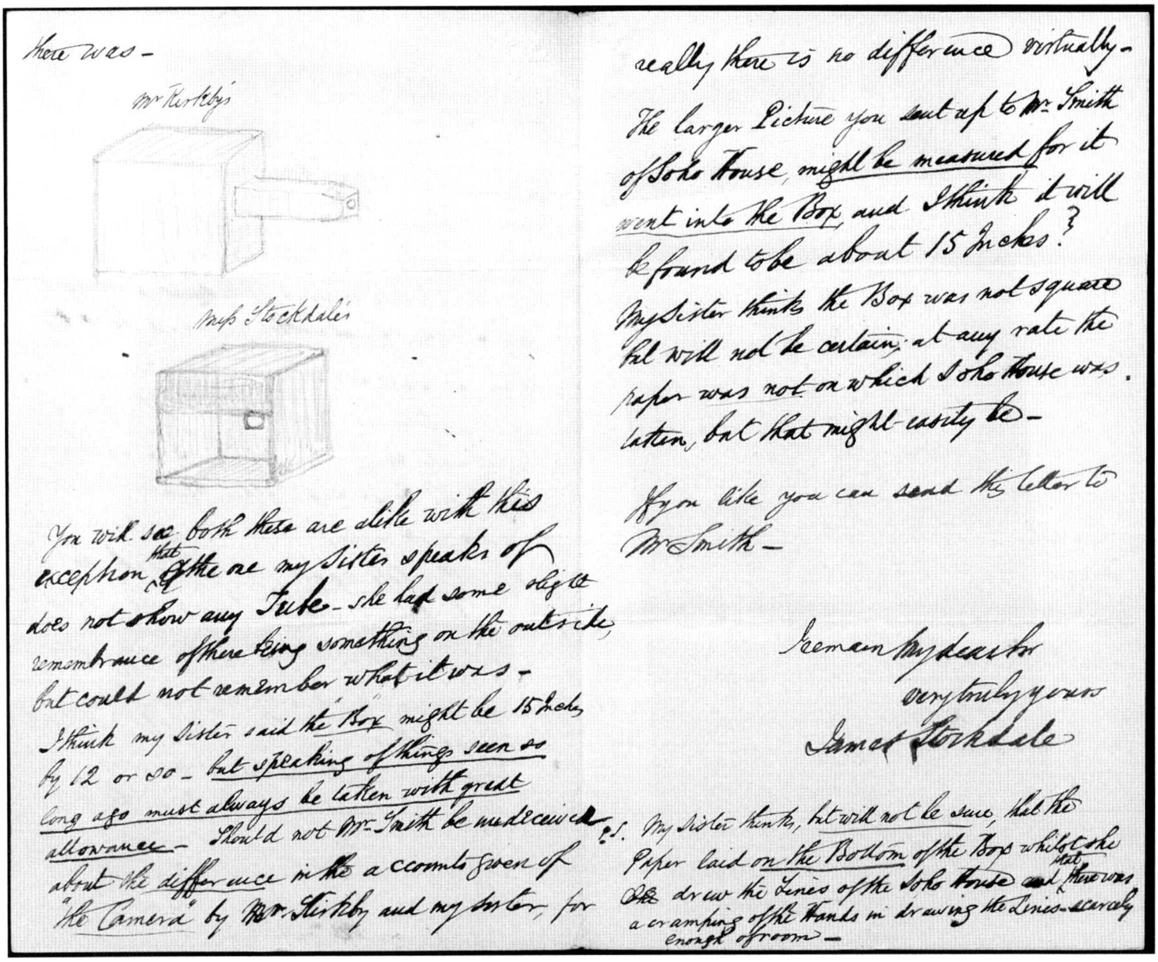

there was -

Mr Kirkby's

Mrs Stockdale's

You will see both these are alike with this exception that the one my Sister speaks of does not show any Tube - she has some slight remembrance of there being something on the outside, but could not remember what it was -

I think my Sister said the Box might be 15 Inches by 12 or so - but speaking of things seen so long ago must always be taken with great allowance - Should not Mr. Smith be undeceived about the difference in the accounts given of "the Camera" by Mr. Kirkby and my Sister, for really there is no difference virtually -

The larger Picture you sent up to Mr. Smith of Soho House, might be measured for it went into the Box, and I think it will be found to be about 15 Inches?

My Sister thinks the Box was not square but will not be certain, at any rate the paper was not on which Soho House was taken, but that might easily be -

If you like you can send this letter to Mr Smith -

I remain My dear Sir
very truly yours
James Stockdale

P.S. My Sister thinks, but will not be sure, that the Paper laid on the Bottom of the Box whilst she drew the Lines of the Soho House and that there was a cramping of the Hands in drawing the Lines - scarcely enough of room -

Figure 4.15 Unknown draftsman's depiction of two cameras in use at Soho in the second decade of the nineteenth century. James Stockdale to Amos Beardsley, January 18, 1864; Science Museum Library & Archives (MS 2117 G 6).

will be found somewhere in the north of Staffordshire nailed against a cottage chimney corner and used as a *Saltbox*."[57]

Although the trail of this camera goes cold, other correspondents come forward to offer their own recollections. Smith learns from one Amos Beardsley how, while taking a sketch of the Soho factory in 1818, a friend had been approached by Matthew Robinson Boulton (1770–1842), the elder Boulton's son and heir at the factory.[58] The younger Boulton had asked: "'Why not use the Camera' and the camera was fetched out of the library & used."[59] A schematic drawing posted to Beardsley exemplifies the depictions sent to Smith from a host of volunteers eager to recall instruments vaguely remembered and practices putatively employed at Soho some half a century earlier (fig. 4.15). Surely, these acts of memorialization speak to a period perception that the industrial present was separated

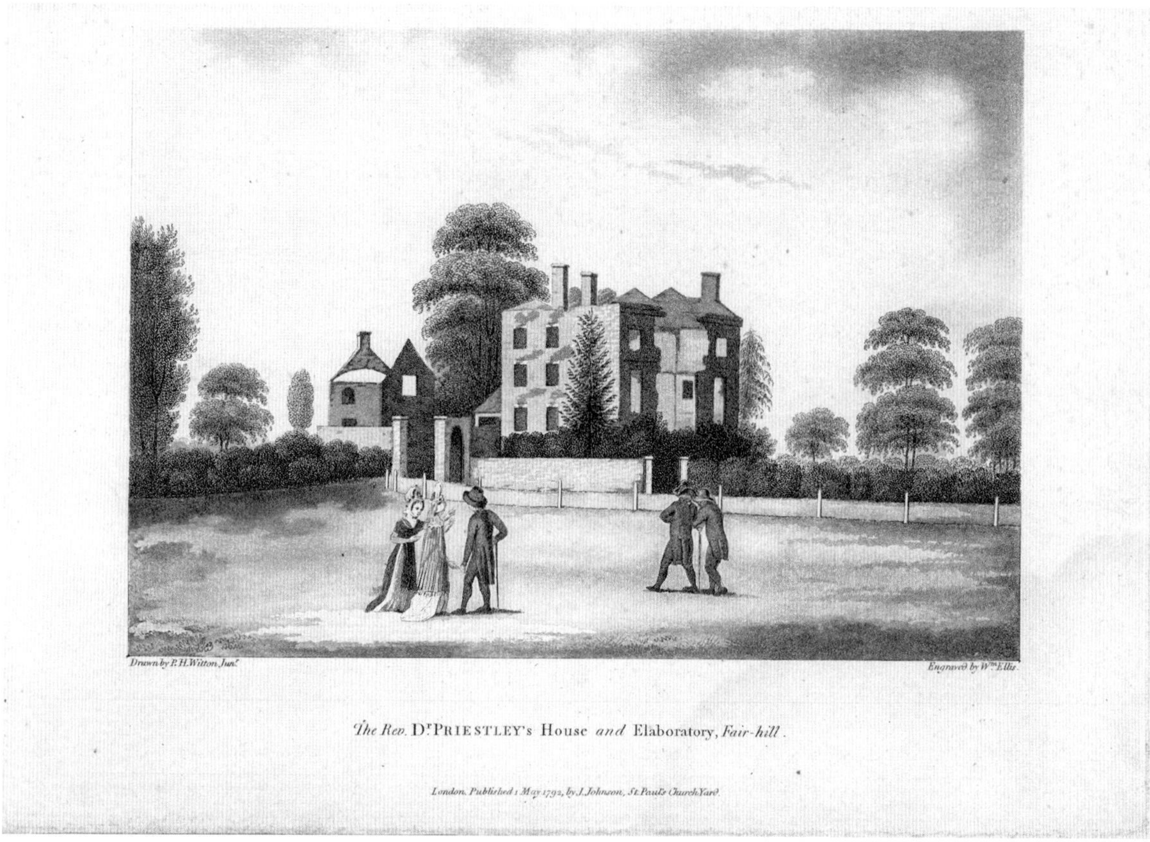

Figure 4.16 William Ellis, engraving after Philip Henry Witton, *Views of the ruins of the principal houses destroyed during the riots at Birmingham* [*Vues des ruines des principaux batiments qui ont souffert dans les emeutes de Birmingham*] ([London: Printed for J. Johnson], 1791 [1792]). Huntington Library (386313).

from its eighteenth-century forebears by a gap only widening in its intensity. By 1833 reformers such as Peter Gaskell were identifying the advent of the steam engine as having imposed the factory system and its pernicious health consequences on workers in merely one human generation.[60] Friedrich Engels traced a comparable time line in 1844 as he anatomized an English proletariat effectively made by rapid transformation of the textile industry around the steam engine, water frame, and other technological innovations. "The industrial revolution is of the same importance for England as the political revolution for France," Engels observed, ". . . and the difference between England in 1760 and in 1844 is at least as great as that between France, under the *ancien régime* and during the revolution of July."[61] Birmingham had felt those revolutions acutely. In 1791, the city had seen reactionary mobs destroy the homes of prominent political progressives including Joseph Priestley and William Russell, William Russell Birch's cousin (fig. 4.16). Birmingham became a site of intensive union organizing after the economic slump of the 1820s and Chartist riots in 1839, even as its city center was remade around Watt's steam-engine technology in the 1840s.[62] Indeed, a key aim for founders of the Birmingham and Midlands Institute (1854) and backers of the city's triennial music festival was to counter Birmingham's dismal reputation as "the birthplace of mediocrity, the home of 'lacquered shams.'"[63]

Writing on the wrong side of the city's Bull Ring riots—with the benevolent paternalism he associated with the first Matthew Boulton replaced by Victorian boosters' bureaucratized schemes—Price saw such changes as willful amnesia. Against Smith's ready appetite for Soho-related artifacts, Price positioned the Birmingham elite and their shoddy treatment of the city's industrial past. To Smith, he claimed:

> The Trustees of our Midland Institute will be awfully riled when they hear of your success, for they are now very busy forming a museum and portrait gallery of all local associations. It serves our conceited Brums right—they have had many opportunities of securing all the relics of Soho, and also the Soho site itself, for some years ago it was offered to the town. But instead of being thankful, the corporation merely insulted Mr. Boulton and then the destruction of this dear old place commenced.[64]

But those descendants of Soho were hardly any better. "It is too bad of Mr. J. W. G. Watt," Price lamented. "He, like Mr. Boulton, don't do anything to the glory of the name."[65] Watt's heirs were wholly occupied with shooting and other leisured pursuits, he wrote; "the present Mr. Boulton takes no interest whatever in these things and would never give himself the trouble to enquire."[66]

If the phrase "Industrial Revolution" was not common in English parlance until the 1880s, the years around 1860 were an active moment for documenting the titans of eighteenth-century industry.[67] Titles such as James Patrick Muirhead's three-volume *Origin and Progress of the Mechanical Inventions of James Watt* (1854) and *Life of James Watt* (1859), and Samuel Smiles's *Lives of Boulton and Watt* (1865) and three-volume *Lives of the Engineers* (1861–1868), were joined by the two-volume *Life of Josiah Wedgwood* (1865–1866), penned by the radically inclined historian Eliza Meteyard (1816–1879).[68] Keen to underscore that Wedgwood was an occasional visitor at the Lunar Society and not a fully fledged member, Meteyard depicted her protagonist as a catalyst to the ebullient, intellectual foment of the 1750s Midlands: crucial infrastructure on which the "Lunar club" would subsequently stand.[69] "The comparatively free and healthy condition of public and private opinion in these busy places, compared to the bovine sloth of the country districts," so Meteyard claims of Wedgwood's learned contacts in Birmingham, Liverpool, and the Warrington Academy of John Aikin and Priestley, "was a remarkable feature of this period of our national industry."[70]

Minimizing her public involvement in Smith's investigation, Meteyard was actively engaged with that project insofar as it had bearing on Thomas Wedgwood's position in the history of photography.[71] In the early fall of 1863, Meteyard could set "the date of Thomas Wedgwood's first trials in Photography . . . to the middle of the year 1791. I find from letters in the handwriting of Alexander Chisholm (Josiah Wedgwood's Chemist) to Burley the cameo and seal setter in Birmingham that lenses and instruments, which would have been only serviceable for a scientific purpose of the kind were to be procured, exchanged, or altered for Mr. Wedgwood."[72] In subsequent correspondence with Smith, she identified two campaigns in Wedgwood's "experiments in Photography"—the first by himself in 1790–1791, the second with Davy after 1797, when they met at Thomas Beddoes's Pneumatic Institution in Bristol—but both yielding images with "a faded appearance, a defect probably due to the quality of the chemical agents employed rather than to a want of manipulation [and] dexterity."[73] Magnifying lenses were required for the artist producing Meteyard's prints to see the target of replication; but no chemical fading would stop her from publishing examples of Wedgwood's early "photographs." The second volume of Meteyard's biography included an engraved view of a tureen flanked by pitchers, tongs, and settings on a legless table (fig. 4.17). It floats against a bare expanse of page, above a caption identifying the figure as "an early photograph by Thomas Wedgwood."[74] Meteyard subsequently reproduced a *different* image as "the earliest known Heliotype, or Sun Picture taken by the Inventor of Photography Thomas Wedgwood in 1791–1793" in her *A Group of Englishmen* (1871). Through etched aquatint, Henry Adlard depicts Wedgwood's rendering of a musician fingering bagpipes as he stares out from the picture plane (fig. 4.18).[75] Apparently made itself after a print of a Savoyard piper cut into copper from a painting by David Teniers II, Wedgwood's fading image was, by Meteyard's account, "clearly a heliotype, or, as we should now say, a photographic copy, of an ordinary book-engraving."[76] By Meteyard's lights, this aquatint after a replication of a print after an oil painting proved that photography had been an eighteenth-century invention.

Edward Price imagined Enlightenment industry and its forays into "this branch of chemical optics" in very different terms.[77] The site neither of "industrial enlightenment" nor the polite entrepreneurship described by recent scholarship, the Soho Manufactory imagined by Price was a font of shadowy forces and stupendous power worthy of Wright of Derby's *Alchymist* (see plate 8).[78] To Smith, Price claims that the room at Soho in which the Lunar Society convened for its monthly meetings "was like a Free Masons lodge. Indeed, I believe the Lunar Society was really a Lodge because when

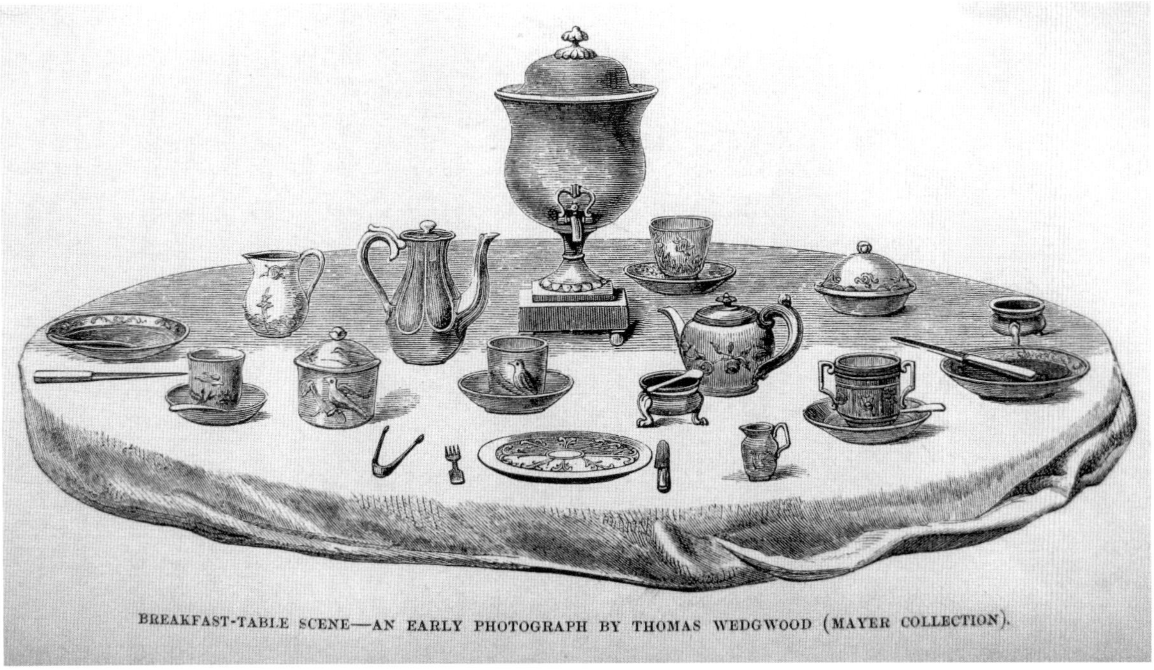

Figure 4.17 Unknown engraver, "Breakfast-Table Scene: An Early Photograph by Thomas Wedgwood," as published in Eliza Meteyard, *The Life of Josiah Wedgwood: From His Private Correspondence and Family Papers . . .* (1865), 2:585.

we pulled down the old Library . . . upon the plaster I found traces of the emblems of masonry. There was (east) delta over C and other signs, but of course you not being a mason I cannot explain more."[79] More extraordinary still were assertions Price relayed from Matthew Boulton's personal manservant, a Mr. Townsend. By that account, the sun pictures had first been disclosed to a wider artistic world in the fall of 1798, when Sir William Beechey (portrait painter to the royal family, who had been knighted earlier that spring) traveled to Handsworth to paint Boulton's portrait (fig. 4.19).[80] "When he was at Soho," Price recounts, "Mr. Boulton explained this invention of taking Sun Pictures. Sir William then went among all the Artists and got up a petition to Matthew Boulton and the Lunar Society begging them to stop because the secret would, if made known, be the means of *shutting up the painters shops*."[81]

Of course, Beechey was one of the artists through whose hands Reynolds's technical notes had passed in the early nineteenth century, per Haydon's transcription. He has also been identified by modern scholarship as a possible source for public exposition of the fraud of the Venetian secret in 1797—a leak putatively made in retaliation against academicians who failed to elect him to their membership that year (see plate 16).[82] If he was thus capable of such conspiratorial plotting, the royal portraitist Beechey certainly also had access to higher powers. "I think Government has some-

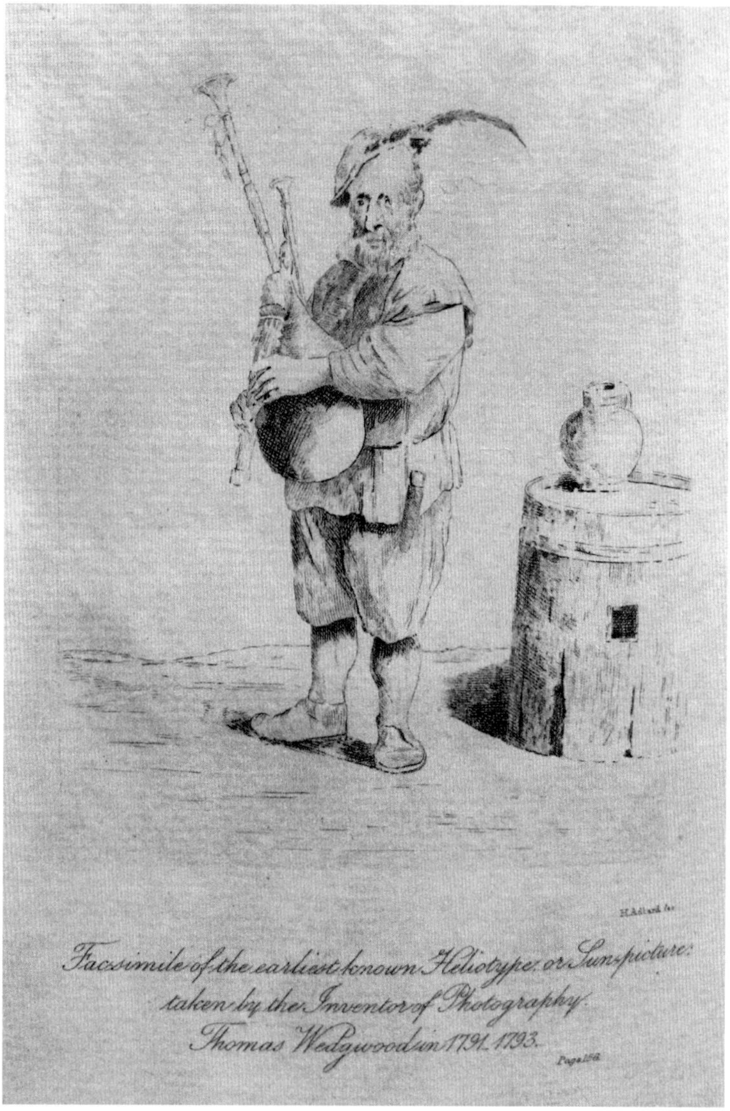

Figure 4.18 Henry Adlard, "Facsimile of the earliest known Heliotype, or Sun Picture taken by the Inventor of Photography Thomas Wedgwood in 1791–1793." Aquatint[?] and etching, published as a frontispiece in Eliza Meteyard, *A Group of Englishmen (1795 to 1815) Being Records of the Younger Wedgwoods and Their Friends . . .* (1871).

thing to do with the suspension of this trade," Price tells Smith, "because the person who held the secret was offered a pension, but Matthew Boulton objected to it."[83] As he and Smith had begun to disclose the secret of the sun pictures anew, Price fears retribution from those dark, highly placed corners. "I don't know how I shall steer my craft thro' all this," he confesses to Smith in 1863, "For I have already raised the ire of many of the Birmingham + Soho men for parting with these curiosities—they blame me very much. Why did they allow that Shrine of Sciences and arts to be razed to the ground and its contents dispersed by the ruthless auctioneers?"[84] Price predicts that what local snobbism and lack of filial piety had not already squandered, a sinister, stealthy establishment would stamp out: "You may depend upon it: this secret was allowed to die out with the deaths of the

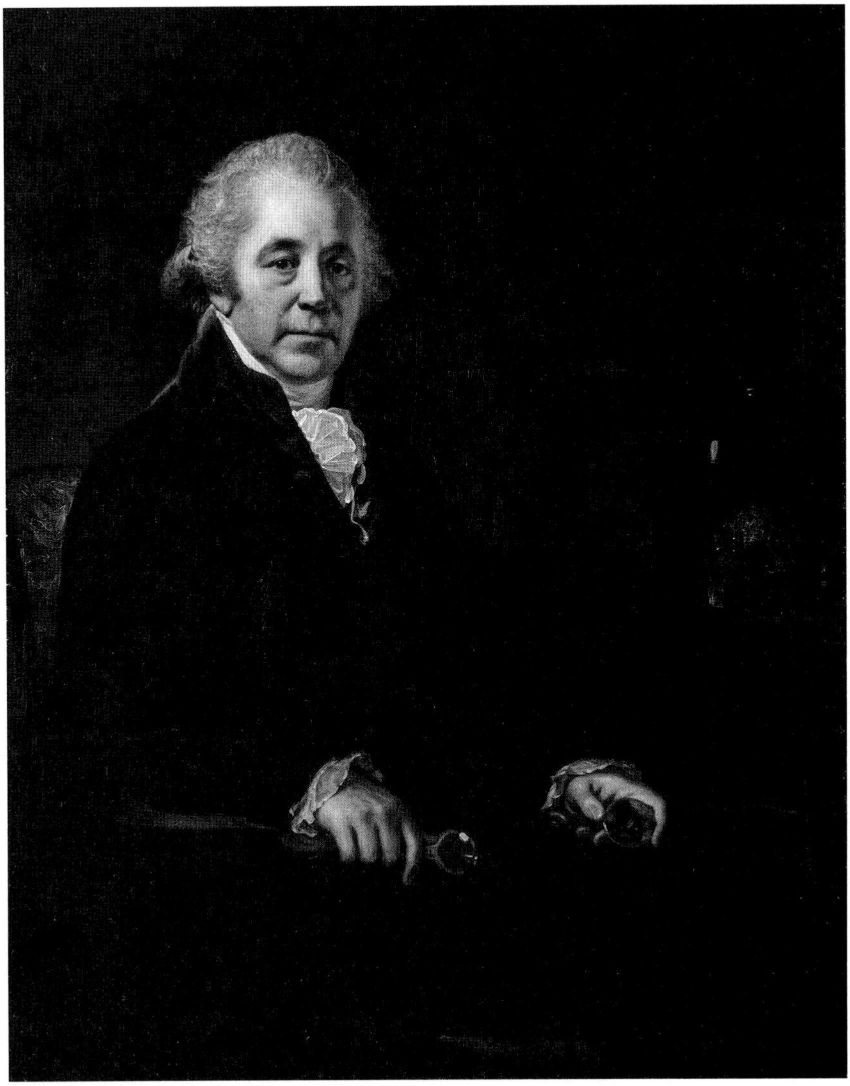

Figure 4.19 Sir William Beechey, *Matthew Boulton (1728–1809)*, ca. 1802–1803. Oil on canvas. Birmingham Museums and Art Gallery, Birmingham, UK.

Lunar Society and all traces of it were destroyed at the instigation of the Royal Academy and some members of the Government."[85]

Delusional as Price sounds in such passages, his fears would prove entirely justified once Smith presented his sun-picture research to a leading body of British photographers, the Photographic Society of London. At King's College on November 3, 1863, Smith read excerpts from his correspondence and presented the material evidence of his findings:

> The pictures . . . were passed round for the examination of Members. Some consisted of reproductions of pictures in monochrome and colour, by Benjamin West, Angelica Kauffman, and other eminent names

(these pictures were on paper, some very large, done in two pieces, and altogether unlike any class of pictures we know); of the two pictures on silver plate, about 7 by 4, and of the two pictures by Wedgwood (one about 8 by 6, a view of a breakfast-table, having much the appearance of a faded silver print; and another similar in appearance, a small reproduction of a drawing).[86]

Smith's claims for the sun pictures as early photographs convinced many. Leading microphotographer and founding member of the Photographic Society, George Shadbolt (1817–1901), was persuaded by the silver specimens: "He had examined them carefully with a lens, and he was assured they were camera images from nature. . . . One of these, it appeared, took back the invention of photography to a period before 1791."[87] Engraver and antiquary Frederick William Fairholt (1813–1866) declared the sun pictures to bear no relation to known printing techniques. "A magnifying-glass would . . . show a grained or engraved surface, which never appears on these," he contended. "It is also very remarkable that, whether in simple tints or in colours, they lie, as it were, on the surface of the paper; they do not sink in as printing-ink or colour would do."[88]

Just as they prompted Eliza Meteyard to new regard for Tom Wedgwood's research, so the sun pictures and their peculiar absorbency called to mind other Enlightenment chemical innovations. Correspondent John J. Cole traced the sun pictures' origins back to the copying machine James Watt had patented in 1780, as noted in chapter 3 (see figs. 3.8–3.9). Discovering old documentation on Watt's device, Cole declares that "the materials for the ink and the paper, and the mention of the substances 'capable of turning black or deep-coloured with solution of iron' are very suggestive."[89] What they *suggested* was that because the Wedgwoods at Etruria—or Watt, Boulton, and Eginton working at Soho—had been capable of making light-sensitive, chemo-mechanical images that changed in time, those Enlightenment savants *must have been* in command of photography.

Other interpreters took exception to such inferences. In her 1857 essay, Lady Eastlake had surveyed the photochemical research of Robert Hunt as she lyricized upon the unnatural pictorial effects yielded by the asymmetrical speeds at which light acts on photographic surfaces. "So impatient have been the blues and violets to perform their task upon the recipient plate," Eastlake writes, "that the very substance of the colour has been lost and dissolved in solar presence; while so laggard have been the reds and yellows . . . that they have hardly kindled into activity before the light has been withdrawn."[90] In late 1863, Hunt himself wrote in to the Photographic Society, stressing the principles of photochemistry that militated against the sun pictures' photographic status.[91] Citing an image like the replica of West's *Antiochus and Stratonice* (see plate 20), Hunt called attention to the even resolution of form across the pictorial field. "The pictures show none of those inequalities," he argued, "which would result had they been copied by means of the camera obscura from *oil paintings* . . . inequalities due to the variations of chemical force in the radiations from surfaces of different colours."[92] Since no photographic process in use in the 1860s—let alone in the eighteenth century—could deliver such results, the sun pictures could not be photographs. More dismissive still was Thomas Augustine Malone (1823–1883), a photographic chemist employed by Talbot and editor of the Liverpool-based journal that became the *British Journal of Photography*.[93] Malone objected that the polygraphic pictures produced by Eginton, Boulton, and Watt bore no connection to "the operation of light, and it was a gratuitous assumption that there were in any sense photographs." Further, the fact that Thomas Wedgwood had been unable to fix his images squarely demonstrated to Malone that there never was any great secret of photography possessed by the Lunar Society; had there been, the younger Wedgwood surely would have received it from his famous father. Worse, Malone charged, Smith simply had no understanding of photography: "If the matter had fallen into the hands of a photographer for examination, they would never have been brought before a photographic meeting or have had such claims made for them."[94]

Nonetheless, Smith's interpretation found warm reception in other quarters. "I need not say how much gratified we all were with the sight of the sun pictures," wrote Thomas Lloyd (1803–1875), a naval engineer who had overseen trials of Smith's screw-propelled *Archimedes* in 1840 and held the post of inspector of steam machinery in the Royal Navy since 1850.[95] Having viewed the pictures in early 1863, "I have not recovered my astonishment. Landseer says they are superior to any he ever saw (I think I am not exaggerating his expression) in respect of semi tints or tones. We know this is the great defect of photography. It is all black or white or at all events approximating too nearly to the two extremes."[96] For Birmingham's local press, meanwhile, the sun pictures offered an opportunity for unabashed boosterism in an age of imperial rivalry. Against the priority awarded to the Frenchman Daguerre, "evidence has now been disclosed which seems to show that the merit really belongs to James Watt, and that consequently Birmingham may claim the honour of being the place in which the art of photography was discovered and first practiced."[97] Less than a week after his presentation at King's College, Smith's findings were summarized in an article entitled "Who Discovered Photography?" in the *Saturday Review of Politics, Literature, Science and Art*.[98] And there, the story was read by Matthew Piers Watt Boulton.

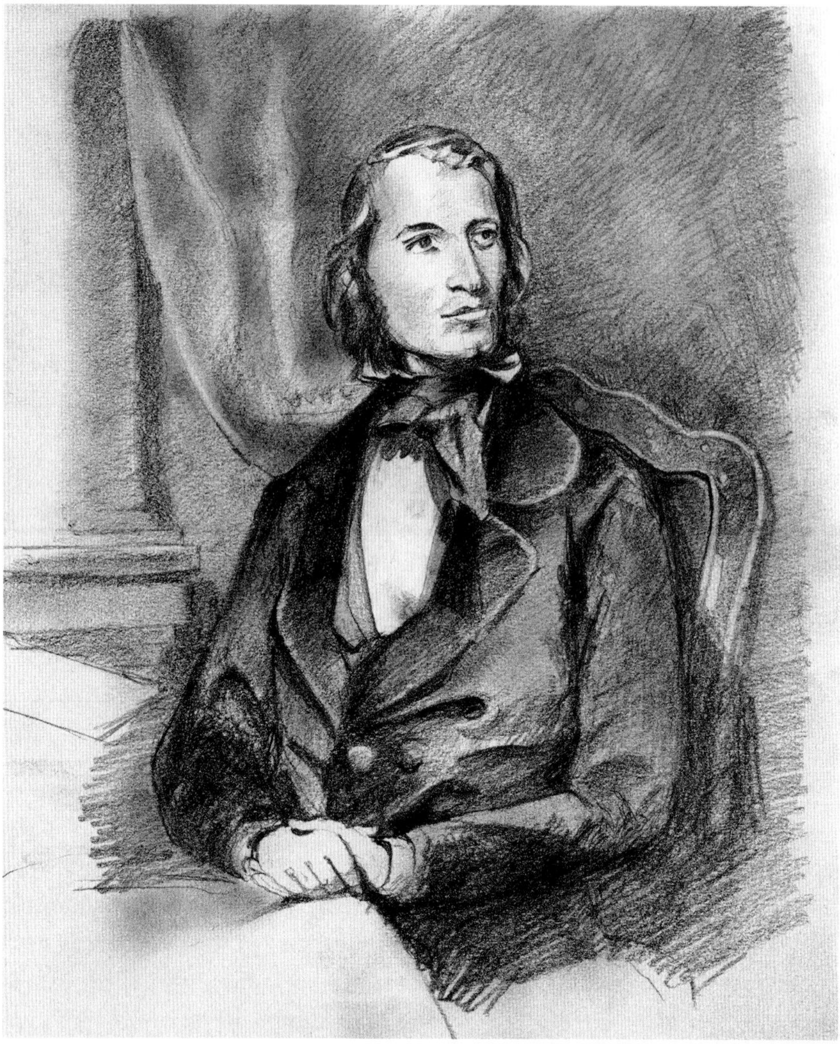

Figure 4.20 Harry Sandoval Jr., after Sir Francis Grant, *M. P. W. Boulton*, ca. 1840.

The Dark Side of the Sun

Raised on an eight-thousand-acre estate in Oxfordshire, educated at Eton and Cambridge, M. P. W. Boulton (1820–1894) would seem to cut the very figure of the Victorian country gentleman (fig. 4.20).[99] Writing some seven letters to Smith in less than ten days of early November 1863 from the resort of St.-Leonards-on-Sea on England's south coast, Boulton was initially sympathetic to the sun pictures. He allowed that they likely were "specimens of the peculiar process which was carried on in my grandfather's time."[100] If doubting the plausibility of "sun pictures" as a period name, Boulton expressed pique at having "never heard one syllable on this subject from Price."[101] By year's end, he had published his position on the matter in a short pamphlet entitled *Remarks on Some Evidence Recently Communicated to the Photographic Society*. Despite the fact that his "intelligent and useful servant" Edward Price had likely misapprehended some of the details, Boulton declared, "The pictures made at Soho by the secret method practised there were produced in some way by the agency of light."[102]

Privately, however, Boulton was taking a darker view. Warning Smith of his "discovery of Price's great dishonesty," Boulton announced systematic irregularities in Soho's accounts; Price, he claimed, had embezzled over one thousand pounds.[103] And five days before Christmas, he dropped the ax. As a frantic Price relayed to Smith, "He has discharged me in toto. Ordered me to leave his house in three weeks. Has taken all my account Books + papers & prohibited me from ever transacting any business whatever for him or in his name. I am left penniless, miserable and wretched."[104] Like the mythical curse of King Tutankhamun's tomb in a later age, the sun pictures were to blame for Price's misfortune: "I feel convinced—nay, I am sure—that had it not been for the annoyance he had experienced at my indiscretion & folly in making myself so busy and bringing his family name before the public, I am sure all other matters would have been amicably settled."[105] Mere months after Smith's presentation to the Photographic Society, Price had fled England with warrants out for his arrest: "I and my wife are poor miserable wretches now."[106] Price's voice thence vanishes from the correspondence entirely.

Having dispensed with Price, Boulton went on a far more aggressive campaign against the sun pictures. If his initial pamphlet had covered some six pages, *Remarks Concerning Certain Photographs Supposed to Be of Early Date* (1864) ballooned to seventy-one. Describing Price as "guilty of great dishonesty, having fraudulently appropriated a large sum of money," Boulton turns his attack on Smith.[107] Not only had Smith been a gullible dupe to Price's plots, but he had been a poor, even immoral, manager of gentlemanly affairs. "If a servant of his own were to deal similarly with his property, Mr. Smith's eyes would be opened. So, too, if I were to accept Mr. Smith's property from a servant of his, believing the servant's statements, however suspicious, and making no communication on the matter to Mr. Smith himself, Mr. Smith would strongly disapprove of my behaviour."[108] Where he had previously allowed that polygraphic picture-making, photography, and sun pictures might all be continuous in harnessing the agency of light, Boulton now followed the course of Thomas Augustine Malone, insisting they be split apart. Quoting the dimensions of Eginton's polygraphic replicas after paintings by Reynolds, Boulton affirms that those images might be called "mechanical."[109] But the technique that made them was "probably widely different from photography" and its products never then described as "sun pictures."[110] Moreover, after delegating agents to interrogate octogenarians surviving in greater Birmingham, Boulton charged that neither was the building depicted in Smith's silver plate Soho House nor was the plate old (see fig. 4.13). Rather than being made by Boulton's grandfather through some mysterious technique in the 1780s, the silver plate was credited to his *aunt*, working in the 1840s with a then-standard Daguerreotype process.[111] Splitting metal and paper apart—dismissing the former as photographic but not ancient, the latter as from the eighteenth century but not photographic—Smith's cache of material was evacuated of its evidentiary value. The sun pictures *qua* photographs had effectively ceased to exist.

In a still-longer iteration of his pamphlet, Boulton names that demotion in his title: *Remarks Concerning Certain Pictures Supposed to Be Photographs of Early Date* (1865–1866). Both the images' date and photographic nature now stand as mere suppositions. Price's villainy is further elaborated; he is discovered "engaged in most dishonest practices, carried on under the cover of gross falsehoods. To escape the punishment of his doings he absconded, and is affirmed to have quitted the country." Boulton also examines new technical evidence on the "perishable nature" of Eginton's mechanical paintings, now made worthy of the mythical Reynolds himself, "supposing the printed colour to have been mixed with gum, honey, and essential oil."[112] Yet, already in early 1864, Francis Pettit Smith had retaliated. In a strongly worded letter, Smith excoriated Boulton for his pamphlets, which served "to throw a very considerable wet blanket on the whole question at issue."[113] Chastening Boulton for his treatment of Price as "harsh in the extreme," Smith turns the charge of poor management back on his adversary. Since he had met Price but briefly on four occasions, the curator claims, "The error in my judgment respecting him is a very excusable one when compared with your own, who had known him so much longer and permitted him to fill a post of trust in which he had of course your fullest confidence."[114]

Assailing Boulton's ability to manage underlings amid what social historians describe as a nineteenth-century shift from the aristocratic retinue to the bourgeois notion of the domestic servant, Smith deploys his sharpest edge when sounding Boulton's inscrutable motives.[115] Smith pulls no punches as he quotes from correspondents on the ground at Soho:

> "The plates being found at Soho House would seem to imply that Boulton [primus] had more to do with them than Mr. Watt. The destruction of the Manufactory seems to have revived much interest in the old associations of that once celebrated spot. You would be horrified at the vandalism of the 'Brums' and the snobbism of the descendants of the Glorious Old Matthew Boulton *whose only aim seems to be to erase everything by which they got their money*."[116]

The relayed accusation is bald. The only reason the sun pictures had been targeted was that they forced into modern visibility a lucrative, industrial heritage that an *arriviste* landowner like M. P. W. Boulton had actively forced into a state of decay—and not a graceful one at that. Since Boulton has insisted on obfuscating the intellectual issues while hindering access to evidence through his self-serving attacks, Smith slams the door. He has "no alternative but to hand you this assurance that I shall take no notice whatever of any further correspondence which you may address to me."[117]

In this light, a down-to-earth interpretation of the sun pictures could be dealt out in the following way. Despite the fervent protests of leading practitioners, photography had been classed at London's 1862 International Exhibition not as a fine art but as a subset of machinery.[118] That labeling was well known to the protagonists in the sun pictures conversation, which too emerged in late 1862. Edward Price could quote the International Exposition's catalogue essay on photography—an essay written by the secretary of the institutional body to which the sun pictures were first made public. From M. P. W. Boulton's point of view, the unearthing of images apparently made by chemo-mechanical techniques in the 1780s *might* have highlighted a dynamic, innovative culture of picture-making whose products were some sixty years ahead of their time. But it also drew attention to spaces at the heart of the Midlands' Industrial Revolution that he himself had recently demolished.[119] Further, by dint of a controversial name that promised their making by natural agency, those "sun pictures" pledged continuity with a polarizing ideology binding photography to a then-leading edge of industrial capitalism (fig. 4.21). There, as photo historian Steve Edwards puts it, the camera was one with "the self-acting machines [that] allowed factory masters to dispense with workers and their skills and thus produce an unceasing flood of commodities."[120] From his genteel haunts of St. Leonards-on-Sea and Thomas's Hotel on Berkeley Square in London's West End, so this earthy account could conclude, M. P. W. Boulton no more wanted to hear of prescient industrial products made in the eighteenth-century factory he had just destroyed than he wanted to see material evidence placing his family in opposition to those working by hand—from Sir William Beechey at 1790s Soho to the Chartists massing in Birmingham's Bull Ring.[121] The sun pictures had to be buried.

Whatever its merits as an interpretation, this earthy account takes on trust the period claim that photography is a vector integral to apprehending the sun pictures. What, then, is a photograph in this conversation? Contesting Smith's presentation at the Photographic Society in early November 1863, Thomas Augustine Malone had contended that photography required the operation of light to produce images.[122] Boulton had shared that view in his 1864 pamphlet. He pledges gratification to Smith if it could be proven that "the pictures in question are veritable photographs, produced by rays of light impinging on a sensitive surface."[123] In the same text, Boulton also embraces a second criterion: the use of optics. "Photography properly so called," he claims, requires "making pictures by means of a lens."[124] And,

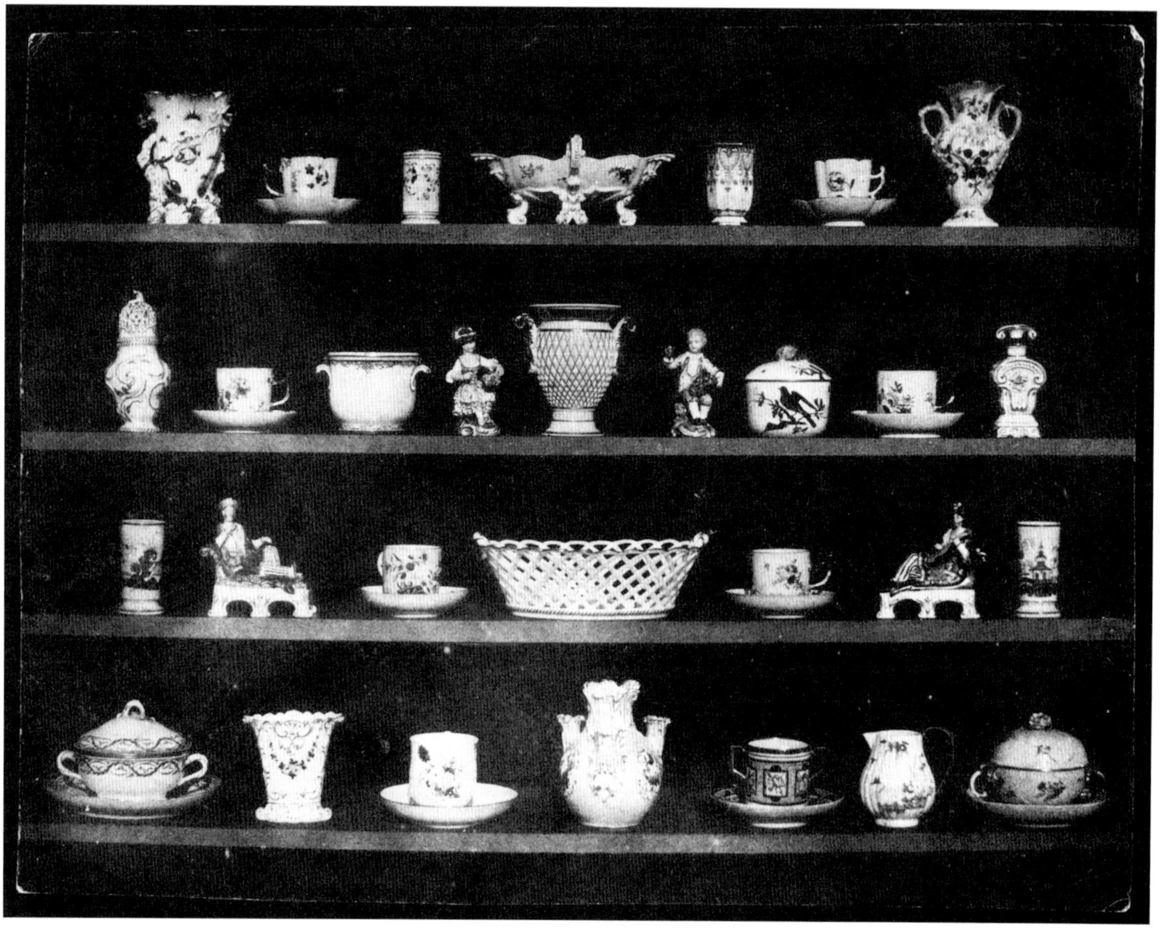

Figure 4.21 William Henry Fox Talbot, *Articles of Porcelain*. 1844. Calotype. Los Angeles County Museum of Art, Los Angeles.

later on in the same text, Boulton avows a third requirement. The sun pictures could be considered photographs, if "the pictures were produced by *fixing* the image formed by a lens."[125] Photo historian Joel Snyder affirms the salience of these criteria—optical instrumentation, the use of light to produce images, chemical means to fix them—as he deflates the speculation (famously encouraged by the work of Helmut and Alison Gernsheim) that photography could have been invented in the eighteenth century.[126] Since the chemicals used for successful fixative solutions were not produced until around 1815, Snyder argues, "the invention of photography so soon after the discovery of some of its basic chemical materials is less mysterious than it is stunning."[127]

But what if we were to release the sun pictures from photography's grasp? What might we learn about the sun pictures, photography, and the trajectory of British Enlightenment painting binding them together in the 1860s (since, of course, most of these images were undisputed replicas after

West and other painters of Reynoldsian stripe) were we to frame all three in a longer, broader matrix of making and thinking with temporally evolving chemical materials?[128] The next section moves toward that view—and to a different reading of both Smith and Boulton—by shifting elemental coordinates, from earthiness to a seemingly outlandish, aerial approach.

Air in an Age of Steam

When J. M. W. Turner painted the burning of the Houses of Parliament in the mid-1830s, he was depicting more than architectural destruction (see plate 19). Possibly imagining divine retribution visited on a seat of government then enacting a controversial amendment to the Poor Law, Turner's cataclysmic renderings of fire and smoke rising over water effectively visualize a new science of matter.[129] So has claimed historian of science Michel Serres. Where Enlightenment science and art reduced space to linear order, Serres proposes, Turner "enters the boiler, the stove and the furnace. He sees matter transformed by fire; the world's new matter at work, with geometry cut down to size. The whole order is turned upside down, materials and paint triumph over drawing, geometry, and form."[130] Ushering the railroad and the steamship into his pictures, Turner not only registers revolutionary applications of James Watt's steam engine but anticipates the new science of heat and energy.[131] Turner is "the introduction of igneous matter into culture; the first real genius of thermodynamics."[132]

A familiar narrative in the history of science traces that thermodynamic lineage back through James Watt himself. Such arguments typically turn on the improvements Watt made to a model of the Newcomen steam engine he was charged to repair in his capacity as instrument maker at University of Glasgow in the early 1760s. In the Newcomen engine's standard operation, water heated into steam by a boiler would rush into a metal cylinder, driving up a plunger. Cold water injected into the cylinder then condensed the steam, creating a vacuum and drawing the plunger down. But this cycle, alternating heating and cooling, wasted steam and fuel. Watt's solution was the introduction of a second cylinder: an independent condenser (fig. 4.22). Connected by valves of ever-greater sophistication as his research progressed, Watt's piston would remain constantly hot, the condenser constantly cold, with near-vacuum conditions obtaining in both to maximize their power and efficiency. Harnessing the expansive power of steam in ruthlessly efficient ways, historian of technology Donald Cardwell has argued, Watt "foreshadowed the progressive improvement of heat-engines and the postulation of Sadi Carnot of a general theory of the motive power of heat. With astonishing insight Watt had laid one of the cornerstones of thermodynamics."[133]

Such proto-thermodynamic interpretation plays a game of Whiggish

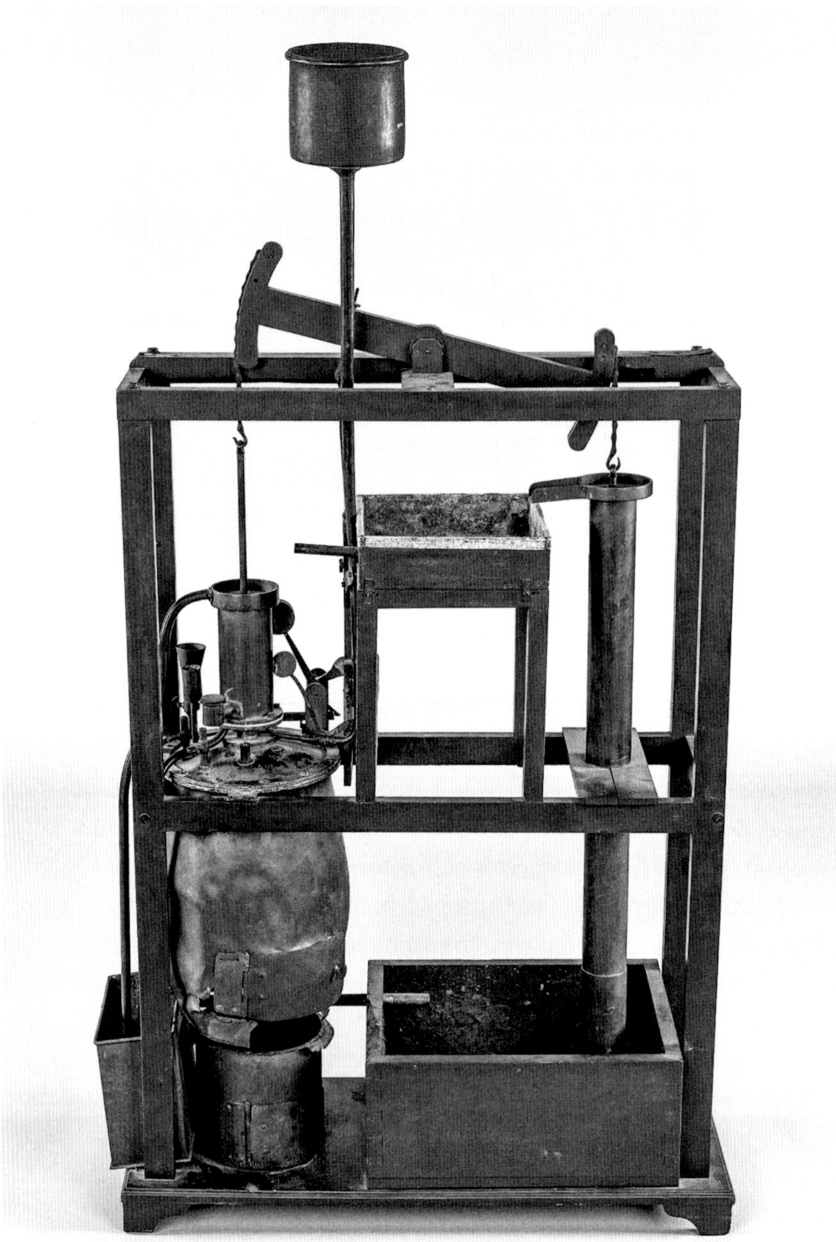

Figure 4.22 Model Newcomen engine, repaired by James Watt. Hunterian Museum & Art Gallery, Glasgow (GLAHM C.29). © The Hunterian, University of Glasgow, 2018.

history. Taking on the terms of the science that would later prove victorious, it obscures the centrality of chemistry to Watt's thinking in the crucial era of his steam-engine research. That is the revisionary argument made by historian of science David Philip Miller. Returning to well-known if controversial relations in early 1760s Glasgow, Miller highlights how the idea of latent heat Watt discussed with chemist Joseph Black in 1764 was stranger than the concept as now understood, construed in terms of kinetic theory.[134] Black's latent heat was not the product of particles in vibratory motion; instead, it was a special state of matter that affected other bodies via chemical attraction. "The cold experienced when water evaporates," per Miller's reconstruction of Black's position, "is the result of the water absorbing sensible heat as the water becomes vapour. The heat is not lost, rather the heat combines chemically with the vapour and it is this that gives the vapour the property of elasticity."[135] Because steam's elastic, expansive properties are products of chemical bonding between water and heat, a different interpretation of the improved steam engine follows. By the account Watt adopted from Black, Miller argues:

> The elasticity of steam was the basis of the force that it exerted in order to drive the piston and was also the property that allowed the separate condenser to work. The degree of elasticity of the steam in the cylinder of the engine depended directly on the amount of latent heat combined in the steam. In short, all key aspects of the engine's working were the product of steam's *chemical* character.[136]

Integral to his patent-making work of the 1760s, Watt's chemistry would expand through Lunar Society contacts in the early 1780s; then, his chemical copier rivaled Erasmus Darwin's "bigrapher" even as Priestley's arrival in Birmingham prompted new collaborations. Watt's chemical reputation resounded more broadly. Litigating against "pirates" in the 1790s, Watt's lawyers argued that the steam-engine patent stood on "chemical principles."[137] Yet, even as Humphry Davy and other leading figures in the early nineteenth century acknowledged him to be a great chemist, research on the nonmaterial nature of heat was making Watt's affiliations with Black's chemistry of latent heat look increasingly precarious.[138] Parsed from its constitutive cosmology into a post-Lavoisier revisionist history promoted by his associates, especially in the decades after 1800, Watt's chemical thinking was actively effaced from view by the 1850s with his backers' unsuccessful bid to claim Watt as the discoverer of water's compound nature.[139] Jettisoning that heritage of Watt's "archaic" chemistry, the Victorian image of Watt as triumphant engineer was born.[140]

We need not subscribe to all of Miller's claims to appreciate the visibility of this chemical Watt among near-contemporaries and the methodological benefits of retrieving it, a point to which I will return momentarily. First, though, let us remember why Francis Pettit Smith had originally gone to Soho in 1862: to acquire a prototype of Watt's steam engine. Smith had no small interest in Watt. Modestly educated, as was Watt, Smith had patented an application of the screw propeller to ship navigation in May 1836, months before his rival claimant John Ericsson did so. Like Watt, Smith had built a cadre of financiers and backers to turn a standing technological possibility into a devastating military advantage.[141] As one testimonial writer advocating for Smith put it, "The conditions of naval warfare have been altogether altered, and in every way to the advantage of this country, by this new application of mechanical power."[142] The analogy between Smith's revolutionary engine at sea and Watt's on land—devices both with precedent in concept, but not previously rendered practical—was keenly noted by contemporaries. Per one editorial pleading his case for greater remuneration in the 1850s, "Mr. Francis Pettit Smith stands in much the same relation to the screw propeller in which Watt stands with regard to the steam engine."[143]

In fact, the Soho heirs took a strong interest in Smith's case. Although divested of connection with Watt's relations by that time, the engineering firm of James Watt & Co., of Soho, Birmingham, donated one hundred pounds to Smith's testimonial fund in 1856.[144] And none other than Matthew Piers Watt Boulton praised Smith's mechanical innovations, in the very sun-picture pamphlets that otherwise assailed the inventor's judgment. "Though Mr. Smith's zeal and impatience of obstacles may render him unfit for the careful weighing of evidence," Boulton wrote in 1865, "it is to be remembered that on other occasions he has displayed these qualities to good purpose. With their aid he achieved the successful introduction of the screw-propeller, in the face of great external disadvantages."[145]

These remarks become even more interesting when we note that, just as he was scuppering the sun pictures in the middle years of the 1860s, Boulton was simultaneously making patent applications of his own: patents on aircraft design that continue to figure in histories of aviation.[146] A diagrammatic detail in Boulton's first aeronautic patent of 1868 shows two "vanes" or ailerons (b and c, in his figure 5) mounted on axles and controlled by a pulley-drawn cable system (fig. 4.23). When the aircraft becomes unstable during flight, ailerons on the "descending side are so inclined that the air impinging upon them exerts a pressure upward" while those on the ascending side are moved down to create downward pressure; the craft is thus balanced.[147] A legitimate challenger to the claims of the Wright brothers nearly half a century later to have invented lateral control mechanisms of contra-acting wing-warping, so contends historian Charles Gibbs-Smith, Boulton's 1868 patent stands as "one of the most remarkable *tours de force* in the history of aviation."[148]

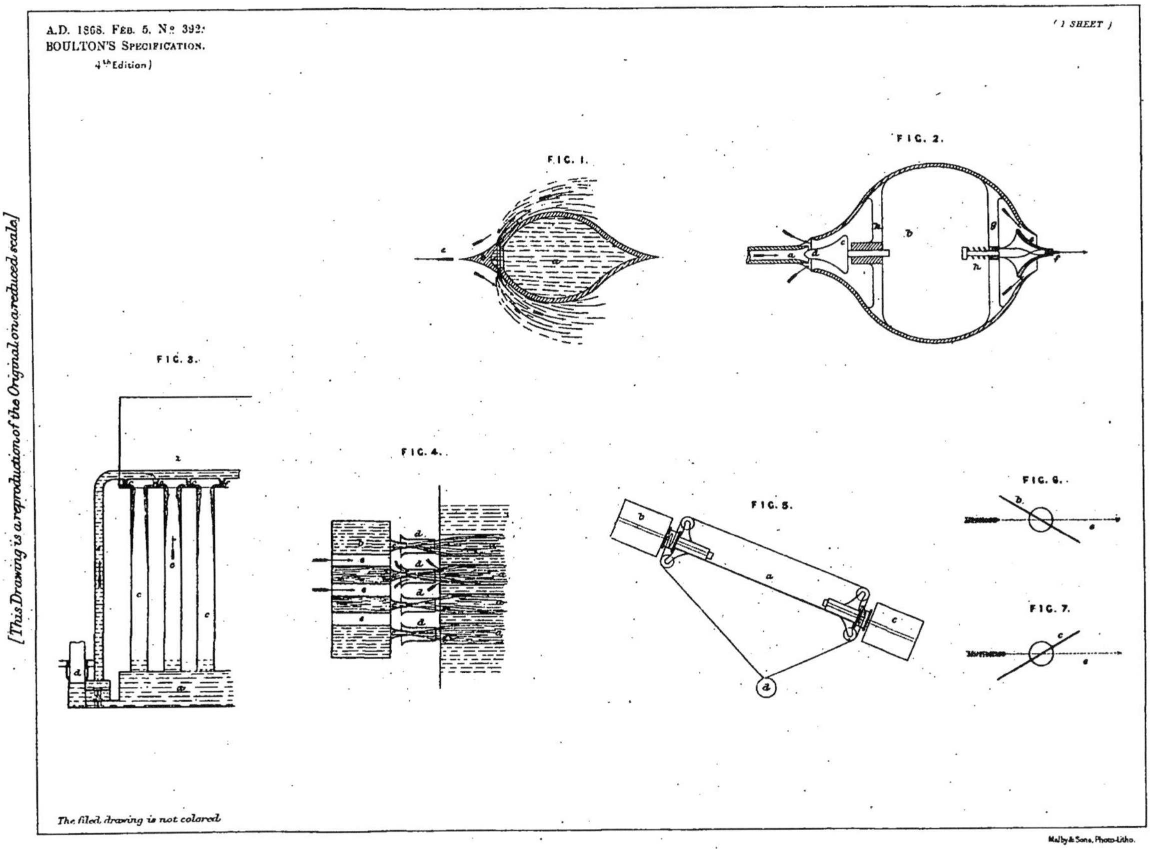

Figure 4.23 Design for ailerons (*lower right*, figs. 5–7) proposed in M. P. W. Boulton, "Improvements in Propulsion and in Aerial Locomotion, and in Apparatus Connected Therewith . . ." (patent no. 392; 1868). Photolithograph by Maltby and Sons.

In 1864 Boulton set out his basic aeronautic principles in a brief text entitled *On Aërial Locomotion*. Appealing directly back to the Lunar Society circles of his grandfather and James Watt, Boulton gestures to the famous prophecies of Erasmus Darwin in *The Botanic Garden* (1791): "Soon shall thy arm, *unconquer'd Steam!* Afar / Drag the slow barge, or drive the rapid car."[149] Although bested by Watt's chemical copying device, Darwin had indeed successfully predicted the use of the steam engine as means of terrestrial and maritime motion. Darwin also anticipated, so Boulton claims as he moves to direct quotation, "that it will 'bear / The flying chariot through the fields of air.'"[150] This is no nostalgia. Forgoing any appeal to the balloons and lighter-than-air craft that had fascinated his grandfather's generation, Boulton describes how his own airplane designs depart radically from predecessors in taking their model from modern technology, not from nature. Earlier aspirants to the art of flight "principally aimed at imitating the wings of birds," he notes. "But since the *use of the screw for the propulsion of steamboats*, the

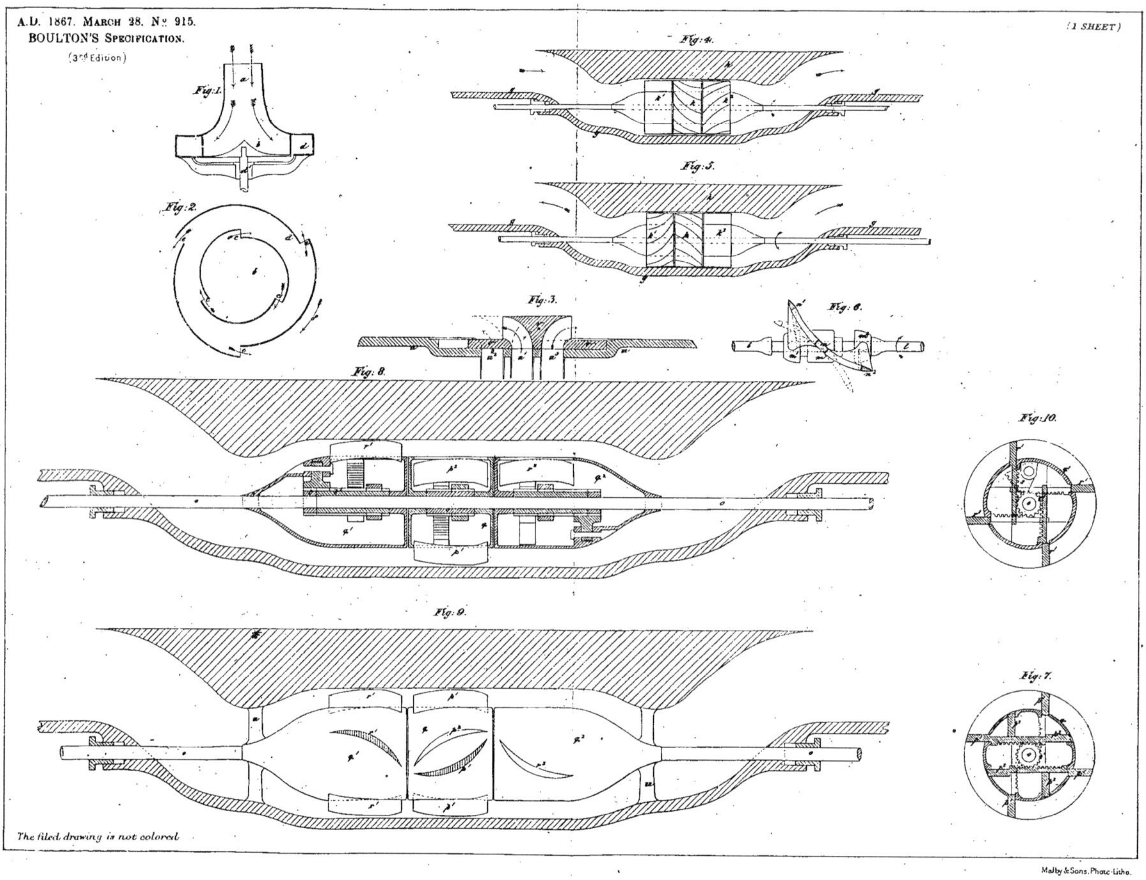

Figure 4.24 Design for turbine-driven aircraft as reproduced in M. P. W. Boulton, "Improvements in Apparatus for Receiving Motion or Energy from Fluids, and for Imparting it to Them, and in the Propulsion of Vessels and Apparatus to be used for that Purpose" (patent no. 915; 1867). Photolithograph by Maltby and Sons.

employment of a similar propeller for aërial locomotion naturally suggests itself."[151] By the end of March 1867, Boulton had succeeded in obtaining a patent on those principles (fig. 4.24). Boulton's diagram here renders a turbine seen from the side and mounted on the hull of an aircraft. Described in an abridgment of the patent specification as a "turbine or screw propeller for ships" applied for aerial propulsion, the engine-driven blades of the turbine would turn on a central shaft to move the craft through the air.[152]

Propelled by the screw mechanism he knew well to be the product of Francis Pettit Smith's hard, unremunerated graft, Boulton's aircraft also faced a crucial challenge. Where could an engine be found that would generate sufficient motive power to lift the craft off the ground without accruing unmanageable weight? Here too, Boulton turns back to his grandfather's partner, James Watt. "The steam-engine first occurs to mind as the possible agent of propulsion," Boulton proposes.[153] Canvassing the inevitable difficulties presented by the steam-driven airship Erasmus Darwin had predicted, he imagines one driven by the combustion of volatile substances.[154] The text then concludes with a plea for chemical innovation: "If the attention of chemists were specially directed to the investigation of explosive compounds, it would be found possible to devise [a substance] which, on ignition, should generate a large volume of gaseous products without excessive velocity of combustion."[155] These too were areas Boulton was actively seeking to claim through the legal instrument of the patent. From the mid-1860s well into the late 1880s, he secured patent after patent on improvements to combustion engines. In 1864 alone, he obtained patent no. 1099 for a "mode of working steam and combustion engines to employ that portion of heat which is generated by the combustion of fuel"; patent no. 1291, "improvements in engines worked by heated air or gases mixed with steam"; patent no. 1636, "improvements in obtaining motive power from aëriform fluids"; and patent no. 3044, "improvements in heating aëriform fluids by injecting some substance in a state of fusion (chlorides. &c.)."[156]

Yet, already in 1864—the year of his second pamphlet on the sun pictures—Boulton had located an extremely promising chemical explosive for use in the engines of his contra-acting, warp-winged, screw-propelled airship. "An engine worked by gun-cotton," Boulton argues,

> might closely resemble the steam-engine. A suitable quantity of the gun-cotton (or other similar substance) being introduced into a chamber near the cylinder, and there ignited, the gas thus generated would rush into the cylinder, and work the piston, just as is now done by steam. A very powerful engine could thus be . . . employed to communicate a rapid rotation to an aërial propeller (or to any desired number of these).[157]

The compound Boulton had called out for special aeronautic attention then stood at a cutting edge of research into chemical explosives. First synthesized in the 1830s, gun-cotton (nitrocellulose) was typically made by suspending purified cotton for two to forty-eight hours in a solution of equal parts nitric and sulphuric acids, washing out the acids, and mordanting the cotton in alkaline baths.[158] Some six times more explosive than gunpowder, the preparation was rapidly taken into British industrial production at the factory of John Hall and Sons in Faversham, Kent. Manufacture was halted by a horrific explosion on July 14, 1847, which killed at least eighteen men, women, and children. "Portions of the limbs of the victims," so the *Illustrated London News* reported, "were found in the fields and adjacent meadows."[159] Only contemporaneously to Boulton's publication was domestic production of gun-cotton resuming following a report on its safety commissioned in 1863 by the British Association for the Advancement of Science.[160]

By then, however, gun-cotton had also become a major component of advanced *photographic* technique. The compound could be mixed with ether and alcohol to form collodion, a cornerstone of the "wet-collodion" process developed by Frederick Scott Archer (1814–1857) in 1851 to significantly decrease the time required for photographic exposure.[161] Combined with silver and potassium iodides, the collodion would be poured over a glass-plate negative, then dipped into a bath of silver nitrate. "It is now placed in the camera obscura to receive the radiant image," so photochemist Robert Hunt had explained in 1854, "which, when the plate is properly prepared, it does in considerably less than a second of time."[162] What Scott Archer and numerous other claimants on the wet-collodion technique had achieved was an exponential decrease from the five to forty minutes of exposure required by Daguerre, let alone the two to three hours needed for Niépce's *héliographs*.

Gun-cotton, like Alois Senefelder's lithography half a century earlier, was a further chemical accelerant (hardly less dangerous in its photographic applications than in its ordnance origins) underpinning that blazing temporality Walter Benjamin would famously ascribe to the modern image.[163] Such an interpretation was also available to contemporaries. In Lady Eastlake's account, the wet-collodion process had perhaps been *too* effective in accelerating photography's speed. The new art now sped away from human time. Fueled by gun-cotton compounds, "the powers of photography . . . are already more nimble than we need," she observed. "Light is made to portray with a celerity only second to that with which it travels; it has been difficult to contrive the machinery of the camera to keep pace with it, and collodion has to be weakened in order to clog its wheels."[164] Boulton too was thinking about the nature of time in those years of the mid-1860s as he sought to apply chemical materials of temporal acceleration to achieve superior means of spatial motion. In 1866 he published pseudonymously an inquiry into metaphysical philosophy, offering a spirited defense of Immanuel Kant's assaults on common sense. By Boulton's account of principles from *Critique of Pure Reason*, the "phenomena" known to human understanding are products of our subjectivity, which is bound by space and time, while the "noumenon" need not be so constrained.[165] Vulgar apprehension inverts these points, thus obstructing philosophical insight:

> Ordinary persons suppose that the real man suffers a change in time; and if they are told that what has undergone change is only a phenomenon which they themselves have created; that the man or thing as it really is may not have changed in time, and may not exist in time at all,—they are puzzled, and regard such a doctrine as contrary to common sense.[166]

Where Lady Eastlake saw the photographic compounds of gun-cotton needing to be slowed to keep pace with ordinary human perception, Boulton casts that phenomenal horizon as "'traumartig,' dream-like, produced to a great extent by ourselves"—capable of being remade if, as it were, collodion's unclogged wheels were allowed to accelerate.

Let us gather these pieces together to see how an approach from the air allows us to apprehend Boulton, the sun pictures, and that compound we now call photography in different terms. As his mid-1860s patents demonstrate, Boulton was seeking chemical and mechanical means to achieve new kinds of motion. Where the earth of Victorian Britain had been transformed by railroads harnessing Watt's steam engine, Boulton imagines new chemicals generating the motive thrust required to move inhabited, mechanical craft through the air—a fusion that would prove, as Boulton puts it, "of great service, especially in warfare."[167] Yet in taking its means of aerial propulsion from Francis Pettit Smith and not from nature, Boulton's design drew on the model of the screw propeller that had *already* revolutionized warfare upon the watery expanses so crucial to imperial Britain. In 1822 Sir Walter Scott had described Watt, then recently deceased, as "this potent commander of the elements—this abridger of time and space."[168] The screw propeller similarly had compressed space and time at sea, reducing the journey from Malta to Britain from three months to ten days.[169] And it had been achieved by a figure, Smith, who contemporaries not only likened to Watt, but who had then trespassed into Boulton's ancestral factory. Searching for a device that accomplished motion through space by chemical means (if we follow Miller's revisionist account), Smith emerged from that factory with a chemo-mechanical assemblage in the sun pictures that not only stood to change in time via their perishable, material composition, but that bent the new time of photographic history back half a century in advance of itself.[170] More to the point, the engines-to-"photographs" arc at the heart of Smith's Soho adventure found its mirror opposite in Boulton's inventive aim. He dreamed of harnessing gun-cotton—the speedy chemical agent of wet-collodion photography—to drive airplane engines.

By these terms, Boulton's apparent campaign of sabotaging his grandfather's invention of photography discloses a different aim: blocking Smith's possession of chemical and mechanical techniques harnessed to achieve motion. Smith, the likeness of Watt; Smith, whose own screw-propeller design Boulton was simultaneously appropriating; this same Smith who had lifted out from under the falling walls of Boulton's property potential chemo-mechanical clues to spatiotemporal movement.[171]

Movement; or, Moving beyond Photography

This aerial view of the sun pictures might seem too strange to yield broader lessons. Yet three intertwined directives offer themselves willingly to the historian of temporally evolving chemical objects. First, Boulton's is certainly not the only crossing with combustion engines to be found among central figures in the early history of photography.[172] Collaborating with his elder brother Claude Niepce, Nicéphore Niépce (the Gernsheims' candidate for "first photographer") obtained a *brevet* of ten years' duration in 1807 for a combustion engine dubbed the Pyréolophore, which was used to propel a boat.[173] Fueled by a mixture of coal dust, resins, and, later, essential oils, the engine turned on a principle whereby aerial gas in a closed chamber could yield a particularly significant amount of energy when violently penetrated by a highly combustible substance finely atomized and evenly distributed through it.[174] Positively reviewed by leading chemists Lazare Carnot (father of thermodynamicist Sadi Carnot) and Claude-Louis Berthollet, the brothers' patent was revisited in 1816–1817. By that time, the chemically minded Nicéphore Niépce had moved from efforts at replicating Senefelder's lithography on pewter plates to heliographic experiments using asphaltum and other grounds employed by etchers.

That research with bitumen of Judaea and other photosensitive resins would lead to the younger Niépce's collaboration with Daguerre and his lionization by generations of photo historians (fig. 4.25).[175] To hear it told by even the strongest advocate in the 1860s for Niépce's photographic priority, however, heliographic research was but a small subsidiary to the brothers' main objective: the perfection of the mechanism and combustible chemicals for the Pyréolophore. Even as he sought to assert Niépce's importance for Daguerre, biographer Victor Fouque acknowledged:

> [The Pyréolophore] was the principal focus of the Niépce brothers' serious labors. Indeed, their letters are filled with descriptions of new devices, attempts and experiments relating to the Pyréolophore, perfections and improvements of this machine; along with the results of their experiments on combustible materials suitable for impressing movement to their motors. Claude and Nicéphore spent all their time and all their intelligence not only on these works but in taking the steps necessary to the advancement of their favored device.
>
> In fact, lithography and, above all, *héliographie* held only a very small place in their vast correspondence.[176]

That interest in engine-driven transit was shared more widely within the highest echelons of Parisian photography in the 1860s. Active in promoting steam-engine-driven aircraft, Nadar (né Gaspard-Félix Tournachon) made

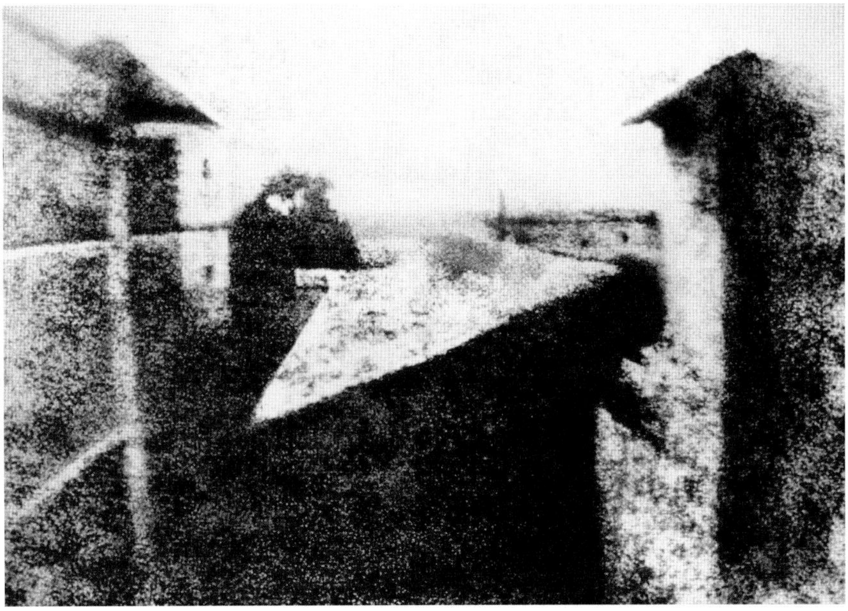

Figure 4.25 Nicephore Niépce, *Point of View Taken from a Window of Le Gras in Saint-Loup-de-Varennes*, ca. 1826. Tinplate "heliograph." Harry Ransom Humanities Research Center, University of Texas, Austin. Enhanced version by Helmut Gernsheim, ca. 1952.

a public display of the helicopter models he photographed in his studio in the summer of 1863.[177]

The most important precedent for Boulton was surely William Henry Fox Talbot himself. From the frantic years of his calotype production in the early 1840s, Talbot began obtaining a sequence of patents on electrical and gas engines.[178] Where his 1840 patent centered on "obtaining motive power by heating expansible liquids or inflammable gases," a 1846–1847 design replaced those volatile liquids with solid combustibles. As visualized in the diagram accompanying his patent (fig. 4.26), solid chemical substances would be packed into tiny cavities evenly spaced in rod R. Ignited by a charge communicated from a galvanic battery by a metal wire (Z), the solid substances would explode to drive up a piston (P), thus turning turn a crank and flywheel. As the piston descends, reciprocating motion advances the rod so the next cavity can fire. Like Boulton in 1864, moreover, Talbot identifies an ideal chemical preparation for packing these combustible cavities: "The substance . . . of most convenient application is that now commonly known by the name of gun cotton, prepared with nitric and sulphuric acids."[179] Seen this way, Talbot's chemical research offers a speculative return to Tom Wedgwood. Incubated amid Etruria's improvements to fired earthenware, Wedgwood's forays into flaming, colored motion and other evolving chemical spectacles promised to unfix time. Those had looked like failures to Talbot in 1844. But even as he visualized products

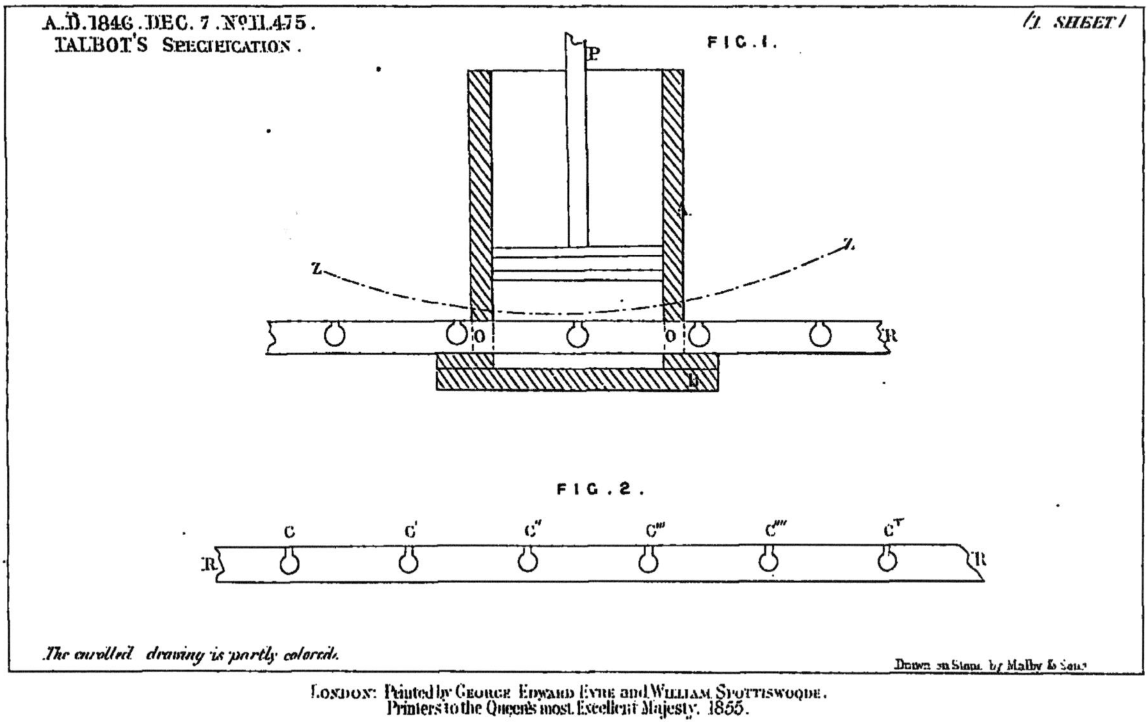

Figure 4.26 Engine specification from W. H. F. Talbot's "Improvements in Obtaining and Applying Motive Power" (patent no. 11,475; 1846). Lithograph by Maltby and Sons.

of that chemically inflected tradition of fired earthenware—the china depicted in the third plate of *The Pencil of Nature*'s first fascicle, for example (see fig. 4.21)—Talbot would soon see the solid, chemical combustibles of his motor engines liquidated and repupurposed as key agent in the wet-collodion process, which he would also try to arrogate under his existing, photographic patents.[180]

To put it in more concrete terms, an aerial view of the sun pictures enables us to see these conjunctions not as curious quirks of Victorian, inventive personality, but as material challenges for expansive historical investigation. In chapter 1, we saw how experiments on vital fermentation, air, and combustion in the early Royal Society shaped thinking about the spectacle of artificial phosphorus' temporal evolutions and the nature of time. Reciprocally, we need to be willing to excavate how products of a Reynoldsian pictorial tradition that embraced temporally evolving objects could then offer clues for thinking about movement and combustion-related problems. Indeed, if this nexus of picture-making, engines, and motion is treated more flexibly, that view might equally take in a figure such as Robert Fulton (1765–1815). An accomplished draftsman, the American-born Fulton arrived in London in 1787 to pursue a painting career through the friendship and tutelage of Benjamin West (himself an artist with no small interest in chemical experimentation); he would go on to become a leading steamship entrepreneur.[181] West also taught American painter Samuel F. B. Morse (1791–1872), the inventor of magnetic telegraphy and an early Daguerreotype practitioner, who lectured on Watt's steam engine as an instructive model for the painter's emulation in the 1820s.[182] Market enthusiast Milton Friedman once described economists' notoriously stylized models as providing "an 'engine' to analyze [the world], not a photographic reproduction of it."[183] A history of temporally evolving chemical objects turns Friedman's terms around. Taking to the air, it challenges us to parse the historical trajectories of chemical research shared between photography, painting, and devices of motive force modeled after the steam engine.

Second, as this aerial view discloses to the point of tedium, patents were a common denominator shared among nearly all the participants in the sun-pictures conversation. Watt had come to the Soho Manufactory in 1774 to join a partnership with Matthew Boulton, which would extend and exploit a steam-engine patent. Like the chief British claimant on photographic invention, Talbot, so Boulton and Watt enforced their patents vociferously, litigating extensively against alleged violators.[184] After obtaining his own patent for screw propulsion of ships in 1836, Francis Pettit Smith had witnessed such disappointment in spite of his invention's fiscal-military advantages that he was awarded as compensation a post at the newly created Patent Museum. There, Smith's superior was Bennet Woodcroft (1803–1879), himself an early patentee of screw-propelled ships, who had authored his own history of steamship invention.[185] Following an abortive professorship of Descriptive Machinery at University College London, Woodcroft became closely involved with Prince Albert's efforts to reform patent law.[186] As the engine behind the Patent Office Library, Woodcroft also closed the circle to Reynolds's experimental milieu of the 1750s when he brought the Society of Arts' collection of models and premium-winning designs under the aegis of the Patent Museum, which opened to the public as part of the larger South Kensington Museum in 1857. Smith discovered the sun pictures while working for the Patent Museum (where the artifacts would eventually be displayed). And, as he contested Smith's interpretation of the sun pictures, M. P. W. Boulton himself took out dozens of patents on aircraft and combustion engine designs. Photo historian Jordan Bear has astutely described the sun pictures debate of the 1860s as "a process of apportioning the rights of property among communities and individuals in order to arrive at a conceptual model for the circulation and commodification of knowledge."[187] Understood in those terms, airplane parts, steam- and combustion engines, photographs, and sun pictures would all be continuous claimants upon legal protection for invention. Painting would be the odd man out, not because of some ontological difference, but because of ethical principle. That is, as exemplified by Nathaniel Hone's charge of alchemical plagiarism leveled against Reynolds in 1775 (as examined in chapter 2; see plate 11), painters had come to assert the moral category of originality as assessed by a public—not inventive privilege as judged by law—as the defining bounds of their art.[188] The aerial perspective of the sun pictures compels us to examine with greater attention when, how, and why the juridico-moral matrices parsing photography from painting would come to be renegotiated.[189]

A third lesson from the aerial view returns us back to Miller's account of the steam engine's chemistry. Academic oil paintings, their polygraphic replicas, calotypes, combustion-driven airplanes, boat engines: that array looks strange if taken as so many contributions to the history (or prehistory) of photography. But the strangeness recedes when, instead of photography, ways of making and knowing with temporally evolving chemical objects become the target of historical analysis. Reviewing Geoffrey Batchen's account of photography's origins, Joel Snyder declared, "There could have been no desire to photograph prior to the means of photographing. Judging by the remarkable differences in the products of the various processes, there is no reason to *assume* that the early experimenters were moved by the same impulse."[190] If nothing else, the arguments of Price, Smith, Meteyard, and their sympathetic interlocutors reveal the violence required to remake the artifacts of Enlightenment chemical experimentation into prototypes for the then-victorious congeries of chemo-mechanical techniques called photography. The aerial view shows how a history of temporally evolving chemical objects can illuminate interactions between trajectories of research, some of which crystallize as artifacts called oil paintings, photographs, or the sun pictures that sit somewhere between them, and do so unburdened by the victor's terms. Cloven as it is by contested, evolving conceptions of chemistry, this tradition of making and thinking with materials valued for changing in time (while affording reflections *on* time) would intercept figures key to the histories of painting, photography, and technology alike. Exemplified by the engines of Niépce, Talbot, Boulton, and even Watt, those ways of making and thinking are not circumscribed to photography; instead, they map a different history of the planetary present and the place of the fine arts therein. Sketching that problematic will be the work of the conclusion.

CONCLUSION

Art History in/as an Age of Combustion

Writing in 2002, chemist Paul J. Crutzen sketched the outlines of a new epoch in planetary history. Several markers defining this era of terrestrial life shaped by human activity dated to the twentieth century: exploding consumption of nitrogen fertilizers, dramatic upticks in production of nitric oxide from combustion of fossil fuels, exponential population growth. Yet evidence of elevated carbon dioxide and methane levels in air bubbles extracted from polar ice also indicated an earlier advent for humanity's impact on the global climate. The "Anthropocene," as Crutzen helped dub this new era, was inseparable from the British Enlightenment examined in this book. Noting how anomalous chemical indicators "coincide with James Watt's design of the steam engine," Crutzen found convenient provenance for the epoch: "The Anthropocene could be said to have started in the latter part of the eighteenth century."[1]

Subsequent research has shifted and complicated the chronology.[2] Investigators have unearthed evidence of far earlier anthropogenic changes—evolutions—in climate history. Analysis of sediments from lakes, peat bogs, and other strata in Scandinavia suggests that, long before Matthew Boulton's Soho and competing manufactories industrialized metal production, lead smelting in medieval Europe had been spreading atmospheric pollution.[3] Citing unnaturally elevated methane and carbon dioxide deposits extracted from Greenland ice samples, some environmental scientists have pushed the onset of the Anthropocene back even to the cultivation of rice and forest clearances in Eurasia some five to eight thousand years ago.[4] In the other direction, the search for a "golden spike" of geosynchronous, stratigraphic signals unambiguously indicating the dawn of a new epoch has recently impelled researchers, including Crutzen, to date the Anthropocene to the detonation of the first nuclear weapons in 1945.[5] Nonetheless, Watt's "chemical" engine and British industrialization remain touchstones in Anthropocene narratives. Writing in 2007, Crutzen (with coauthors Will Steffen and John R. McNeill) envisioned that British-led period inaugurated between 1800 and 1850 as a conflagration of "carbon stored from millions of years of photosynthesis: a massive energy subsidy from the deep past to modern society, upon which a great deal of our modern wealth depends."[6] The consumption of coal in the steam-engine era would change the earth's future by setting ancient time on fire.

Evidence of anthropogenic climate change continues to meet resistance from some interested quarters, of course. And at the other end of the political spectrum, the Anthropocene's terms and narrative are hardly the only framework in which to grapple with the larger realities of global warming made by capitalism.[7] But it is instructive to note that perceptive observers from within the era of this book's relay were keenly aware of these unequal, man-made effects as they began to accelerate. George Perkins Marsh (1801–1882), lawyer, naturalist, and President Abraham Lincoln's envoy to Italy from 1861 to 1882, is an instructive voice.[8] Publishing in the year of M. P. W. Boulton's second sun-pictures pamphlet, his airplane treatise, and several engine patents, Marsh sought to demonstrate how the earth's lands, seas, and skies had been slowly altered by generations of human action.[9] If there could be little doubt that human engineering had transformed the planet's surface and the life-forms growing on it, so argues *Man and Nature; or, Physical Geography as Modified by Human Action* (1864), "operations of rural husbandry and industrial art have tended to produce great changes in the hygrometric, thermometric, electric, and chemical condition of the atmosphere."[10] Marsh underscored the extent of human destructiveness even as he allowed that not all of these "comprehensive mutations" were negative. Nature "has left it within the power of man irreparably to derange the combinations of inorganic matter and of organic life, which through the night of aeons she has been proportioning and balancing."[11] But destruction on a global scale was less some ontological fable than it was a profoundly political history. Following the course of William Godwin (Tom Wedgwood's friend and Mary Shelley's father), *Man and Nature* stressed the deleterious force of tyrants on humans and the planet alike. From a golden age of agricultural plenty, republican Rome had degenerated into empire and the crushing feudalism of Europe's middle ages. A savvy art collector familiar with J. J. Winckelmann's identification of ancient Greek political freedom with aesthetic greatness, Marsh details that "host of temporal and spiritual tyrannies which she [i.e., Rome] left as her dying curse to all her wide dominion, and which, in some form of violence or of fraud, still brood over almost every soil subdued by the Roman legions."[12] As fat bishops hoarded lands—as a nobility of the blood invented rights to first fruits—oppressed peasants devastated once-fertile territories in their desperate exploitation of diminishing resources. Natural history and human history were inseparable.

The foregoing chapters have argued that a relay of temporally evolving chemical objects and their material lives in British Enlightenment culture miscarries when skewed into replays of familiar dates and objects associated with what has come to be called photography. Tracing its ignition through the early Royal Society, this narrative has mapped a tradition of making and thinking with chemical materials changing visibly in time (and affording conceptual purchase on time) into the heart of British painting through the figure of Joshua Reynolds. Recapturing that "chemical Reynolds" and generations of its votaries, this relay has unearthed Reynoldsian chemical homunculi haunting grand-manner art. It has pursued chemical replicas of those chemically unstable, unctuous lives as they morph under ingenious hands, shape-shifting into materials binding early photographs to engine fuels. But, placed alongside Marsh's nightmarish story, might that relay's conjunction of artificial phosphorus, oil paintings, chemo-mechanical replicas, boat and airplane engines, along with photographs, be seen instead as so much dreaming in the materials then *making* a heated planet? Could the reckonings with time via artificial phosphorus in the hands of Robert Hooke or Tom Wedgwood offer episodes in what Marshall Sahlins calls a "native anthropology of Western society, [or] the indigenous conceptions of human existence" as it struggled to understand its position in the Anthropocene's new time?[13] Should the combined legs of this relay all be seen, that is, as so many moments in a history of combustion?

From a certain point of view, an increasing imprint of chemical industry on the cultural and physical landscapes of modernity is hard to escape. Lured by the demands for soaps and bleaches of the textile mills that enchanted Alois Senefelder and inspired Joseph Booth, industrialists founded alkali works at Prestonpans, St. Helens, and other centers of the British northwest, shipping their wares out via steam-powered rail lines established by the 1830s. Chemical syntheses of the coal ignited to power those steam engines, in turn, underpinned the artificial dyes yielding the kaleidoscopic range of colors new to Victorian textiles made on that alkali-bleached cotton, spun on steam-powered machinery.[14] By the early twentieth century, laudatory narratives such as *British Chemical Industry: Its Rise and Development* (1938) stressed the importance of the heavy chemicals necessary to those intertwining enterprises and the inventors who improved them. Yet, beyond the terrifying modernity announced by chemicals' weaponized release above the trenches of World War I, the privations of the Great War also revealed the pervasive penetration of "fine chemicals"—perfumes, photographic preparations, disinfectants, a capacious array of pharmaceuticals—as much into British economic interests as in the fabric of everyday life.[15]

The impact of this chemical modernity has not been lost on the history of art. No longer ground in studios by James Northcote and his ilk, pigments were increasingly prepared by consolidating, fine-chemical industrialists, sold to artists "ready-made" from shops such as that set up at Bloomsbury's 38 Rathbone Place by William Winsor and Henry Charles Newton in 1832.[16] Pressured by the camera's mechanization of depiction on one side and legacies of industrialized color's "new chemical aesthetic" on the other, the painter (at least per Thierry de Duve's clever reading) was left with but one move: the readymade.[17] Eschewing craft, Marcel Duchamp's readymades redefined art as choosing—as selecting among the porcelain urinals and other prefabricated consumer goods that now count as so many "technofossils" in the Anthropocene's sedimentary layers.[18] The unnamed elephant in this room is, of course, the petrochemical industry that emerged in the mid-nineteenth century, which has done so much to compound the conditions of global warming even while becoming key financial and material infrastructure for the visual arts. Big oil, as Ross Barrett and Daniel Worden trenchantly observe, is now a central player in contemporary art "as an underwriter of nearly every major museum . . . whose petroleum products are used to make film, ink, paint, and countless other tools used to produce art."[19]

Productive as it would be to better understand how, when, and where chemo-industrial and petrochemical interests have come to intercept the making and display of art, I recognize that placing environmental destruction (or one influential reading of it called the Anthropocene) at the end of a relay of temporally evolving chemical objects can hardly be any less teleological than the dominant stories about photography's coming this book has sought to disarm. Surely, the urgency of the environmental situation warrants it; but a more expressly art-historical point also merits reinforcement. For, however its historical relay ends, this book's ultimate intervention aims to be a methodological one. *Painting with Fire* has proposed interpretation of the art (broadly conceived) of industrializing society via technical channels; it has highlighted the importance of chemo-material factors impinging upon the making of that art, even though traces of those chemical supports and processes may have now become invisible as its formal properties. But all of those are so many elementary errors against which the intellectual edifice of academic art history once was built. In *Stilfragen* (1893), Alois Riegl railed against the pernicious consequences of technological adulation run rampant in the study of art. Just as Darwinists of all stripes had flattened and overextended the insights of Erasmus Darwin's grandson, Riegl charged, so histories of art had suffered by undue application of the thought of architect and critic Gottfried Semper (1803–1879). Hardly would Semper have assented to the claims made in his name; he "would never have agreed to exchanging free and creative artistic impulse [*Kunstwollen*] for an essentially mechanical and materialist drive."[20] Moving into and inspiring a consolidating infrastructure of university-taught art history, Riegl envisioned instead a forensic analysis of stylistic propagations propelled not by materials and techniques, but by what he called an "all-pervasive law of causality" driving artistic form.[21]

The kinds of problems Riegl articulated through his ingenious analyses of form (shifts from ancient privileging of the plane as organizing field to the perception of body in deep space; movements from haptic to optic apprehension; the liberation of the artistic beholder) continue to carry substantial truck among practitioners of academic art history. True, it is no longer fashionable to speak as Riegl did about art's "constant progress," but the mandate to excavate, rather than impugn, aleatory desires departing from canonical taste has inspired some of the discipline's most ambitious writing. For Riegl, the alternative to avowing a positive *Kunstwollen* making such formal otherness meant subjecting art's history to the crudest exigencies and doing so with the grossest deference to present taste. Riegl positioned his *Die spätrömische Kunstindustrie* (1901) as an emphatic rejoinder to the view voiced by Marsh and many others that art had not changed but *declined* amid the collapse of imperial Rome. "To destroy this prejudice," Riegl wrote, "is the principle object of all the studies contained in this book."[22]

Ingenious though they may be, recent materialist-inclined contributions to the art historian's "coy science" have tended to abide Riegl's admonitions, privileging the perceptible properties of high-status objects as endpoints of analysis. But as makers from Hooke to Niepce, Talbot to M. P. W. Boulton and, perhaps above all, Tom Wedgwood knew only too well, resolutions of matter into visible form hardly exhaust the potentials of chemical objects evolving in time. My view is that histories of Anglo-American art stand significantly to gain by seeing themselves reflected in the gasoline rainbow of that relay, which has made our chemical ontology and our precarious planetary condition. And this point brings me back around to Jeff Wall's ideas about liquid intelligence. Veining through the preceding pages has been recourse to a language of the elements. Appeal to the elemental has pulled us down from speculation on the "allegorical" meanings of Reynolds's attraction to unstable paint chemistries. Where an elemental chemistry of fire, water, and air released Shelley's homunculus back to the earth, a stolid, down-to-earth analysis guided one way of reading Boulton's attacks on the sun pictures. Aerial pursuits—flights of apparent fancy—

then revealed a more complicated (and compelling) picture. Fire abounds, of course, making and remaking this story's artifacts and their poetics. Implicitly present in steam perhaps, water (the fourth of the Empedoclean elements, at least according to the familiar way of parsing them) has not borne the same argumentative charge.[23] Yet returning to Wall on a methodological plane is suggestive. "For me," Wall writes, "water . . . embodies a memory-trace of very ancient production processes—of washing, bleaching, dissolving and so on, which are connected to the origin of *technè*."[24] Focused on a historical moment in which "technology" was busily being invented, *Painting with Fire* and its elementary errors per Riegl can also suggest contours for an *elemental art history*.[25]

What might that mean? With all the perversity of Reynolds himself, such an approach would insist on thinking the fine and the industrial arts together, and doing so with one and the same artifact. It gives equal footing to form and materials, to the visible and the invisible, to the realm of the hand (whether manual or mechanical) and its lesser-known interlocutor, the chemical penumbra. When it amasses evidence, this elemental approach examines perceptible features of artifacts, but it also gives ear to scientific analyses of paint films and to material reenactments of their chemical components. It studies condition reports of artifacts in museums and, indeed, interrogates the conditions of possibility (logistical, actuarial, or otherwise) whereby artifacts are transported from their native elements. This approach does not require us to deny differences between an oil painting by Reynolds and a motor oil (to adapt Duchamp's plans for a Rembrandt). What it does mean is that we are willing to pursue historical understanding of why, when, and how Reynolds's replicas, sun pictures, and wet-collodion photographs could look like they offered clues to improvement of combustion engines, as a figure such as M. P. W. Boulton was able to hold in the mid-1860s. It challenges us, moreover, to grapple with a slippery world of dreams wherein chemicals that prompted reflection on the nature of time in the long eighteenth century now count for geological stratigraphers of the Anthropocene among the artificial markers of a new planetary time.

Exemplary work in these directions is, of course, well under way, and *Painting with Fire* has sought to learn and employ its lessons. What remains in the end is not want of intelligence. If anything, we need a simpler, more elemental way of apprehending and of showing how the passions and the politics—the material drives—that have made the arts of our world are also the making of a world on fire.

ACKNOWLEDGMENTS

This book was generated under circumstances, by turns, idyllic and devastating. In the former mood, I give much love to Susan M. Bielstein, editor extraordinaire, who has been an ideal interlocutor and tireless advocate for this project from the very start. At that inception, the research was also supported by generous, multiyear grants from Fonds de recherche du Québec–Société et Culture (FRQSC) and the Social Sciences and Humanities Research Council (SSHRC), along with fellowships from the Paul Mellon Centre for Studies in British Art, the Yale Center for British Art, and the Huntington Library. I am very grateful to Tim Campbell who so generously extended an invitation to coteach a seminar related to this project in its early gestation through the Center for Disciplinary Innovation at University of Chicago's Franke Center for the Humanities in 2015. Much of the core writing was then done at the Huntington Library, where I was a National Endowment for the Humanities Fellow in 2016–2017. I wish to acknowledge my enduring gratitude to Steve Hindle and his colleagues at the Huntington Library and Art Gallery, along with Jon Mee, John Demos, Tawrin Baker, and other members of our Long-Term Fellows Working Group who made that research environment so productive. I have learned much from the book's two anonymous referees; I am very grateful for their capacious insights and analytical acumen.

Along the way, I had the good fortune to share and expand arguments presented here with critical audiences at the following venues: the Courtauld/UCL Early Modern Symposium; the Royal Society; SUNY Binghamton; Yale Center for British Art; Max Planck Institute for the History of Science; Reed College; the Getty Research Institute; UCLA's History of Science Colloquium; University of Cambridge's Centre for Research in the Arts, Social Sciences and Humanities (CRASSH); California Institute of Technology; University of Toronto; University of Southern California; University of Chicago; University of Toronto's Jackman Humanities Institute; Clark Art Institute; Huntington Library/USC Early Modern Studies Institute; UCLA's Department of Architecture and Urban Design; Harvard Art Museums; University College London; Montréal's Séminaire des Nouveaux Modernes; Washington University; University of Sydney's Power Institute; Massachusetts Institute of Technology; Brown University; and Columbia University's Society of Fellows. I thank the organizers for these wonderful opportunities and the audiences for their brilliant feedback. I also acknowledge invitations to present related work at the University of Minnesota and the Wallace Collection, which were thwarted by adverse circumstances.

As noted in the introduction, an early catalyst for this book was Jeff Wall's writing on "liquid intelligence." I extend virtual thanks to Wall, as well as to the students at McGill University and the University of Chicago who participated in my seminars on that topic. Additional thanks to Michael Cole, Rebecca Zorach, and Richard Neer, who generously presented their work to our meetings. In 2013 the seminar expanded to a conference, "Liquid Intelligence and the Aesthetics of Fluidity," held in Montreal. I am delighted to thank the following presenters for their stunning contributions: Caroline Arscott, Fabio Barry, Yukio Lippit, Jeffrey Moser, Alexander Nemerov, Jennifer L. Roberts, and Itay Sapir. Thanks as well to Sarah Hamill for providing a culminating comment, and to moderators Nick Dew, Byron Hamann, Amelia Jones, and Jonathan Sachs. I am very grateful to Jeffrey Moser and Jennifer L. Roberts, along with John Harwood, for their superb contributions to "Liquid Intelligence," a special issue of that project I edited as *Grey Room* 69 in 2017.

Some of the material presented in these pages has appeared in earlier iterations. Elements explored in the book's introduction and second chapter surfaced in my article "Joshua Reynolds's 'Nice Chymistry': Art and Accident in the 1770s," *Art Bulletin* 97, no. 1 (March 2015): 58–76. Reynolds's camera obscura, an object I discuss in the introduction, showed up at the end of "'Mr. Hooke's Reflecting Box': Modeling the Projected Image in the Early Royal Society," in "Curiously Drawn: Early Modern Science as a Visual Pursuit," a special issue of *Huntington Library Quarterly* 72, no. 2 (Summer 2015): 301–28. Earlier thoughts on Reynolds's work at the Society of Arts, discussed here in chapter 2, were presented in "Reynolds's Science of Experiment in Practice and Theory," in *Joshua Reynolds: Experiments in Paint* (London: Wallace Collection/Paul Holberton Publishing, 2015), 100–111. Finally, I first tried to reckon with Reynolds's life in chemical replication, a key concern for the third chapter, in "Did Joshua Reynolds Paint His Pictures? The Trans-Atlantic Work of Picturing in an Age of Chymical Reproduction," in *Picturing* (Chicago: University of Chicago Press/Terra Foundation for American Art, 2016), 44–80. I thank, respectively, Kirk T. Ambrose; Felicity Henderson, Sachiko Kusukawa, and Alex Marr; Lucy Davis and Mark Hallett; and Rachael Z. DeLue for their encouragement and excellent editorial insights.

From the start, this book has taken me far beyond my comfort zones. In such a thrilling ride, I have turned to and learned from the generosity of colleagues who have kindly indulged my forays into their areas of expertise. While all remaining errors and infelicities are, of course, mine alone, I wish to thank the following scholars for their insightful comments on individual chapters: Fredrik Albritton Jonsson, Tim Barringer, Jennifer Chuong, Noam M. Elcott, Charles Ford, Michael Gaudio, Mark Hallett, John Harwood, Ashley Inglehart, Melissa Lo, Jon Mee, Omar Nasim, Martin Postle, Amy Knight Powell, Katie Scott, and Leslie Tomory. John Brewer, Jason Lafountain, Mark Ledbury, and Jennifer L. Roberts generously read and shared critiques on several chapters, while Byron Hamann plowed through multiple iterations of the entire manuscript with characteristic and indomitable brio. A joy of this project has come from informal conversations with investigators beyond the humanist's usual stomping grounds: with chemists Tony Jia, Janek Szychowski, and David Tirrell at the California Institute of Technology (along with Moti Feingold and Jonathan Katz, who encouraged those conversations); painters Ross Chisholm and Titus Kaphar; and painting conservators Mark Aronson, Alex Gent, Alan Fenix, Rica Jones, and David Saunders. Along similar lines, I thank Rick Brettell for encouraging me many years ago to consider how art historians might make more robust use of the museum as knowledge-producing infrastructure, and Melinda McCurdy and James Glisson at the Huntington Art Gallery for their support as I put that strategy to work. For their generosity and wisdom, I thank the archivists and librarians to whom this project is indebted, especially Rupert Baker and Keith Moore (Royal Society), Lucy Lead (Victoria & Albert Museum, Wedgwood Collection), Ross MacFarlane (Wellcome Collection), Mark Pomeroy (Royal Academy), and Eve Watson (Royal Society of Arts). My thanks to Joel Score for his expert copyediting, June Sawyers for indexing, and James Toftness for his patience and guidance. I am pleased to acknowledge the photographic talents of Sara Tarter, along with the research assistance of the following students who worked on the project: in chronological order, Elisabeth Niemczyk, Maryse Ouellet, Rebecca McEwen, Jennifer Mueller, Jemma Israelson, Nic Holt, and Rachel Sutherland.

Conversations over many years with friends and colleagues have kept this project going through the darkness. Thanks especially to Zeynep Çelik Alexander, Caroline Arscott, Fredrik Albritton Jonsson, Jordan Bear, David Brafman, John Brewer, Bill Brown, Jim Chandler, Catherine E. Clark and Brian Jacobson, Susan Dackerman, Nick Dew, Emily Doucet, Anne Dunlop, Sven Dupré, Noam M. Elcott, Charles Gagnon, Michael Gaudio, Becky Givan, Florence Grant, Charles Ford, Mark Hallett, John Harwood, Chriscinda Henry, Karin Leonhard, Catherine Lock, Alex Marr, Jon Mee, Jeff Moser, Michael Osman, Andrew Perchuk, Nick Robbins, Jennifer L. Roberts, Itay Sapir, Katie Scott, Joel Snyder, Eva Struhal, Richard Taws, Mary Terrell and Ted Porter, Leslie Tomory, and Jennifer Tucker. I am grateful to my colleagues at McGill's Department of Art History and Communication Studies, especially the Hard Red Spring/Tactical Unit of Rock collective (Charles Acland, Darin Barney, Jenny Burman, Carrie Rentschler, and Jonathan Sterne). I am honored by the constant challenge of attempting to keep up with fellow editors of *Grey Room*—the smartest group of people I know: Zeynep Çelik Alexander, Lucia Allais, Eric de Bruyn, Noam M. Elcott, Byron Hamann, John Harwood, and Ben Young. On a personal level, I am grateful to the Kouretas/Alivizatos family, the family of Patrick W. Hunter and Laura Melbin, the Henry/Lange family (Cinda, Fabian, and Elsa), and all those who helped make the death of my mother, Susan W. Hunter (1948–2015), as dignified as it could be. I dedicated my first book to my wonderful wife, Daphne Kouretas, who deserves much more and much better than that. If there is a third one, I will dedicate it to our second daughter, Zoë. But this book is for Lucia, who had to live through the worst of it, and was—is—the best of it. *Lux in tenebris.*

NOTES

Introduction

1. On Coutts's various sittings with Reynolds, see David Mannings and Martin Postle, *Sir Joshua Reynolds: A Complete Catalogue of His Paintings* (New Haven, CT: Yale University Press, 2000), 1:148–49. On the Coutts bank, see Albert Boime, *Art in an Age of Revolution, 1750–1800* (Chicago: University of Chicago Press, 1987), 267.

2. Reproduced in Sir Joshua Reynolds, *Discourses on Art*, ed. Robert R. Wark (New Haven, CT: Yale University Press, 1997), 291. Cf. Richard Wendorf, *Sir Joshua Reynolds: The Painter in Society* (London: National Gallery, 1996); David H. Solkin, "Great Pictures or Great Men? Reynolds, Male Portraiture, and the Power of Art," *Oxford Art Journal* 9, no. 2 (1986): 42–49; John Barrell, *The Political Theory of Painting from Reynolds to Hazlitt: "The Body of the Public"* (New Haven, CT: Yale University Press, 1986).

3. Louis Simond, *Journal of a Tour and Residence in Great Britain during the Years 1810 and 1811*, 2nd ed. (Edinburgh: J. Ballantyne, 1817), 1:49.

4. James Northcote, "A Disappointed Genius," *Artist* 19 (July 18, 1807): 13; 14. On Northcote's artistic and literary careers, see Mark Ledbury, *James Northcote, History Painting, and the Fables* (New Haven, CT: Yale University Press, 2014).

5. Northcote, "Disappointed Genius," 15.

6. John T. Smith, *Nollekens and His Times* (London: Henry Colburn, 1828), 2:291–92.

7. Important contributions to this conservation literature include M. Kirby Talley, "'All Good Pictures Crack': Sir Joshua Reynolds's Practice and Studio," in *Reynolds*, ed. Nicholas Penny (New York: Harry N. Abrams, 1986), 55–70; Stephen Hackney, Rica Jones, and Joyce Townsend, eds., *Paint and Purpose: A Study of Technique in British Art* (London: Tate Gallery, 1999); Ashok Roy et al., eds., "Joshua Reynolds in the National

Gallery and Wallace Collection," special issue, *National Gallery Technical Bulletin* 35 (2015); and Lucy Davis and Mark Hallett, eds., *Joshua Reynolds: Experiments in Paint* (London: Wallace Collection/Paul Holberton Publishing, 2015).

8. Francis Charteris, Lord Elcho, to unidentified recipient, June 9, 1859; Scottish National Gallery, object file NG 338, ff3_v–4_r, f2_{r-v}.

9. Joseph Booth, *A Treatise Explanatory of the Nature and Properties of Pollaplasiasmos; or, the Original Invention of Multiplying Pictures in Oil Colours . . .* (London: J. Rozea, 1784), 9; [Joseph Booth], *An Address to the Public on the Polygraphic Art* (London: J. Walter, 1788[?]), 4.

10. Booth, *Address to the Public*, 5.

11. William Birch, "The Life and Anecdotes of William Russell Birch, Enamel Painter" (ca. 1815–1834); reproduced in Emily T. Cooperman and Lea Carson Sherk, *William Birch: Picturing the American Scene* (Philadelphia: University of Pennsylvania Press, 2011), 169–70; 172.

12. For other recent histories of photography that give substantial attention to Reynolds, cf. Joel Snyder, "Res Ipsa Loquitur," in *Things That Talk: Object Lessons from Art and Science*, ed. Lorraine Daston (New York: Zone Books, 2004), 195–221; Steve Edwards, *The Making of English Photography: Allegories* (University Park: Pennsylvania State University Press, 2006), esp. 136–46; and Josh Ellenbogen, *Reasoned and Unreasoned Images: The Photography of Bertillon, Galton, and Marey* (University Park: Pennsylvania State University Press, 2012).

13. Robin Kelsey, *Photography and the Art of Chance* (Cambridge, MA: Belknap Press/Harvard University Press, 2015), 22.

14. The variable nature of argument in the *Discourses* and their apparent distance from his own practice are classic topics in Reynolds scholarship. Cf. Charles Mitchell, "Three Phases of Reynolds's Method," *Burlington Magazine* 80, no. 467 (1942): 35–40; E. H. Gombrich, "Reynolds's Theory and Practice of Imitation," *Burlington Magazine* 80, no. 467 (1942): 40–45; Robert R. Wark, "Introduction," in Reynolds, *Discourses on Art*, xiii–xxxiii.

15. Kelsey, *Art of Chance*, 17, 23.

16. Reynolds, "Discourse XIII" (1786), in *Discourses on Art*, 243.

17. Kelsey, *Art of Chance*, 23. For Alpers's account of Reynolds on the camera obscura as foil to her analysis of an early modern Dutch pictorial tradition allied with photography, see Svetlana Alpers, *The Art of Describing: Dutch Art in the Seventeenth Century* (Chicago: University of Chicago Press, 1983), xvii–xxvii, 26–71.

18. Horace Walpole to William Mason, September 21, 1777, in *Horace Walpole's Correspondence with William Mason*, ed. W. S. Lewis, G. Cronin Jr., and C. H. Bennett (New Haven, CT: Yale University Press, 1955), 1:328–29. On the use and provenance of Reynolds's camera obscura, see Matthew C. Hunter, "Mr. Hooke's Reflecting Box: Modeling the Camera Obscura in the Early Royal Society of London," *Huntington Library Quarterly* 78, no. 2 (2015): esp. 322–28. On the larger interest in optical devices in the late eighteenth-century Academy (centered instructively around Reynolds's rival Thomas Gainsborough), see Ann Bermingham, *Sensation and Sensibility: Viewing Gainsborough's Cottage Door* (New Haven, CT: Yale Center for British Art, 2005); Ann Bermingham, ed., "Technologies of Illusion: The Art of Special Effects in Eighteenth-Century Britain," special issue, *Huntington Library Quarterly* 70, no. 2 (2007).

19. For the provenance of this canvas and its eventual sale in 1878 for fifty pounds sterling, see correspondence of George Barker Jr. to the president and council of the Royal Academy, October 29, 1877, and January 29, 1878; Royal Academy of Arts, object file 03/576.

20. *The Diary of Benjamin Robert Haydon*, ed. Willard Bissell Pope (Cambridge, MA: Harvard University Press, 1963), 5:574.

21. Henry Peach Robinson, *Pictorial Effect in Photography: Being Hints on Composition and Chiaroscuro for Photographers* (London: Piper & Carter, 1869), 11.

22. Robinson, *Pictorial Effect*, 83. Thanks to Noam M. Elcott.

23. Walter Benn Michaels, "Photographs and Fossils," in *Photography Theory*, ed. James Elkins (New York: Routledge, 2007), 444.

24. This literature is vast, but some key statements include Roland Barthes, *Camera Lucida: Reflections on Photography* (1980), trans. Richard Howard (New York: Hill and Wang, 1981); Stanley Cavell, *The World Viewed: Reflections on the Ontology of Film* (Cambridge, MA: Harvard University Press, 1979); Rosalind Krauss, "Notes on the Index: Seventies Art in America," *October*, no. 3 (1977): 68–81, and "Notes on the Index: Seventies Art in America, Part 2," *October*, no. 4 (1977): 58–67; and Kaja Silverman, *The Miracle of Analogy; or, The History of Photography, Part 1* (Stanford, CA: Stanford University Press, 2015).

25. For this view of Reynolds, cf. Celina Fox, *The Arts of Industry in the Age of Enlightenment* (New Haven, CT: Yale University Press, 2009), 216; Martin Kemp, "True to Their Natures: Sir Joshua Reynolds and Dr. William Hunter at the Royal Academy of Arts," *Notes and Records of the Royal Society of London* 46, no. 1 (1992): 77–88.

26. See [William Cole], "A Letter from Mr. William Cole of Bristol, to the Philosophical Society of Oxford, containing his Observations on the Purple Fish," *Philosophical Transactions* 15, no. 178 (1685): 1285–86. On Cole, Cf. A. J. Turner, "A Forgotten Naturalist of the Seventeenth Century: William Cole of Bristol and His Collections," *Archives of Natural History* 11, no. 1 (1982): 27–41; Matthew C. Hunter, *Wicked Intelligence: Visual Art and the Science of Experiment in Restoration London* (Chicago: University of Chicago Press, 2013), 131–38.

27. A century ago, these swatches were available to historian R. T. Gunther, who replicated them as an indigo insert between pages 100 and 101 in his *Early Science in Oxford*, vol. 4, *The Philosophical Society* (Oxford, 1925). Despite several searches and correspondence with numerous institutions, the swatches are presently unlocated.

28. William Cole to Robert Plot, February 5, 1685, in *Early Science at Oxford*, vol. 12, *Dr. Plot and the Correspondence of the Philosophical Society of Oxford*, ed. R. T. Gunther (Oxford, 1939), 263–64.

29. See Leonard Trengrove, "Chemistry at the Royal Society of London in the Eighteenth Century: IV, Dyes," *Annals of Science* 26, no. 4 (1970): 331–53. Pliny relayed how, in preparing the dye, the murex were steeped in salt for three days, boiled in water for nine more, and then paired with a different sea-snail to stain wool "the colour of congealed blood (*sanguinis concreti*), blackish at first but gleaming when held up to the light"; Pliny the Elder, *Natural History*, vol. 3, *Libri VIII–XI*, trans. H. Rackham (Cambridge, MA: Harvard University Press, 1940), book 9, LXII: 135, pp. 254–55.

30. Boyle explains how white paper dyed blue by a syrup of violets turns red on contact with acidic lemon juice; but if the same paper is moistened by "a little Oyl of Tartar *per Deliquium* [potassium carbonate that has absorbed atmospheric moisture], or the like quantity of Solution of Potashes . . . you shall find the Blew Colour of the Syrrup turn'd in a moment into a perfect Green"; Robert Boyle, *Experiments and Considerations Touching Colours* (1664), in *The Works of Robert Boyle*, vol. 4, *Colours and Cold, 1664–5*, ed. Michael Hunter and Edward B. Davis (London: Pickering & Chatto, 1999), 124–25. Cf. William Eamon, "Robert Boyle and the Discovery of Colour Indicators," *Ambix* 27, no. 3 (1980): 204–9.

31. William Cole to Robert Plot, May 21, 1685, in Gunther, *Early Science at Oxford*, 12:293.

32. Cole, "Observations on the Purple Fish," 1281.

33. Cole to Plot, February 5, 1685, 264. For Cole's timing of "less than 2 minutes" required for the initial white to green shift, a "2 or 3 minutes" for green to turn deep

blue, and "3 or 4 minutes" for that blue to become dark red, see William Cole to Robert Plot, October 31, 1684, in Gunther, *Early Science at Oxford*, 12:233–35.

34. Ann Bermingham, *Learning to Draw: Studies in the Cultural History of a Polite and Useful Art* (New Haven, CT: Yale University Press, 2000), 67.

35. See Meghan C. Doherty, "Discovering the 'True Form': Hooke's *Micrographia* and the Visual Vocabulary of Engraved Portraits," *Notes and Records of the Royal Society* 66, no. 3 (2012): 211–34; Craig Ashley Hanson, *The English Virtuoso: Art, Medicine, and Antiquarianism in the Age of Empiricism* (Chicago: University of Chicago Press, 2009), esp. chaps. 2–3; Ken Arnold, *Cabinets for the Curious: Looking Back at Early English Museums* (Aldershot, UK: Ashgate, 2006); Carol Gibson-Wood, "Picture Consumption in London at the End of the Seventeenth Century," *Art Bulletin* 84, no. 3 (2002): 491–500; *Printed Images in Early Modern Britain: Essays in Interpretation*, ed. Michael Hunter (Farnham: Ashgate, 2010); and Felicity Henderson, Sachiko Kusukawa, and Alexander Marr, eds., "Curiously Drawn: Early Modern Science as a Visual Pursuit," special issue, *Huntington Library Quarterly* 78, no. 2 (2015).

36. Fox, *Arts of Industry*, esp. 13–43; Kathleen H. Ochs, "The Royal Society of London's History of Trades Programme: An Early Episode in Applied Science," *Notes and Records of the Royal Society of London* 39, no. 2 (1985): 129–58; Sachiko Kusukawa, "Richard Waller's Table of Colours (1686)," in *Colour Histories: Science, Art, and Technology in the 17th and 18th Centuries*, ed. Magdalena Bushart and Friedrich Steinle (Boston: De Gruyter, 2015), 3–21; Anna Marie Roos, "The Saline Chymistry of Color in Seventeenth-Century English Natural History," *Early Science and Medicine* 20 (2015): 562–88; Hunter, *Wicked Intelligence*; and Carol Gibson-Wood, *Jonathan Richardson: Art Theorist of the English Enlightenment* (New Haven, CT: Yale University Press, 2000).

37. Hanson, *English Virtuoso*, 3.

38. M. A. R. Cooper, "Robert Hooke (1635–1703): Proto-Photogrammetrist," *Photogrammetric Record* 15, no. 87 (1996): 411; Edward Eigen, "On Purple and the Genesis of Photography or the Natural History of an Exposure," in *Ocean Flowers: Impressions from Nature*, ed. Carol Armstrong and M. Catherine Zegher (New York: Drawing Center, 2004), 272.

39. Geoffrey Batchen, *Burning with Desire: The Conception of Photography* (Cambridge, MA: MIT Press, 1997).

40. Robert Hooke, "Cometa; or, Remarks about Comets," in *Lectures and Collections made by Robert Hooke, Secretary of the Royal Society* (London: J. Martyn, 1678), 31–32. For the longer trajectory of chemical analyses of the heavens, see Craig Martin, *Renaissance Meteorology: Pomponazzi to Descartes* (Baltimore: Johns Hopkins University Press, 2011), esp. 80–105. As here, I adopt the convention of recent histories of early modern chymistry and retain period spellings of chemical materials, only adopting modern terms when explicitly sanctioned by the agents.

41. John Evelyn, *Sculptura; or, The History, and Art of Chalcography and Engraving in Copper*... (London: J. C. for T. Beedle, 1662), 9.

42. I thank Dr. Szychowski, along with David Tirrell, Moti Feingold, Jonathan Katz, and Tony Jia, for their generous assistance in this project, which was conducted between November 2009 and March 2010.

43. We rounded up Hooke's three-quarters of a quart of water (709 milliliters) to 750 milliliters. A quarter pound (113 grams) of sulfuric acid (H_2SO_4, density 1.83 g/mL) is roughly 62 milliliters.

44. Nehemiah Grew, *The Anatomy of Plants; With an Idea of a Philosophical History of Plants, and Several Other Lectures, Read before the Royal Society* (London: Printed by W. Rawlins, 1682), 244.

45. For an earlier analysis of this model as cognitive device, see Matthew C. Hunter, "Experiment, Theory, Representation: Robert Hooke's Material Models," in

Beyond Mimesis and Convention: Representation in Art and Science, ed. Roman Frigg and Matthew Hunter (New York: Springer Verlag, 2010), 193–219.

46. On period geology and Hooke's role in it, compare Paolo Rossi, *The Dark Abyss of Time: The History of the Earth and the History of Nations from Hooke to Vico* (1979), trans. Lydia G. Cochrane (Chicago: University of Chicago Press, 1984); Ellen Tan Drake, *Restless Genius: Robert Hooke and His Earthly Thoughts* (New York: Oxford University Press, 1996).

47. On making and testing evidence through reenactments, see Peter Heering, "The Enlightened Microscope: Re-Enactment and Analysis of Projections with Eighteenth-Century Solar Microscopes," *British Journal for the History of Science* 41, no. 3 (2008): 345–67; and Elizabeth Cavicchi, "A Witness Account of Solar Microscope Projections: Collective Acts Integrating across Personal and Historical Memory," *British Journal for the History of Science* 41, no. 3 (2008): 369–83. For more ambitious pursuits of chymical reenactment, see "The Chymistry of Isaac Newton" (http://webapp1.dlib.indiana.edu/newton/reference/chemLab.do), headed by William R. Newman, and "The Making and Knowing Project" (http://www.makingandknowing.org), led by Pamela H. Smith. A synthetic statement of this approach is Pamela H. Smith, "In the Workshop of History: Making, Writing, and Meaning," *West 86th* 19, no. 1 (2012): 4–31. On the place of reenactment (albeit understood in a different way) in art-historical method, see Michael Baxandall, *Patterns of Intention: On the Historical Explanation of Pictures* (New Haven, CT: Yale University Press, 1985), esp. 32–40.

48. At the very least, the iron-covered ball has to be stylized imaginatively to approximate the comet's core (stylized because, contrary to the superficial reaction of the recipe's wax ball, Hooke theorized comets to be rent by massive internal instabilities). The maker also has to posit that the bath of diluted acid can stand in for the fluid "aether" through which a comet moves, that principles of pressure governing the movement of bodies in fluids act as laws of celestial motion, and that the force of terrestrial gravity in relation to the laboratory apparatus approximates the force of solar gravitation operating on a comet visible from earth. This fit between laboratory situation and heavenly blaze remains provisional (the recipe doesn't produce light, as Hooke took comets to do, for example). "In some things analogous to the one, and somewhat to the other, though not exactly the same with either," so Hooke aptly noted of his approach to representing comets; see Hooke, *Lectures and Collections*, 11–12, 47.

49. On the history of figure-ground research, see J. L. Pind, *Edgar Rubin and Psychology in Denmark: Figure and Ground* (Cham: Springer, 2014), esp. 90–109.

50. An exemplary work here is Theresa Fairbanks Harris and Scott Wilcox, *Papermaking and the Art of Watercolor in Eighteenth-Century Britain: Paul Sandby and the Whatman Paper Mill* (New Haven, CT: Yale University Press, 2006).

51. For different paths in this direction, cf. James Elkins, *What Painting Is: How to Think about Oil Painting Using the Language of Alchemy* (New York: Routledge, 1999), esp. 117–46; Leslie Carlyle, "The Artist's Anticipation of Change as Discussed in British Nineteenth Century Instruction Books on Oil Painting," in *Appearance, Opinion, Change: Evaluating the Look of Paintings*, ed. Victoria Todd (London: United Kingdom Institute for Conservation, 1990), 62–67.

52. Michael Baxandall, *The Limewood Sculptors of Renaissance Germany* (New Haven, CT: Yale University Press, 1980), 36. The literatures impinging on time in the long eighteenth century are vast, but path-marks in from art history, history of science and technology, social history, and historiography include Michael Fried, *Absorption and Theatricality: Painting and Beholder in the Age of Diderot* (Chicago: University of Chicago Press, 1980); Nina Dubin, *Futures and Ruins: Eighteenth-Century Paris and the Art of Hubert Robert* (Los Angeles: Getty Research Institute, 2010); Richard Taws, *The Politics of the Provisional: Art and Ephemera in Revolutionary France* (University Park: Pennsylvania State

University Press, 2013); Martin J. S. Rudwick, *Bursting the Limits of Time: The Reconstruction of Geohistory in the Age of Revolution* (Chicago: University of Chicago Press, 2005); E. P. Thompson, "Time, Work-Discipline, and Industrial Capitalism," *Past and Present* 38, no. 1 (1967): 56–97; Paul Glennie and Nigel J. Thrift, *Shaping the Day: A History of Timekeeping in England and Wales 1300–1800* (Oxford: Oxford University Press, 2009); Stephen Bann, *The Clothing of Clio: The Representation of History in Nineteenth-Century Britain and France* (Cambridge: Cambridge University Press, 1984); J. G. A. Pocock, *Virtue, Commerce, and History: Essays on Political Thought and History, Chiefly in the Eighteenth Century* (Cambridge: Cambridge University Press, 1995); and Reinhart Koselleck, *Futures Past: On the Semantics of Historical Time* (1985), trans. Keith Tribe (New York: Columbia University Press, 2004).

53. Cf. Andrew McClellan, *Inventing the Louvre: Art, Politics, and the Origins of the Modern Museum in Eighteenth-Century Paris* (New York: Cambridge University Press, 1994); Danielle Rice, "The Fire of the Ancients: The Encaustic Painting Revival, 1755 to 1812" (PhD diss., Yale University, 1979).

54. Theo. Swift, "Epigram on the Paintings of Sir Joshua Reynolds," *Whitehall Evening Post*, no. 5695 (September 21–24, 1782): n.p.

55. For a period conception of patina as a "physico-chemico-economic" matter, see Gio Maironi da Ponte, *Sul Verderame: Memoria Fisico-Chimico-Economica* (Bergamo: Per Francesco Locatelli, 1784). More broadly, see Randolph Starn, "Three Ages of 'Patina' in Painting," *Representations* 78, no. 1 (2002): 86–115.

56. Cf. George Kubler, *The Shape of Time: Remarks on the History of Things* (New Haven, CT: Yale University Press, 1962); Georges Didi-Huberman, *Devant le temps: Histoire de l'art et anachronism des images* (Paris: Editions de Minuit, 2005); Alexander Nagel and Christopher S. Wood, *Anachronic Renaissance* (New York: Zone Books, 2010); and Keith P. F. Moxey, *Visual Time: The Image in History* (Durham, NC: Duke University Press, 2013).

57. Darwinian evolution, according to one prominent biologist, "is best referred to as a theory of variational evolution. According to this theory, an enormous amount of genetic variation is produced in every generation, but only a few individuals of the vast number of offspring will survive to produce the next generation. The theory postulates that those individuals with the highest probability of surviving and reproducing successfully are the ones best adapted, owing to their possession of a particular combination of attributes.... Since all changes take place in populations of genetically unique individuals, evolution is necessarily a gradual and continual process"; Ernst Mayr, *What Evolution Is* (New York: Basic Books, 2001), 85–86. On the crossings between Darwin and the visual arts, cf. Diana Donald and Jane Munro, eds., *Endless Forms: Charles Darwin, Natural Science and the Visual Arts* (Cambridge: Fitzwilliam Museum, 2009); Phillip Prodger, *Darwin's Camera: Art and Photography in the Theory of Evolution* (Oxford: Oxford University Press, 2009).

58. William Henry Fox Talbot, "Obtaining Motive Power" (patent no. 8650; 1840), 3–4.

59. Erasmus Darwin, *Zoonomia; or The Laws of Organic Life* (London: J. Johnson, 1796), 1:538.

60. Thomas Wedgwood to James Mackintosh, ca. January 10, 1801 (postmarked December 18, 1801); Victoria & Albert Museum, Wedgwood Collection, MS E40 28489 a, $f2_r$. For Erasmus Darwin's theory of biological evolution, cf. James Harrison, "Erasmus Darwin's View of Evolution," *Journal of the History of Ideas* 32, no. 2 (1971): 247–64; Roy Porter, "Erasmus Darwin, Doctor of Evolution?" in *History, Humanity and Evolution: Essays for John C. Greene*, ed. James Moore (New York: Cambridge University Press, 1989), 39–69. In quotations, I use angle brackets to signal passages where a writer has made manual insertions into a manuscript text. For this convention of transcription,

see Michael Hunter, "How to Edit a Seventeenth-Century Manuscript: Principles and Practice," *Seventeenth Century* 10, no. 2 (1995): 277–310, esp. 294.

61. On Riegl's links to evolutionary thought, see Jas' Elsner, "From Empirical Evidence to the Big Picture: Some Reflections on Riegl's Concept of *Kunstwollen*," *Critical Inquiry* 32, no. 4 (2006): 741–66; Alois Riegl, *Problems of Style: Foundations for a History of Ornament* (1893), trans. Evelyn M. Kain (Princeton, NJ: Princeton University Press, 1992), 4.

62. Cf. Marco Beretta, *Imaging a Career in Science: The Iconography of Antoine Laurent Lavoisier* (Canton, MA: Science History Publications, 2001).

63. See Denis I. Duveen, "Madame Lavoisier 1758–1836," *Chymia* 4 (1953): 13–29. As but one of many art-historical readings of David's modernism, see T. J. Clark, *Farewell to an Idea: Episodes from a History of Modernism* (New Haven, CT: Yale University Press, 1999).

64. Lawrence M. Principe, "A Revolution Nobody Noticed? Changes in Early Eighteenth-Century Chymistry," in *New Narratives in Eighteenth-Century Chemistry*, ed. Lawrence M. Principe (Dordrecht: Springer, 2007), 2.

65. William R. Newman and Lawrence M. Principe, "Alchemy vs. Chemistry: The Etymological Origins of a Historiographic Mistake," *Early Science and Medicine* 3, no. 1 (1998): 32–65.

66. Cf. Ursula Klein and Wolfgang Lefebvre, *Materials in Eighteenth-Century Science: A Historical Ontology* (Cambridge, MA: MIT Press, 2007).

67. For longer histories of art/alchemy intersections, see William R. Newman, *Promethean Ambitions: Alchemy and the Quest to Perfect Nature* (Chicago: University of Chicago Press, 2004); Pamela H. Smith, *The Body of the Artisan: Art and Experience in the Scientific Revolution* (Chicago: University of Chicago Press, 2004); Spike Bucklow, *The Alchemy of Paint: Art, Science, and Secrets from the Middle Ages* (London: Marion Boyars, 2009).

68. See Gerald J. Gruman, "A History of Ideas about the Prolongation of Life: The Evolution of Prolongevity Hypotheses to 1800," *Transactions of the American Philosophical Society* 56, no. 9 (1966): 1–102, esp. 62–68.

69. A classic study remains Charles Webster, *The Great Instauration: Science, Medicine and Reform, 1626–1660* (New York: Holmes & Meier, 1976).

70. Cf. Kevin C. Knox, "'The Deplorable Frenzy': The Slow Legitimisation of Chemical Practice at Cambridge University," in *The 1702 Chair of Chemistry at Cambridge: Transformation and Change*, ed. Mary D. Archer and Christopher D. Haley (New York: Cambridge University Press, 2005), 1–30; Roy Porter, *Doctor of Society: Thomas Beddoes and the Sick Trade in Late-Enlightenment England* (London: Routledge, 1992); Mike Jay, *The Atmosphere of Heaven: The Unnatural Experiments of Dr. Beddoes and His Sons of Genius* (New Haven, CT: Yale University Press, 2009).

71. Philo, "For the Morning Post," *Morning Post and Daily Advertiser*, no. 571 (August 26, 1774), n.p.

72. Archibald Clow and Nan L. Clow, "Vitriol in the Industrial Revolution," *Economic History Review* 15, nos. 1–2 (1945): 44.

73. Cf. Beverly Lemire, *Fashion's Favourite: The Cotton Trade and the Consumer in Britain, 1660–1800* (Oxford: Pasold Research Fund, 1991); Jan de Vries, *The Industrious Revolution: Consumer Behavior and the Household Economy, 1650 to the Present* (Cambridge: Cambridge University Press, 2008); John Styles, *The Dress of the People: Everyday Fashion in Eighteenth-Century England* (New Haven, CT: Yale University Press, 2007), esp. 109–33; and Andreas Malm, *Fossil Capital: The Rise of Steam Power and the Roots of Global Warming* (London: Verso, 2016).

74. F. W. Gibbs, "Robert Dossie (1717–1777) and the Society of Arts," *Annals of Science* 7, no. 2 (1951): 149.

75. Samuel Johnson quoted the following passage from Boerhaave when defining chemistry in the second edition of his dictionary: "'An art whereby sensible bodies contained in vessels . . . are so changed, by means of certain instruments, and principally fire, that their several powers and virtues are thereby discovered, with a view to philosophy, or medicine"; Samuel Johnson, *A Dictionary of the English Language . . .*, 2nd ed. (London: J. and P. Knapton et al., 1755), vol. 1, page headed CIC–CIN.

76. See Samuel Johnson, "Life of Boerhaave," *Gentleman's Magazine* 9 (1739): 37–38, 72–73, 114–16, 172–76.

77. James Northcote, *Supplement to the Memoirs of the Life, Writings, Discourses, and Professional Works of Sir Joshua Reynolds . . .* (London: Henry Colburn, 1815), vi.

78. I adopt this phrase from the "medical penumbra" proposed by L. W. B. Brockliss and Colin Jones, *The Medical World of Early Modern France* (Oxford: Clarendon, 1997), esp. 236–37. I take this approach to move with the spirit of Jan Golinski's account of eighteenth-century Britain's public culture of chemistry, wherein theoretical disputes were discouraged in favor of consensual agreements about experimental matters of fact; Jan Golinski, *Science as Public Culture: Chemistry and Enlightenment in Britain, 1760–1820* (New York: Cambridge University Press, 1992).

79. Many exceptions could be cited, but particularly important to the development of this project has been Jennifer L. Roberts, *Transporting Visions: The Movement of Images in Early America* (Berkeley: University of California Press, 2014). Of course, the contention that photographs are neither pictures nor objects but indexes has been a leading theoretical contention for the past half century. A useful compendium of these debates is James Elkins, ed., *Photography Theory* (New York: Routledge, 2007).

80. Some historians of science have used the term "materials" to highlight the artificially processed salts, acids, and other compounds that became targets of eighteenth-century chemists' study. I resist this term since my story focuses primarily on chemical products shaped (however fugitively) by makers interested in materials but most frequently pursuing targets other than the materials themselves; see Klein and Lefebvre, *Materials in Eighteenth-Century Science*.

81. Critique of the ontology advanced by talk of objects and subjects is key to the twentieth century's most prominent philosophical meditation on time: Martin Heidegger's *Sein und Zeit* (1927). In turn, Michael Fried has made "the *object*-beholder (one is tempted so say object-*subject*) relationship" a target of French critics of the Rococo from the 1750s onward; Michael Fried, "Toward a Supreme Fiction: Genre and Beholder in the Art Criticism of Diderot and His Contemporaries," *New Literary History* 6, no. 3 (1975): 581. For the larger arc of Fried's arguments about objects and subjects and alternatives to them within a Heideggerian ambit, cf. Walter Benn Michaels, *The Shape of the Signifier: 1967 to the End of History* (Princeton, NJ: Princeton University Press, 2004); Bill Brown, ed., *Things* (Chicago: University of Chicago Press, 2004).

82. Lorraine Daston and Peter Galison, *Objectivity* (New York: Zone Books, 2007), 22.

83. Roy Porter, "The Enlightenment in England," in *The Enlightenment in National Context*, ed. Roy Porter and Mikuláš Teich (New York: Cambridge University Press, 1981), 6, 7.

84. See James Northcote to Samuel Northcote Jr., August 15, 1776; Royal Academy of Arts, MS NOR 28_r.

85. On Northcote's friendships with the Yonge and Holmes families, see Ledbury, *James Northcote*, 12–13, 26. For Hooke's contacts with the Holmes family, see Lisa Jardine, *The Curious Life of Robert Hooke: The Man Who Measured London* (New York: HarperCollins, 2003), 256–57.

86. Laurence Sterne, *The Life and Opinions of Tristram Shandy, Gentleman* (1759–1766), ed. Ian C. Ross (New York: Oxford University Press, 1983), vol. 1, chap. 22, p. 58.

87. Bernhard Siegert has claimed that "Classical/Romantic authorship was established by anthropomorphizing the seventeenth-century postal *raison d'état*. Eighteenth-century poetry was simultaneously a cover-up of the postal service and a delivery of the mail in the production and communication of knowledge"; Bernhard Siegert, *Relays: Literature as an Epoch of the Postal System* (1993), trans. Kevin Repp (Stanford, CA: Stanford University Press, 1999), 13.

88. Esther Leslie, *Synthetic Worlds: Nature, Art and the Chemical Industry* (London: Reaktion, 2005), 10.

89. Jeff Wall, "Photography and Liquid Intelligence," in *Jeff Wall*, ed. Thierry de Duve et al. (London: Phaidon, 1996), 90.

90. Cf. Hunter, *Wicked Intelligence*; Matthew C. Hunter and Francesco Lucchini, eds., *The Clever Object* (Hoboken, NJ: John Wiley & Sons, 2013); Matthew C. Hunter, ed., "Liquid Intelligence," special issue, *Grey Room* 69 (Fall 2017).

91. John Fowles, *A Maggot* (London: Jonathan Cape, 1985), 5.

92. A powerful endorsement of the historical mandate is Andreas Malm and Alf Hornborg, "The Geology of Mankind? A Critique of the Anthropocene Narrative," *Anthropocene Review* 1, no. 1 (2014): 62–69.

93. Fowles, *Maggot*, 5.

Chapter 1

1. On this celebrated purchase, I draw from Michael I. Wilson, *Nicholas Lanier: Master of the King's Musick* (Aldershot, UK: Scolar, 1994), 83–132; Jerry Brotton, *The Sale of the Late King's Goods: Charles I and His Art Collection* (London: Macmillan, 2006), 113–29; Lucy Whitaker and Martin Clayton, with contributions by Aislinn Loconte, *The Art of Italy in the Royal Collection: Renaissance & Baroque* (London: Royal Collections Publications, 2007); Alessandro Luzio, *La Galleria dei Gonzaga: Venduta All'inghilterra nel 1627–28* (Rome: Bardi, 1974); and Sir Isaac Wake to Secretary of State Edward Conway, April 18, 1628; reproduced in *Original Unpublished Papers Illustrative of the Life of Sir Peter Paul Rubens*, ed. W. Noël Sainsbury (London: Bradbury & Evans, 1859), 326–27.

2. Mayerne describes the pictures' movement "dans un nave recharge de raisins de Chorinte ou estoyent plusieurs barils pleins de Mercure sublimé, dont la vapeur excitee par la chaleur des raisins noircit comme encre touts les tableaux tant a huile qu'a destrempe"; Sir Theodore de Mayerne, "Pictoria, sculptoria et quae subalternarum artium" (the "Mayerne manuscript"), British Library, MS 2052, f14v.

3. "Il fault donque user ou d'eau fort ordinaire ou d'une faite du sel & d'alum. Ou d'une Espirite du sel du Vitriol, ou du Soulphre"; Mayerne manuscript, British Library, MS 2052, f14v.

4. See Whitaker and Clayton, *Art of Italy*, 224–26.

5. "Pour ceste vapeur du sublimé (que se faict avec du sel & du vitriol, je croy qu'un espirit acide tiré du ces deux eut esté la meilleur. Parce que simile agit in simile"; Mayerne manuscript, British Library, MS 2052, f14v. For more on Mayerne's chemistry, see Hugh Trevor-Roper, *Europe's Physician: The Various Life of Sir Theodore De Mayerne* (New Haven, CT: Yale University Press, 2006).

6. See entries in the Royal Society's minutes for May 31, June 7, and June 28, 1665, in Thomas Birch, *History of the Royal-Society of London, for Improving of Natural Knowledge* (London: A. Millar, 1756), 2:52, 55, 60.

7. For Mayerne's research into colors at the Royal Society, see Vera Keller, "Scarlet Letters: Sir Theodore de Mayerne and the Early Stuart Color World in the Royal Society," in *Archival Afterlives: Life, Death, and Knowledge-Making in Early Modern British Scientific and Medical Archives*, ed. Vera Keller, Anna Marie Roos, and Elizabeth Yale (Leiden: Brill, 2018), 72–119. For subsequent gleanings from Mayerne's research, see

July 30, 1668; July 17, 1679; May 8, 1680; May 18, 1681; Birch, *History of the Royal-Society*, 2:311, 3:496, 4:37, 4:87.

8. Walter E. Houghton, "The English Virtuoso in the Seventeenth Century: Part II," *Journal of the History of Ideas* 3, no. 2 (1942): 205 (190–219).

9. An extensive history is E. Newton Harvey, *A History of Luminescence from the Earliest Times until 1900* (Philadelphia: American Philosophical Society, 1957).

10. [Marcello Malpighi], "An Improvement of the Bononian Stone, shining in the dark," *Philosophical Transactions* 134 (1677): 842. For the full letter, which unfortunately contains no further details on this interesting meeting of art and phosphorus, see Marcello Malpighi to Henry Oldenburg, February 27, 1676/7, in *The Correspondence of Henry Oldenburg*, vol. 13, *July 1676–July 1681*, ed. and trans. A. Rupert Hall and Marie Boas Hall (London: Taylor and Francis, 1986), 222–23. An inventory of the glowing collection is reproduced in Raffaella Morselli, *Italian Inventories 3: Collezioni e quadrerie nella Bologna del Seicento: Inventari 1640–1707*, ed. Anna Cera Sones (Los Angeles: Provenance Index of the Getty Information Institute, 1998), 425–30.

11. Editor William Derham included a recipe for this phosphorus preparation (which calls for taking "the Crust, which comes off the Stones, and beat it small, fearcing it as before; then when you have made your Figure, or Image, wet it with the White of an Egg, and sprinkle upon it your fine Powder") in Robert Hooke, *Philosophical Experiments and Observations of the Late Eminent Dr. Robert Hooke, S.R.S. and Geom. Prof. Gresh., and Other Eminent Virtuoso's in His Time*, ed. William Derham (London: Printed by W. and J. Innys, 1726), 176. I have not been able to locate the recipe in Hooke's surviving manuscripts.

12. See, for example, Roland Barthes, "The World as Object," in *Calligram: Essays in New Art History from France*, ed. Norman Bryson (New York: Cambridge University Press, 1988), 106–15; Krista Thompson, "The Sound of Light: Reflections on Art History in the Visual Culture of Hip-Hop," *Art Bulletin* 91, no. 4 (2009): 481–505.

13. Paraphrasing the subtitle of Hans Belting's *Bild und Kult* in this way, my point is of course not to deny that there was art in Britain prior to the eighteenth century. For a recent survey thereof, see David H. Solkin, *Art in Britain 1660–1815* (New Haven, CT: Yale University Press, 2015).

14. Robert G. Frank Jr., *Harvey and the Oxford Physiologists: Scientific Ideas and Social Interaction* (Berkeley: University of California Press, 1980).

15. An exhaustive account of research around phosphorus and combustion in the 1670s Oxford-associated networks would need to engage with John Mayow's *Tractatus Quinque* (1674). Instructively, Mayow's writings would be ushered back into publication in 1790 by Thomas Beddoes, a catalyst in the chemical tradition under investigation here; see Douglas McKie, "Fire and the Flamma Vitalis: Boyle, Hooke, and Mayow," in *Science, Medicine and History: Essays on the Evolution of Scientific Thought and Medical Practice Written in Honour of Charles Singer*, ed. E. Ashworth Underwood (New York: Geoffrey Cumberlege/Oxford University Press, 1953), 1:469–88; Mike Jay, *The Atmosphere of Heaven: The Unnatural Experiments of Dr. Beddoes and His Sons of Genius* (New Haven, CT: Yale University Press, 2009), 30–31.

16. I thank Chriscinda Henry for pushing me on this point.

17. [John Beale], "Two Instances of Something Remarkable in Shining Flesh, from Dr. J. Beal of Yeavel in Somersetshire," *Philosophical Transactions of the Royal Society of London*, no. 125 (May 22, 1676), 600.

18. See [Robert Boyle], "Some Observations about Shining Flesh, made by the Honourable Robert Boyle . . . ," *Philosophical Transactions*, no. 89 (1672), 5108–16.

19. Beale, "Two Instances," 602.

20. For more on Boyle's exchanges with various German-speaking adepts, see Law-

rence M. Principe, *The Aspiring Adept: Robert Boyle and His Alchemical Quest* (Princeton, NJ: Princeton University Press, 1998), esp. 134–36.

21. [Robert Boyle], "A Short Memorial of some Observations made upon an Artificial Substance, that shines without any precedent Illustration," in *Lectures and Collections made by Robert Hooke, Secretary of the Royal Society* (London: J. Martyn, 1678), 60.

22. [Frederick Slare], "An account of several Experiments made with the shining substance of the liquid and of the solid *Phosphorous. . . ,*" *Philosophical Collections*, no. 3 (1681), 48.

23. Pamela H. Smith, *The Business of Alchemy: Science and Culture in the Holy Roman Empire* (Princeton, NJ: Princeton University Press, 1994), 248.

24. See Jan Golinski, "A Noble Spectacle: Phosphorus and the Public Culture of Science in the Early Royal Society," *Isis* 80 (1989): 11–39, esp. 15–17.

25. Simon Werrett, *Fireworks: Pyrotechnic Arts and Sciences in European History* (Chicago: University of Chicago Press, 2010), 81–82.

26. [Anonymous], "An Account of Four Sorts of Factitious Shining Substances . . . ," *Philosophical Transactions*, no. 135 (May 26, 1677), 867.

27. Slare, "Account of several Experiments," 50. For the extensive use of urine in Boyle's contemporaneous phosphorus preparations, see "A Paper of the Honourable Robert Boyl's despoited with the Secretaries of the Royal Society, October 14, 1680. and opened since his Death; being an Account of his making the Phosphorous, &c.," *Philosophical Transactions*, no. 196 (1693), 583–84.

28. Thomas Willis, *Two Discourses concerning The Souls of Brutes* (1672), trans. S. Pordage (London: Thomas Dring, Ch. Harper and John Leigh, 1683), 22.

29. Thomas Willis, "Of Fermentation; or, The Inorganical Motion of Natural Bodies" (ca. 1656), in *Dr. Willis's Practice of Physick, Being the Whole Works of That Renowned and Famous Physician* (London: T. Dring, C. Harper, and J. Leigh, 1684), 32.

30. Where post-Lavoisierian chemistry could turn to the concept of oxidation to explain both metabolism and combustion, Frank argues, Willis's notion of fermentation provided to the Oxford physiologists "a conceptual system within which the generation of heat, even in such palpably different manifestations as animal heat and fire, could be seen as resulting from a common chemical action"; Frank, *Harvey and the Oxford Physiologists*, 164.

31. See Antonio Clericuzio, "Carneades and the Chemists: A Study of *The Sceptical Chymist* and its Impact on Seventeenth-Century Chemistry," in *Robert Boyle Reconsidered*, ed. Michael C. W. Hunter (Cambridge: Cambridge University Press, 1994), 79–90; Principe, *Aspiring Adept*, 27–62; and Ku-Ming Chang, "Fermentation, Phlogiston and Matter Theory: Chemistry and Natural Philosophy in Georg Ernst Stahl's *Zymotechnia Fundamentalis*," *Early Science and Medicine* 7, no. 1 (2002): 31–64.

32. Robert Boyle, *The Aerial Noctiluca* (1680), in *The Works of Robert Boyle*, vol. 9, *Publications of 1678–83*, ed. Michael Hunter and Edward B. Davis (London: Pickering & Chatto, 2000), 278.

33. Boyle, *Aerial Noctiluca*, 280.

34. Boyle's much-discussed experiment would be taken by contemporary critics such as French physician Samuel Cottereau Duclos as exemplifying the corpuscularians' unjustified reliance on invisible particles and other theoretical concepts, with insufficient attention to chemical practice and literature; see Victor D. Boantza, *Matter and Method in the Long Chemical Revolution: Laws of Another Order* (Farnham, Surrey: Ashgate, 2013), esp. 48–56.

35. Robert Boyle, "A Physico-Chymical Essay, Containing An Experiment with some Considerations touching the differing Parts and Redintegration of Salt-Petre," in *Certain Physiological Essays* (1661), in *The Works of Robert Boyle*, vol. 2, *The Sceptical Chymist*

and other Publications of 1661, ed. Michael Hunter and Edward B. Davis (London: Pickering & Chatto, 1999), 108.

36. On the importance of this essay to Boyle's broader project, see Michael Hunter, *Boyle: Between God and Science* (New Haven, CT: Yale University Press, 2009), 116–18.

37. See Boyle, "Physico-Chymical Essay," 106. For a reenactment of an experiment Boyle proposed in support of Christ's "Redintegration," see https://alexwraggemorley.wordpress.com/2015/01/03/robert-boyles-experimental-proof-of-the-possibility-of-the-resurrection/ (accessed February 21, 2017).

38. I draw from D. J. Lysacht, "Hooke's Theory of Combustion," *Ambix* 1, no. 2 (1937): 93–105; H. D. Turner, "Robert Hooke and Theories of Combustion," *Centaurus* 4, no. 4 (1956): 297–310; McKie, "Fire and the Flamma Vitalis"; and Frank, *Harvey and the Oxford Physiologists*, esp. 137–39.

39. March 1, 1665; Birch, *History of the Royal-Society*, 2:19.

40. This, of course, is to overlook the problem of the air pump's notorious leaks, which are explored at length in Steven Shapin and Simon Schaffer, *The Leviathan and the Air-Pump: Hobbes, Boyle, and the Experimental Life* (Princeton, NJ: Princeton University Press, 1985).

41. Robert Hooke, *Micrographia; or, Some Physiological Descriptions of Minute Bodies made by Magnifying Glasses* (London: John Martyn and James Allestry, 1665), 103.

42. Hooke, *Micrographia*, 105. Hooke's interpretation of combustion was sufficiently compelling to guide Thomas Sprat's definition of fire as "the Act of the dissolution of heated *Sulphureous Bodies*, by the *Air* as a *Menstruum*, much after the same manner, as *Aqua Fortis* . . . on dissoluble Bodies, as Iron, *Tin*, *Copper*: that heat, and light are two inseparable effects of this dissolution"; Thomas Sprat, *The History of the Royal-Society of London* (London: T. R. for J. Martyn, 1667), 215.

43. See Golinski, "Noble Spectacle," 21–23.

44. Robert Hooke, "Lectures of Light: Sect. IV" (ca. May 1681), in *The Posthumous Works of Robert Hooke*, ed. Richard Waller (London: Sam Smith and Benjamin Walford, 1705), 112. On Balduin, see Vera Keller, "Hermetic Atomism: Christian Adolph Balduin (1632–1682), *Aurum Aurae*, and the 1674 Phosphor," *Ambix* 61, no. 4 (2014): 366–84.

45. Hooke, "Sect. IV," 112.

46. Hooke, "Sect. IV," 108.

47. On the role of phosphorus in Hooke's last "Lecture of Light," see Douwe Draaisma, *Metaphors of Memory: A History of Ideas about the Mind* (1995), trans. Paul Vincent (New York: Cambridge University Press, 2000), 49–67. Other recent readings of this important lecture include Nick Wilding, "Graphic Technologies," in *Robert Hooke: Tercentennial Studies*, ed. M. A. R. Cooper and Michael Hunter (Aldershot, UK: Ashgate, 2006), 123–34; Matthew C. Hunter, *Wicked Intelligence: Visual Art and the Science of Experiment in Restoration London* (Chicago: University of Chicago Press, 2013), esp. 177–87; and Sean Silver, *The Mind Is a Collection: Case Studies in Eighteenth-Century Thought* (Philadelphia: University of Pennsylvania Press, 2015), esp. 114–24, 157–59.

48. Hooke, "Lectures of Light: Sect. VII" (ca. June 1682), in *Posthumous Works*, 139.

49. Hooke, "Sect. VII," 139.

50. For this assessment, see John Aubrey, *Brief Lives*, ed. Oliver Lawson Dick (London: Secker and Warburg, 1949), 165. Cf. Jim Bennett, "The Mechanics' Philosophy and the Mechanical Philosophy," *History of Science* 24 (1986): 1–28.

51. Cf. Betty J. T. Dobbs, *The Foundations of Newton's Alchemy; or, "The Hunting of the Greene Lyon"* (Cambridge: Cambridge University Press, 1975); Richard S. Westfall, *Never at Rest: A Biography of Isaac Newton* (Cambridge: Cambridge University Press, 1980); and William R. Newman, *Atoms and Alchemy: Chymistry and the Experimental Origins of the Scientific Revolution* (Chicago: University of Chicago Press, 2006).

52. On Hooke's insistence on the difference between impressions and ideas, see

Matthew C. Hunter, "The Theory of the Impression According to Robert Hooke," in *Printed Images in Early Modern Britain: Essays in Interpretation*, ed. Michael Hunter (Burlington: Ashgate, 2010), 167–90.

53. For his claim that ideas are "material and bulky, that is, . . . Bodies of determinate bigness, and impregnated with determinate Motions, and to be in themselves distinct," see Hooke, "Sect. VII," 142. For the timing, see Hooke, "Sect. VII," 143.

54. Hooke, "Sect. VII," 141.

55. Cf. Michael Wright, "Hooke's Longitude Timekeeper," in *Robert Hooke: New Studies*, ed. Michael C. W. Hunter and Simon Schaffer (Woodbridge, Suffolk: Boydell Press, 1989), 63–118; Jim Bennett, "Hooke's Instruments," in *London's Leonardo: The Life and Work of Robert Hooke*, ed. Jim Bennett et al. (Oxford: Oxford University Press, 2003), 63–104.

56. On Huygens's designs, see J. H. Leopold, "The Longitude Timekeepers of Christiaan Huygens," in *The Quest for Longitude*, ed. William J. H. Andrewes (Cambridge, MA: Collection of Historical Scientific Instruments, Harvard University, 1996), 102–14. For a concise summary of the controversy, see Lisa Jardine, *The Curious Life of Robert Hooke: The Man Who Measured London* (New York: HarperCollins, 2003), 177–213.

57. June 28, 1682; Birch, *History of the Royal-Society*, 2:154.

58. See Martin J. S. Rudwick, *The Meaning of Fossils: Episodes in the History of Paleontology* (New York: American Elsevier, 1972), esp. 34–35; Robert Hooke, "Lectures and Discourses of Earthquakes and Subterraneous Eruptions" (1668–ca. 1700), in *Posthumous Works*, 310.

59. Historian of science Paolo Rossi puts it this way: "Slow, uniform changes, unforeseen catastrophes, the destruction and emergence of continents, the mutations within species: All spoke to Hooke of a history that is like a river of unknown course, which might event contain, in its past, treasures of wisdom not lost forever"; Paolo Rossi, *The Dark Abyss of Time: The History of the Earth and the History of Nations from Hooke to Vico* (1979), trans. Lydia G. Cochrane (Chicago: University of Chicago Press, 1984), 15.

60. Hooke, "Earthquakes," 318.

61. Hooke, "Sect. VII," 141.

62. See [Robert Hooke], "Review of Marc Antonio Cellio's *Il Fosforo overo la Pietra Bolognese . . .* (Roma: 1680)," in *Philosophical Collections* 3 (1681), 77–79.

63. For phosphorus in the repository, see Nehemiah Grew, *Musaeum Regalis Societatis; or, A Catalogue & Description of the Natural and Artificial Rarities belonging to the Royal Society and preserved at Gresham Colledge* (London: W. Rawlins, 1681), 353–57.

64. Hooke, "Sect. VII," 141.

65. On this decreasing luminescence, see Lawrence M. Principe, "Chymical Exotica in the Seventeenth Century; or, How to Make the Bologna Stone," *Ambix* 63, no. 2 (2016): 118–44, esp. 123–24.

66. Nehemiah Grew, *The Anatomy of Plants; With an Idea of a Philosophical History of Plants, and Several Other Lectures, Read before the Royal Society* (London: Printed by W. Rawlins, 1682), 158.

67. Grew, *Musaeum*, 254. Cf. Anna Marie Roos, "Nehemiah Grew (1641–1712) and the Saline Chymistry of Plants," *Ambix* 54, no. 1 (2007): 51–68; Anna Marie Roos, "The Saline Chymistry of Color in Seventeenth-Century English Natural History," *Early Science and Medicine* 20 (2015): 562–88.

68. Grew, *Musaeum*, 268.

69. See H. W. Janson, "The 'Image Made by Chance' in Renaissance Thought," in *16 Studies* (New York: Harry N. Abrams, 1974), 53–69; H. V. S. Ogden and Margaret S. Ogden, *English Taste in Landscape in the Seventeenth Century* (Ann Arbor: University of Michigan Press, 1955).

70. Grew, *Musaeum*, 253.

71. For other cases of rocks, stones, and trees confusing the arts, see Walter Benn Michaels, "Prehistoricism," in *The Shape of the Signifier: 1967 to the End of History* (Princeton, NJ: Princeton University Press, 2004), 83–128.

72. Grew, *Musaeum*, 268. A classic art-historical encounter with Kepler's camera-obscura experiments is Svetlana Alpers, *The Art of Describing: Dutch Art in the Seventeenth Century* (Chicago: University of Chicago Press, 1983), esp. chap. 2.

73. Grew, *Musaeum*, 253.

74. Josef Maria Eder, *History of Photography*, trans. Edward Epstean (New York: Columbia University Press, 1945), 62.

75. Eder, *History of Photography*, 61, 63, 62.

76. See Hooke, "Sect. VII," 143.

77. [Henry Oldenburg], "An Experiment of a Way of Preparing a Liquor, That Shall Sink into, and Colour the Whole Body of Marble, Causing a Picture, Drawn on a Surface, to Appear Also in the Inmost Parts of the Stone," *Philosophical Transactions of the Royal Society of London*, no. 7 (1665), 126. More broadly, see Fabio Barry, "Painting in Stone: The Symbolism of Colored Marbles in the Visual Arts and Literature from Antiquity until the Enlightenment" (PhD diss., Columbia University, 2011).

78. Here, I adapt a phrase from Lynda Nead, "Velocities of the Image *c.* 1900," *Art History* 27, no. 5 (2004): 745–69.

79. Cf. *Hiroshi Sugimoto: History of History* (New York: Japan Society, 2005); Walter Benn Michaels, "Photographs and Fossils," in *Photography Theory*, ed. James Elkins (New York: Routledge, 2007), 431–50.

Chapter 2

1. Arie Wallert and Willem de Ridder, "About Skin: Technical Examination of Paintings by Adriaen van der Werff," in *Studying Old Master Paintings: Technology and Practice*, ed. M. Spring et al. (London: Archetype, 2011), 183.

2. See Dániel Margócsy, *Commercial Visions: Science, Trade, and Visual Culture in the Dutch Golden Age* (Chicago: University of Chicago Press, 2014), esp. 183–84.

3. J. C. Le Blon, *Coloritto; or, The Harmony of Colouring in Painting* (London, 1725), 8.

4. On this theme, see Mechthild Fend, *Fleshing Out Surfaces: Skin in French Art and Medicine, 1650–1850* (Manchester: Manchester University Press, 2017), esp. 35–39.

5. See William Hogarth, *The Analysis of Beauty* (1753), ed. Ronald Paulson (New Haven, CT: Yale University Press, 1997), esp. 8–10.

6. See Ronald Paulson, *Hogarth: His Life, Art, and Times* (New Haven, CT: Yale University Press, 1971), 1:43–54. On the storied relations between goldsmithing and alchemy, see Michael Cole, *Cellini and the Principles of Sculpture* (Cambridge: Cambridge University Press, 2002).

7. Hogarth, *Analysis of Beauty*, 88. For period uses of isinglass in image transfer, see Benjamin Martin, *Typographia Naturalis; or, The Art of Printing . . .* (London: Printed for the author, at No 171, in Fleet-Street, 1772).

8. For a brilliant reading of print as liquid transfer, see Jennifer L. Roberts, "The Veins of Pennsylvania: Benjamin Franklin's Nature-Print Currency," *Grey Room* 69 (2017): 50–79.

9. Hogarth, *Analysis of Beauty*, 89, 90.

10. See Malcolm Baker, "Attending to Marble in Eighteenth-Century Britain," in *The Aesthetics of Marble: From Late Antiquity to the Present*, ed. Dario Gamboni and Gerhard Wolf (Munich: Hirmer Verlag, forthcoming). For more on Chambers's "peice of Marble with the several colours used, in it, like a Painters pallet, being greatly saturated with Aquafortis at different times for 20 hours, tho the polish of the marble was quite

effaced, yet there was not the least discharge of any of the colours, nor were they any wise dulled," see Emmanuel Mendes da Costa to Anthony Keck, March 21, 1759; Royal Society of Arts, RSA/PR/GE/110/7/47.

11. Hogarth, *Analysis of Beauty*, 91.

12. For analysis of Hogarth's careful pictorial facture, see David Bomford and Ashok Roy, "Hogarth's 'Marriage à la Mode,'" *National Gallery Technical Bulletin* 6 (1982): 44–67; Rica Jones, "William Hogarth, *Mrs. Salter*," in *Paint and Purpose: A Study in Technique in British Art*, ed. Stephen Hackney, Rica Jones, and Joyce Townsend (London: Tate Gallery, 1999), 42–47.

13. Hogarth, *Analysis of Beauty*, 91.

14. For Hogarth's expanded critique of patina, see Hogarth, *Analysis of Beauty*, 91–92.

15. Hogarth, *Analysis of Beauty*, 91.

16. E. H. Gombrich, "Reynolds's Theory and Practice of Imitation," *Burlington Magazine* 80, no. 467 (1942): 45.

17. John Barrell, *The Political Theory of Painting from Reynolds to Hazlitt: "The Body of the Public"* (New Haven, CT: Yale University Press, 1986), 80. See also Richard Wendorf, *Sir Joshua Reynolds: The Painter in Society* (London: National Gallery, 1996).

18. As interpreters have stressed, Reynolds represents Keppel at a moment of defeat—when his ship, HMS *Maidstone*, had sunk in combat off France's Atlantic coast in 1747. Cf. David H. Solkin, "Great Pictures or Great Men? Reynolds, Male Portraiture, and the Power of Art," *Oxford Art Journal* 9, no. 2 (1986): 42–49; Douglas Fordham, *British Art and the Seven Years' War: Allegiance and Autonomy* (Philadelphia: University of Pennsylvania Press, 2010), esp. 65–72; Daniel O'Quinn, "Facing Past and Future Empires: Joshua Reynolds's Portraits of Augustus Keppel," in *The Culture of the Seven Years' War: Empire, Identity, and the Arts in the Eighteenth-Century Atlantic World*, ed. Frans De Bruyn and Shaun Regan (Toronto: University of Toronto Press, 2014), 307–38; Mark Hallett, *Reynolds: Portraiture in Action* (New Haven, CT: Yale University Press, 2014), 98–105.

19. For this assessment, see Allan Cunningham, *The Lives of the Most Eminent British Painters, Sculptors, and Architects* (London: J. Murray, 1830), 244; February 16, 1757, Minutes of the Society, 1754–1757 (Royal Society of Arts, RSA/AD/MA/100/12/01/01), 227. For protocols on testing verdigris, see Robert Dossie, *The Handmaid to the Arts* (London: Printed for J. Nourse, 1764), 112–15.

20. Important contributions to the vast conservation literature on Reynolds include M. Kirby Talley, "'All Good Pictures Crack': Sir Joshua Reynolds's Practice and Studio," in *Reynolds*, ed. Nicholas Penny (New York: Harry N. Abrams, 1986), 55–70; Hélène Dubois, "'Use a little wax with your colours, but don't tell anybody': Joshua Reynolds's Painting Experiments with Wax and His Sources," *Hamilton Kerr Institute Bulletin* 3 (2000): 97–106; Roy Ashok et al., "Joshua Reynolds in the National Gallery and Wallace Collection," special issue, *National Gallery Technical Bulletin* 35 (2015): 1–111; and Lucy Davis and Mark Hallett, eds., *Joshua Reynolds: Experiments in Paint* (London: Wallace Collection/Paul Holberton Publishing, 2015).

21. Joyce Townsend, Technical Report, August 1998; Tate Britain, object file 4505.

22. For this picture, see David Mannings and Martin Postle, *Sir Joshua Reynolds: A Complete Catalogue of His Paintings* (New Haven, CT: Yale University Press, 2000), 1:485–86.

23. "Report before Cleaning," November 1947; Tate Britain, object file 5750.

24. "Report on the Cleaning," January-February 1948; Tate Britain, object file 5750. On the supposed frailties of Reynolds's drawing, see Luke Hermann, "The Drawings by Sir Joshua Reynolds in the Herschel Album," *Burlington Magazine* 110 (1968): 650–58.

25. "Summary of Technical Examination," July 3, 1968, 5; Tate Britain, object file 79.

26. "Sir Joshua Reynolds," *General Evening Post*, February 25, 1792, n.p.

27. Joseph Farington, *Memoirs of the Life of Sir Joshua Reynolds with Observations on his Talents and Character* (London: Strahan and Spottiswoode, for T. Cadell & W. Davies, 1819), 89.

28. Martin Archer Shee, *The Commemoration of Reynolds, in Two Parts, with Notes, and Other Poems* (London: J. Murray, 1814), 14. This exhibition has been digitally reconstructed in Janine Barchas's noted project *What Jane Saw*, http://whatjanesaw.org/ (accessed August 29, 2018).

29. Shee, *Commemoration of Reynolds*, 13.

30. [Joshua Reynolds], *A discourse delivered to the students of the Royal Academy, on the distribution of the prizes, Dec. the 10th, 1774. By the president.* (London: Thomas Davies, 1775), 28.

31. Reynolds knew that Vasarian legend of van Eyck's alchemical pursuits. Studying van Eyck's *Virgin and Child with Canon Van der Paele* (1436; now in the Groeningemuseum, Bruges) on June 28, 1781, Reynolds wrote the following in his notebook: "This Picture claims perhaps more attention from its being the work of the man who has been said to be the first inventor of the art of Painting in oil, than from any intrinsic merit in the work itself [,] however this mistake which was first published by Vasari and from his authority propagated in the world, has been lately <rectified> by the learned antiquarian Mr. Raspe who has proved beyond all contradiction that this art was invented <and practiced> many ages before Van Eyck was born"; Yale Center for British Art, MS Reynolds 38, n.p. This is an unambiguous reference to the erudite study published that year by R. E. Raspe as *A Critical Essay on Oil-Painting; Proving that the Art of Painting in Oil was Known before the Pretended Discovery of John and Hubert Van Eyck* (London: H. Goldney, 1781). For different visions of the extended, global crossings between visual art and alchemy, see William R. Newman, *Promethean Ambitions: Alchemy and the Quest to Perfect Nature* (Chicago: University of Chicago Press, 2004); Pamela H. Smith, *The Body of the Artisan: Art and Experience in the Scientific Revolution* (Chicago: University of Chicago Press, 2004); and David Brafman's exhibition *The Art of Alchemy* (Getty Research Institute, October 11, 2016–February 12, 2017).

32. See Frederick W. Hilles, *The Literary Career of Sir Joshua Reynolds* (Cambridge: Cambridge University Press, 1936), 112–27. For the auction catalogue of Reynolds's library, where only thirty-five (of over two thousand) lots are devoted to books, see H. Phillips, *Unique Collection of Drawings & Prints: A Catalogue of All the Great and Valuable Collection Ancient Drawings, Scarce Prints, and Books of Prints, Which Belonged to Sir Joshua Reynolds, Deceased, Late President of the Royal Academy* . . . (London: 1798), 89–91.

33. For this negative assessment, see Samuel Edgerton Jr., "Galileo, Florentine 'Disegno,' and the 'Strange-Spottednesse' of the Moon," *Art Journal* 44, no. 3 (Fall 1984): 228. On Reynolds's influence, see Neil Harris, *The Artist in American Society: The Formative Years 1790–1860* (New York: George Braziller, 1966), 11. A different view of Reynolds's contacts to the alchemical tradition is David Bjelajac, *Washington Allston, Secret Societies and the Alchemy of Anglo-American Painting* (Cambridge: Cambridge University Press, 1997).

34. See, for example, Martin Kemp, "True to Their Natures: Sir Joshua Reynolds and Dr. William Hunter at the Royal Academy of Arts," *Notes and Records of the Royal Society of London* 46, no. 1 (1992): 77–88.

35. Incorrectly attributed to Thomas Hudson, the Irons portrait was loaned to a 1979 exhibition in London under a pseudonym for the picture's real owner, Peter Cecil Wilson (1913–1984), the chairman of Sothebys. Wilson died in France in 1984, and the location of the picture is now unknown. I thank William Schupbach for this information. For more on Wilson, see Frank Herrmann, "Wilson, Peter Cecil (1913–1984)," in

Oxford Dictionary of National Biography (Oxford University Press, 2004), http://www.oxforddnb.com.huntington.idm.oclc.org/view/article/31845 (accessed October 3, 2016).

36. See Craig Ashley Hanson, *The English Virtuoso: Art, Medicine, and Antiquarianism in the Age of Empiricism* (Chicago: University of Chicago Press, 2009), 164.

37. John Huxham, MD, *An Essay on Fevers . . .*, 5th ed. (London: J. Hinton, 1767), v.

38. Joshua Reynolds, *Discourses on Art*, ed. Robert R. Wark (New Haven, CT: Yale University Press, 1997), 29.

39. This literature is vast, but I have found the following especially instructive: Thomas E. Crow, *Emulation: Making Artists for Revolutionary France* (New Haven, CT: Yale University Press, 1995); Richard T. Neer, "Poussin, Titian, and Tradition: *The Birth of Bacchus* and the Genealogy of Images," *Word & Image* 18, no. 3 (2002): 267–81; Maria H. Loh, "New and Improved: Repetition as Originality in Italian Baroque Practice and Theory," *Art Bulletin* 86, no. 3 (2004): 477–504; Smith, *Body of the Artisan*.

40. For Huxham's election, see Royal Society Library, EC/1739/02.

41. William Schupbach, "The Fame and Notoriety of Dr. John Huxham," *Medical History* 25 (1981): 420. For more on Huxham's reputation, see Norman Moore, "Huxham, John (c. 1692–1768)," rev. Richard Hankins, in *Oxford Dictionary of National Biography* (Oxford University Press, 2004), http://www.oxforddnb.com.proxy3.library.mcgill.ca/view/article/14319 (accessed September 19, 2016).

42. Yale Center for British Art, MS Reynolds 33, f4.

43. See Martin Postle, "Reynolds, Sir Joshua (1723–1792)," in *Oxford Dictionary of National Biography* (Oxford University Press, 2004), http://www.oxforddnb.com/view/article/23429 (accessed January 8, 2013).

44. See Sam Smiles, ed., *Sir Joshua Reynolds: The Acquisition of Genius* (Bristol: Sansom & Company, 2009); Fordham, *British Art and the Seven Years' War*, esp. 65–72.

45. For more on Baker, cf. James Northcote, *The Life of Sir Joshua Reynolds* (London: Henry Colburn, 1819), 1:6–7; *Biographia Britannica; or, Lives of the Most Eminent Persons who have flourished in Great Britain and Ireland* (1747; Hildesheim: Georg Olms Verlag, 1973), 1:422. For Wadham College in the era of the Civil War, see Charles Webster, *The Great Instauration: Science, Medicine, and Reform, 1626–1660* (New York: Holmes & Meier, 1976). On the amateur scientific interests of Reynolds's father, see Hilles, *Literary Career*, 4.

46. The Rev. Samuel Reynolds to Charles Cutcliff in Bideford, March 17, 1740; Royal Academy of Arts, MS REY 3/95.

47. Charles Robert Leslie and Tom Taylor, *Life and Times of Sir Joshua Reynolds* (London: John Murray, 1865), 1:23.

48. A label affixed to the back of the portrait erroneously identifies the fictively affixed male face as that of Thomas Mudge (ca. 1715–1794); for transcription of this label and correction of its errors, see Schupbach, "Fame and Notoriety," 417–21.

49. See Stamford R. Flint, *Mudge Memoirs: Being a Record of Zachariah Mudge, and Some Members of His Family* (Truro: Netheron and Worth, 1883), 4–5.

50. See *A Sermon on Liberty: By Z. Mudge* (London: Sold by F. Knight and Son, 1790).

51. Cited in Edmond Malone, "Some Account of the Life and Writings of Sir Joshua Reynolds," in *The Literary Works of Sir Joshua Reynolds*, 5th ed. (London: T. Cadell & W. Davies, 1819), 1:xcviii. On Burke's reservations about science and abstraction especially in the 1790s, see Maurice Crosland, "The Image of Science as a Threat: Burke versus Priestley and the 'Philosophic Revolution,'" *British Journal for the History of Science* 20, no. 3 (1987): 277–307.

52. Samuel Johnson, *A Dictionary of the English Language . . .*, 2nd ed. (London: J. and P. Knapton et al., 1755), vol. 2, n.p.

53. For discussion of Mudge's acrimonious relations with Astronomer Royal

Nevil Maskelyne, see David Penney, "Thomas Mudge and the Longitude: A Reason to Excel," in *The Quest for Longitude*, ed. William J. H. Andrewes (Cambridge, MA: Collection of Historical Scientific Instruments, Harvard University, 1996), 293–310. In the 1780s, Elizabeth Johnson, one of Reynolds's sisters, also composed a tract claiming to resolve the longitude problem by spiritual illumination; see *The Geography and Astronomy of the Created World, and of Course the Longitude: Being the Fourth Book by the Author of the Explanation of the Vision to Ezekiel* (Oxford: Printed for the author, 1785). For the attribution to Johnson, see http://cudl.lib.cam.ac.uk/view/MS-RGO-00014-00053/13 (accessed March 23, 2017).

54. For this assessment, see David Landes, *Revolution in Time: Clocks and the Making of the Modern World*, rev. ed. (Cambridge, MA: Belknap Press/Harvard University Press, 2000), 152.

55. See Mannings and Postle, *Complete Catalogue*, 1:345–46.

56. For Reynolds's election, see Royal Society Library, EC/1760/14; for Mudge's, Royal Society Library, EC/1777/08. Reynolds participated in nominations for several candidates to fellowship in the Royal Society (one in the 1760s, two in the 1770s, four in the 1780s, two in the 1790s), and had at least some cognizance of the institution's broader activities. In 1770 he could write to Sir William Hamilton congratulating him "on the honour you have acquired by the account you have given to the Royal Society of Vesuvius and Aetna"; Joshua Reynolds to Sir William Hamilton, June 17, 1770, in *The Letters of Sir Joshua Reynolds*, ed. John Ingamells and John Edgcumbe (New Haven, CT: Yale University Press, 2000), 33.

57. For more on Newton's own reflecting telescopes, see A. A. Mills and P. J. Turvey, "Newton's Telescope: An Examination of the Reflecting Telescope Attributed to Sir Isaac Newton in the Possession of the Royal Society," *Notes and Records of the Royal Society of London* 33, no. 2 (1979): 133–55.

58. John Mudge, *Directions for Making the Best Composition for the Metals of Reflecting Telescopes* (London: W. Bowyer and J. Nichols, 1777), 23.

59. See Robert Smith, *A Compleat System of Opticks in Four Books; viz. A Popular, a Mathematical, a Mechanical, and a Philosophical Treatise . . .* (Cambridge: Printed for the author, 1738), 2:301–12.

60. Mudge, *Directions*, 7.

61. See Mudge, *Directions*, 8–9.

62. The attribution to Hudson appears in *Thomas Hudson, 1701–1779, Portrait Painter and Collector: A Bicentenary Exhibition* (London: Greater London Council, 1979), no. 66. The image is also reproduced with a questioned attribution to Hudson and no further comment in Barrell, *Political Theory of Painting*, 67.

63. Ellis Waterhouse, *Painting in Britain: 1530 to 1790* (Baltimore: Penguin, 1953), 150.

64. Ellen G. Miles, "Thomas Hudson (1701–1779): Portraitist to the British Establishment" (PhD diss., Yale University, 1976), 8. For a more sympathetic reading, see Hallett, *Portraiture in Action*, 25–49.

65. For this connection between Reynell and Reynolds, see Smiles, *Reynolds: Acquisition of Genius*, 18.

66. James Northcote, *Supplement to the Memoirs of the Life, Writings, Discourses, and Professional Works of Sir Joshua Reynolds . . .* (London: Henry Colburn, 1815), vi.

67. For a telling featuring the elder Edgcumbe, see Northcote, *Life of Sir Joshua Reynolds*, 1:25–26.

68. See Samuel Reynolds, after Joshua Reynolds, *The Rev. Mr. Thomas Smart* (mezzotint, 1822; British Museum, London).

69. William Cotton and John Burnet, eds., *Sir Joshua Reynolds, and His Works: Glean-*

ings from His Diary, Unpublished Manuscripts, and from Other Sources (London: Longman, Brown, Green, Longmans, and Roberts, 1856), 30–31. For more on this portrait, see Matthew C. Hunter, "The Cunning of Sir Sloshua: Reynolds, the Sea, and Risk," *Grey Room* 69 (2017): 80–107.

70. For these influential stories, see David H. Solkin, *Painting for Money: The Visual Arts and the Public Sphere in Eighteenth-Century England* (New Haven, CT: Yale University Press, 1992), chaps. 4–5; Ilaria Bignamini, "George Vertue, Art Historian and Art Institutions in London, 1689–1768: A Study of Clubs and Academies," *Walpole Society* 54 (1988): 1–148; Joan Coutu, "'A Very Grand and Seigneurial Design': The Duke of Richmond's Academy in Whitehall," *British Art Journal* 1, no. 2 (2000): 47–54; Louise Lippincott, *Selling Art in Georgian London: The Rise of Arthur Pond* (New Haven, CT: Yale University Press, 1983).

71. Bignamini, "George Vertue," 124; Derek Hudson and Kenneth W. Luckhurst, *The Royal Society of Arts, 1754–1954* (London: Murray, 1954), 37.

72. *Premiums by the Society, Established at London, for the Encouragement of Arts, Manufactures, and Commerce* (London: Printed by order of the Society, 1759), 13.

73. For a discussion of Ramsay's own unusual method of painting portraits with a ground layer of red hues that would establish a tonal layer underneath the "epidermis" of paint, see Rica Jones, "Painting a Face All Red at the First Sitting: Ramsay's Technique for Portraits, 1735–1760," in Mungo Campbell et al., *Allan Ramsay: Portraits of the Enlightenment* (New York: Prestel, 2013), 111–15.

74. Minutes of the Society, 1759–1760 (Royal Society of Arts, RSA/AD/MA/100/12/01/04), 27. Minutes of Committees, 1758–1760 (Royal Society of Arts, RSA/PR/GE/112/12/1), show that Reynolds also served on committees for drawing in March 1759; for "polite and liberal arts" and history painting in early 1760 (pt. 2:15, 21); and for the public exhibition of contemporary arts (pt. 3:63).

75. December 3, 1755; Royal Society of Arts, RSA/AD/MA/100/12/01/01, 68.

76. February 8, 1758; Minutes of the Society, March 1757–February 1758 (Royal Society of Arts, RSA/AD/MA/100/12/01/02), 146. For more on Gardner, see Maxine Berg, "New Commodities, Luxuries and Their Consumers in Eighteenth-Century England," in *Consumers and Luxury: Consumer Culture in Europe 1650–1850*, ed. Maxine Berg and Helen Clifford (New York: Manchester University Press, 1999), 63–85, esp. 78–79.

77. Jean-Henri Müntz, *Encaustic; or, Count Caylus's Method of Painting in the Manner of the Ancients . . .* (London: Printed for the author, and A. Webley, at the Bible and Crown near Chancery Lane, Holborn, 1760), 19.

78. J. H. Müntz to the Society of Arts, May 7, 1760; Royal Society of Arts, RSA/PR/GE/110/8/145, 7/5/1760, 1.

79. See Minutes of the Society, 1760 (Royal Society of Arts, RSA/AD/MA/100/12/01/05), 91.

80. See Dubois, "Use a little wax"; Rica Jones, "Sir Joshua Reynolds (1723–1792): George IV when Prince of Wales 1785," in Hackney, Jones, and Townsend, *Paint and Purpose*, 146–51; Alexandra Gent, "Reynolds, Paint, and Painting," in Davis and Hallett, *Experiments in Paint*, 42–53.

81. "During the early years of the eighteenth century, there was little evidence of any general interest in chemistry—the most fundamental science of the industrial arts"; F. W. Gibbs, "Robert Dossie (1717–1777) and the Society of Arts," *Annals of Science* 7, no. 2 (1951): 149. For recent, bullish views, cf. Lawrence M. Principe, ed., *New Narratives in Eighteenth-Century Chemistry* (Dordrecht: Springer, 2007); Mary D. Archer and Christopher D. Haley, eds., *The 1702 Chair of Chemistry at Cambridge: Transformation and Change* (New York: Cambridge University Press, 2005).

82. See Arthur L. Donovan, *Philosophical Chemistry in the Scottish Enlightenment: The Doctrines and Discoveries of William Cullen and Joseph Black* (Edinburgh: Edinburgh University Press, 1975); Jan Golinski, *Science as Public Culture: Chemistry and Enlightenment in Britain, 1760–1820* (New York: Cambridge University Press, 1992).

83. For the proposed list of works to be translated from December 15, 1758, see Royal Society of Arts, Minutes of Committees, 1758–1760, RSA/PR/GE/112/12/1, f24. For an example of the advertisement of prizes and the financial incentives, see "Continuation of the Premiums offered by the Society for the Encouragement of Arts, Manufactures and Commerce," *London Evening Post*, no. 4920 (May 17–19, 1759), n.p.

84. See, respectively, April 18, 1760 (Royal Society of Arts, RSA/PR/GE/112/12/1, pt. 3:157); April 8, 1761 (Minutes of the Society, 1760–1761 [RSA/AD/MA/100/12/01/06], 219); November 26, 1760 (RSA/AD/MA/100/12/01/06, 62); and November 26, 1760 (RSA/AD/MA/100/12/01/06, 63).

85. In the assessment of one historian, it was "by far the best treatise on the practice of the industrial arts that had yet appeared in the [English] language and addressed specifically to that class of progressive tradesmen through whom the Society was aiming to effect country-wide improvements"; Gibbs, "Dossie," 154.

86. For period critique of Dossie's hasty willingness to reduce nature to first principles, see A. Z., *Remarks on Mr. Robert Dossie's Institutes of Experimental Chemistry* (London: S. Hooper, 1760). On the polymathic approach to image-making in the virtuosic tradition, see Hanson, *English Virtuoso*, esp. 108–21.

87. Dossie, *Handmaid to the Arts*, xiii, ix.

88. On Lewis, see F. W. Gibbs, "William Lewis, M.B., F.R.S. (1708–1781)," *Annals of Science* 8, no. 2 (1952): 122–51; F. W. Gibbs, "A Notebook of William Lewis and Alexander Chisholm," *Annals of Science* 8, no. 3 (1952): 202–20; Larry Stewart, "Assistants to Enlightenment: William Lewis, Alexander Chisholm, and Invisible Technicians in the Industrial Revolution," *Notes and Records of the Royal Society of London* 62, no. 1 (2008): 17–29.

89. In December 1736, Lewis offered a "Course of Chemistry, with a View to the Improvement of Pharmacy, Trades, and the Art itself" from a laboratory off Fetter Lane in the London ward of Faringdon; by 1747 his courses had expanded to five parts, including fifteen lectures on "The Art of Colours." See his advertisements in *London Evening Post*, no. 1421 (December 23–25, 1736); *Daily Post*, no. 7694 (May 2, 1744); and *Whitehall Evening Post or London Intelligencer*, no. 269 (October 31–November 3, 1747).

90. William Lewis, *Commercium Philosophico-Technicum; or, The Philosophical Commerce of Arts . . .* (London: H. Baldwin, 1763), iv.

91. On wresting chemical autonomy from Newtonian models, see Lawrence M. Principe, *The Aspiring Adept: Robert Boyle and His Alchemical Quest* (Princeton, NJ: Princeton University Press, 2000), 15.

92. On Hales's Newtonian chemistry, see Donovan, *Philosophical Chemistry*, 20–25. In his commonplace book, Reynolds wrote the following definitions: "Στατικη is the Doctrine of weighing of Bodies [;] Hydrostatics is the Doctrine of weighing of Bodies in water [;] Hidraulics, the Doctrine of the motion of fluids"; Yale Center for British Art, MS Reynolds 33, 3v.

93. William Lewis to Stephen Hales, February 16, 1760; RSA/PR/GE/110/25/1.

94. Lewis to Hales, RSA/PR/GE/110/25/1.

95. For the identification of Chisholm as the carrier, see RSA/PR/GE/110/25/1.

96. See, for example, Derek Hudson, *Sir Joshua Reynolds: A Personal Study* (London: G. Bles, 1958), 88; see also 60–62.

97. Hogarth had resigned from the Society of Arts in late 1757; see Michael Kitson, "Hogarth's 'Apology for Painters,'" *Walpole Society* 41 (1966–1968): 46–111, esp. 60–62.

98. Cf. Matthew Hargraves, *"Candidates for Fame": The Society of Artists of Great Britain, 1760–1791* (New Haven, CT: Yale University Press, 2005), esp. 27–28, 102–3; Holger Hoock, *The King's Artists: The Royal Academy of Arts and the Politics of British Culture 1760–1840* (New York: Oxford University Press, 2003), 52–62.

99. For a synthetic reading of Northcote as painter and author, see Mark Ledbury, *James Northcote, History Painting, and the Fables* (New Haven, CT: Yale University Press, 2014).

100. "I have receiv'd the Box with the brass patterns and have them all cast except the mettals which could not be cast 'till the man had more to cast them with but you may be sure I shall send them by the first opportunity"; James Northcote to Samuel Northcote, April 8, 1772; Royal Academy of Arts, MS NOR 9 i_r.

101. "I had some difficulty to get a case for the Spectacles. I first desired Mr. Dollond [i.e., noted optical instrument maker Peter Dollond (1731–1820)] to gett one . . . and got that which I think will do exactly but in trying them into so many cases the frames are a little scratch'd and I think it would not be amiss to polish them a little before Mr Mudge has them"; James Northcote to Samuel Northcote, June 25, 1771; Royal Academy of Arts, MS NOR 1.

102. "Sir Joshua is now building a very beautiful house at Richmond where he intends to go often in the summer[;] from it there is a very fine prospect. Now he has not got any telescope and very probably has never look'd through a reflex lens[?]. If you at particular times now and then . . . would fit up one of any size even small and in the plainest manner as he might be told the whole excellence was in the metals and give it him as your own making and I will always consider myself in your debt for it 'till it is in my power to make a return"; James Northcote to Samuel Northcote, December 21, 1771; Royal Academy of Arts, MS NOR 7 i_v.

103. James Northcote to Samuel Northcote, August 23, 1771; Royal Academy of Arts, MS NOR 4 i_v.

104. James Northcote to Samuel Northcote, September 21, 1771; Royal Academy of Arts, MS NOR 5 ii_v.

105. James Northcote to Samuel Northcote, July 27, 1772; Royal Academy of Arts, MS NOR 11 i_v–ii_r.

106. James Northcote to Samuel Northcote, September 21, 1771; Royal Academy of Arts, MS NOR 5 i_r.

107. *A Catalogue of the Pictures, Sculptures, Models, Designs in Architecture, Drawings, Prints, &c. Exhibited at the Great Room, in Spring-Gardens, Charing-Cross, April the Twenty-Sixth, 1771, by the Royal Incorporated Society of Artists of Great-Britain* (London: Printed for the Society, 1771), 16.

108. Joseph Priestley, *The History and Present State of Discoveries relating to Vision, Light and Colour* (London: J. Johnson, 1772), 365. The source on which Priestley draws for his claims about the invention of artificial phosphorus tells an older, more complicated story; see G. G. L., "Historia inventionis Phosphori," *Miscellanea Berolinensia ad Incrementum Scientarum* (Berlin: Johan. Christ. Papenii, 1710), 91–98. On the importance of accidental discovery to Priestley, see Golinski, *Science as Public Culture*, esp. 80.

109. Cf. F. W. Shurlock, "The Scientific Pictures of Joseph Wright," *Scientific Progress in the Twentieth Century* 17, no. 67 (1923): 432–38, esp. 436; Benedict Nicolson, *Joseph Wright of Derby: Painter of Light* (New York: Pantheon, 1968), 1:119.

110. [Robert Baker], *Observations on the Pictures Now in Exhibition at the Royal Academy, Spring Gardens, and Mr. Christie's by the Author of the Remarks on the English Language* (London: Printed for John Bell, 1771), 26.

111. Jenny Uglow, *The Lunar Men: Five Friends Whose Curiosity Changed the World* (New York: Farrar, Strauss and Giroux, 2002); Rica Jones, "Wright of Derby's Techniques

of Painting," in *Wright of Derby*, ed. Judy Egerton (New York: Metropolitan Museum of Art, 1990), 263–71. The former painting is now in the Derby Museum and Art Gallery, the latter at the National Gallery, London,

112. On this point, see Hargraves, *"Candidates for Fame,"* esp. 102–3.

113. Critic Baker concluded, "The pictures he has exhibited have . . . but a very small share of merit. We may reasonably attribute this to his having too much business"; Baker, *Observations on the Pictures*, 11.

114. *The Diary of Benjamin Robert Haydon*, ed. Willard Bissell Pope (Cambridge, MA: Harvard University Press, 1963), 5:574.

115. For a brief contextualization and transcription of the ledgers, see Malcolm Cormack, "The Ledgers of Sir Joshua Reynolds," *Walpole Society* 42 (1968–1970): 105–69.

116. On this bilingual practice, see Talley, "All Good Pictures Crack," 57.

117. Joshua Reynolds, ledger book (Fitzwilliam Museum, Cambridge, MS 2.1916), $f178_r$.

118. Haydon, *Diary*, 5:574.

119. For this provenance, see George Barker to the president and council of the Royal Academy, October 29, 1877; Royal Academy of Arts, object file 03/576. On the Barkers as picture restorers, see http://www.npg.org.uk/research/programmes/directory-of-british-picture-restorers/british-picture-restorers-1600-1950-b.php. For Lawrence's sale, "Sir Joshua Reynolds.—Two Canvasses, containing the artist's various experiments on colours, with his written memoranda, very curious"; *A Catalogue of the Valuable Collection of Paintings by Ancient and Modern Masters of Sir Thomas Lawrence* (London: Catalogues had at Mr. Christie's Office, 1830), 4 (lot 29). As reproduced in figure 2.12, a copy of this catalogue at the Huntington Library (acc. no. 40883) is annotated in pencil with the price of five pounds five shillings at which the canvases apparently sold. An ovular self-portrait by Reynolds (measuring 30 by 20 inches, p. 7) sold in the same sale for 210 pounds.

120. Charles Lock Eastlake, *Materials for a History of Oil Painting* (London: Longman, Brown, Green & Longmans, 1847), 444.

121. See Matthew C. Hunter, "Reynolds's Science of Experiment in Practice and Theory," in Davis and Hallett, *Experiments in Paint*, 100–111.

122. Northcote, *Life of Sir Joshua Reynolds*, 2:21.

123. Horace Walpole to George Montagu, May 31, 1763, in *Horace Walpole's Correspondence*, ed. W. S. Lewis and Ralph S. Brown Jr. (New Haven, CT: Yale University Press, 1941), 10:79. On this theme more broadly, see Aimee Marcereau Degalan, "Dangerous Beauty: Painted Canvases and Painted Faces in Eighteenth-Century Britain" (PhD diss., Case Western Reserve University, 2007).

124. [Unknown writer], *Letters Concerning the Present State of England: Particularly Respecting the Politics, Arts, Manners, and Literature of the Times* (London: Printed for J. Almon, 1772), 253.

125. [Unknown writer], "The Painter's Mirror for 1774: Number 1," *Public Advertiser* (London), no. 13013 (May 3, 1774), n.p. For details on the dating of this picture, now in a private collection, see Mannings and Postle, *Complete Catalogue*, 1:370 (#1414, fig. 1007). For an account of Reynolds's full submission to the 1774 show, see Hallett, *Portraiture in Action*, 1–21.

126. "Painter's Mirror," n.p. For details on the dating of this picture, now in the Royal Collection, see Mannings and Postle, *Complete Catalogue*, 1:476 (#1902, fig. 1106). On the third fugitive picture noted in 1774, a portrait of Thomas Newton, Bishop of Bristol, see Mannings and Postle, *Complete Catalogue*, 1:351–52 (#1338, fig. 1084).

127. [William Combe], *A Poetical Epistle to Sir Joshua Reynolds, Knt.* (London: Fielding and Walker, 1777), 5.

128. See, for example, Mark Hallett, "Reynolds, Celebrity, and the Exhibition

Space," in *Joshua Reynolds: The Creation of Celebrity*, ed. Martin Postle (London: Tate, 2005), 34–47, esp. 46–47.

129. Edward Young, *Conjectures on Original Composition: In a Letter to the Author of Sir Charles Grandison* (London: Printed for A. Millar, 1759), 12.

130. Young, *Original Composition*, 12.

131. Alexander Gerard, *An Essay on Genius* (London: W. Strahan, T. Cadell, 1774), 9; 75–76.

132. Gerard, *Genius*, 84; Peter Kivy, "Genius and the Creative Imagination," in *The Oxford Handbook of British Philosophy in the Eighteenth Century*, ed. James A. Harris (New York: Oxford University Press, 2013), 480 (467–87).

133. Gerard, *Genius*, 88. For this managerial assessment, see Timothy M. Costelloe, *The British Aesthetic Tradition: From Shaftesbury to Wittgenstein* (New York: Cambridge University Press, 2013), 88.

134. Critic F. W. Hilles aligned this fragment with Reynolds's seventh discourse rather than the sixth; see Hilles, *Literary Career*, 228–29.

135. Royal Academy of Arts, MS REY 3/32. Compare Alexander Gerard, *An Essay on Taste . . .* , 2nd ed. (Edinburgh: Printed for A. Millar, London, 1764), 240. In another fragment, Reynolds asks "what objection can ~~there~~ be made for reason to come afterwards ~~to select and <to correct what>~~ seperate the Gold from the dross and determine what is to be preserved and what is to be thrown away"; Royal Academy of Arts, MS REY 3/55.

136. Royal Academy of Arts, MS REY 3/9.

137. A Committee on Manures had been established at the Society of Arts by the end of 1759, where compositions of artificial fertilizers made through ash, urine, flower of sulphur, and other chemical substances were debated. For a recipe presented on December 14, 1759 ("Strewing & sifting dry ashes on the Ground 1 pd. Assa Foetide—one peck of Wood or Coal Sut—Two pounds of Garlick Cloves moderately bruised one pd of Flower of Sulphur, infused in stale Chamberly and sprinkle it with a Wisk Brush on the Plants when they just appear above Ground"), see Royal Society of Arts, RSA/PR/GE/112/12/1, pt. 3:15. On manuring in the period more broadly, see Anna M. Foy, "The Convention of Georgic Circumlocution and the Proper Use of Human Dung in Samuel Martin's *Essay upon Plantership*," *Eighteenth-Century Studies* 49, no. 4 (2016): 475–506. For the Society's use of assaying furnaces and other techniques of refining metals, see, for example, Nicolas Crisp, "Some Thoughts Offered to the Society for the Encouragement of Arts, Manufactures and Commerce on the Subject of Assaying Furnaces," Guard Book, vol. 5, item 76; Royal Society of Arts, RSA/PR/G/110/5/76.

138. Reynolds, *Discourse Delivered*, 12.

139. Reynolds, *Discourse Delivered*, 3.

140. Reynolds, *Discourse Delivered*, 25.

141. Reynolds, *Discourse Delivered*, 28. For different view of Reynolds's position on intention, see Robin Kelsey, *Photography and the Art of Chance* (Cambridge, MA: Belknap Press/Harvard University Press, 2015), 128–29.

142. For modern studies stressing the important role of dung in blue pigments including verdigris, see Mary V. Orna, J. D. L. Manfred, and Norbert S. Baer, "Synthetic Blue Pigments: Ninth to Sixteenth Centuries, I: Literature," *Studies in Conservation* 25, no. 2 (1980): 53–63; and Mary V. Orna, J. D. L. Manfred, and Maureen M. Julian, "Synthetic Blue Pigments: Ninth to Sixteenth Centuries, II: Silver Blue,'" *Studies in Conservation* 30, no. 4 (1985): 155–60.

143. Reynolds, *Discourse Delivered*, 25.

144. "Hoc Casus miscuit Corintho, cum caperetur, incensa"; Pliny, *Natural History*, vol. 9, *Libri XXXIII–XXXV*, trans. H. Rackham (Cambridge, MA: Harvard University Library, 1952), 34:6, pp. 130–31.

145. For ancient sources and modern reconstruction of Corinthian bronze's chemistry, cf. D. M. Jacobson and M. P. Weitzman, "What Was Corinthian Bronze?" *American Journal of Archaeology* 96, no. 2 (1992): 237–47; and David M. Jacobson, "Corinthian Bronze and the Gold of the Alchemists," *Gold Bulletin* 33, no. 2 (2000): 60–66.

146. Petronius, *Satyricon*, trans. Michael Heseltine (Cambridge, MA: Harvard University Press, 1969), 50, p. 103.

147. Reynolds, *Discourse Delivered*, 25.

148. See Sean Silver, "The Prehistory of Serendipity, from Bacon to Walpole," *Isis* 106, no. 2 (2015): 235–56.

149. Reynolds, "Discourse XIII," 243. If Northcote is to be trusted, the painter also implemented such principles in the studio, declaring: "A painter should have two pictures in hand . . . and should work on them alternately, by which means, if chance produced a lucky hit, as it often does, then instead of working upon the same piece, and perhaps by that means destroy the beauty which chance had given, he should go on to the other and improve upon that"; Northcote, *Life of Sir Joshua Reynolds*, 2:7.

150. Horace Walpole, *Anecdotes of Painting in England* (Strawberry-Hill: Thomas Kirgate, 1771), 4:vi.

151. Walpole, *Anecdotes*, 4:vii.

152. See, for example, John Newman, "Reynolds and Hone: 'The Conjuror' Unmasked," in *Reynolds*, ed. Nicholas Penny (New York: Harry N. Abrams, 1986), 344–354.

153. "A Lover of Wit," *Public Advertiser* (May 15, 1775), n.p.

154. "Lover of Wit," n.p.

155. Farington, *Memoirs*, 93.

156. For the broader circumstances of this *furore*, compare Martin Butlin, "An Eighteenth-Century Art Scandal: Nathaniel Hone's 'The Conjuror,'" *Connoisseur* 174, no. 699 (1970): 1–9; Konstantinos Stefanis, "Nathaniel Hone's 1775 Exhibition: The First Single-Artist Retrospective," *Visual Culture in Britain* 14, no. 2 (2013): 131–53.

157. For more on the iconography of alchemy in the northern European tradition, see Lawrence Principe and Lloyd DeWitt, *Transmutations—Alchemy in Art: Selected Works from the Eddleman and Fisher Collections at the Chemical Heritage Foundation* (Philadelphia: Chemical Heritage Foundation, 2002).

158. Cf. Mary Webster, *Johann Zoffany 1733–1810* (New Haven, CT: Yale University Press, 2011), 206–10; Martin Postle, "Johann Zoffany: An Artist Abroad," in *Johann Zoffany RA: Society Observed* (New Haven, CT: Yale University Press, 2011), 25. For dates of the play, see "Theatrical Register," *St. James's Chronicle or the British Evening Post* (December 3–6, 1774), no. 2155, n.p.

159. *Morning Herald and Daily Advertiser* (March 2, 1785), n.p.

160. *Morning Herald and Daily Advertiser* (March 2, 1785), n.p.

161. Baker, *Observations on the Pictures*, 26. For an ingenious treatment of this theme, see Adam Smith, "Of the Nature of that Imitation which takes Place in What are called the Imitative Arts," in *Essays on Philosophical Subjects*, ed. Dugald Stewart (Dublin: Printed for Messrs. Wogan, Byrne, J. Moore, Colbert, Rice W. Jones, Porter, and Folingsby, 1795), 133–84.

162. Peter Perez Burdett to Joseph Wright, February 4, 1771; reproduced in Nicolson, *Wright of Derby*, 1:119.

163. David H. Solkin, "Joseph Wright of Derby and the Sublime Art of Labor," *Representations* 83, no. 1 (2003): 179; Laura Baudot, "An Air of History: Joseph Wright's and Robert Boyle's Air Pump Narratives," *Eighteenth-Century Studies* 46, no. 1 (2012): 1–28.

164. Talley, "All Good Pictures Crack," 55.

165. Rica Jones, "Introduction," in Hackney, Jones, and Townsend, *Paint and Pur-*

pose, 11. More broadly, see Alan Borg, *The History of the Worshipful Company of Painters, Otherwise Painter-Stainers* (Huddersfield: Jeremy Mills, 2005).

166. For these readings, see Neil McKendrick, John Brewer, and J. H. Plumb, *The Birth of a Consumer Society: The Commercialization of Eighteenth-Century England* (Bloomington: Indiana University Press, 1982); Berg and Clifford, *Consumers and Luxury*.

167. William Mason, "Anecdotes of Sir Joshua Reynolds, Chiefly Relating to His Manner of Coloring," in *Sir Joshua Reynolds' Notes and Observations on Pictures*, ed. William Cotton (London: J. R. Smith, 1859), 54.

168. Mason, "Anecdotes," 54.

169. Neil De Marchi and Hans J. Van Miegroet, "Ingenuity, Preference, and the Pricing of Pictures: The Smith-Reynolds Connection," in *Economic Engagements with Art*, ed. Neil De Marchi and Craufurd Goodwin (Durham, NC: Duke University Press, 1999), 401.

170. See John Chu, "High Art and High Stakes: The 3rd Duke of Dorset's Gamble on Reynolds," *British Art Studies*, no. 2, http://dx.doi.org/10.17658/issn.2058-5462/issue-02/jchu (accessed September 8, 2016).

171. J. G. A. Pocock, *Virtue, Commerce, and History: Essays on Political Thought and History, Chiefly in the Eighteenth Century* (New York: Cambridge University Press, 1985), 98.

172. Michel Foucault, "24 January 1979," in *The Birth of Biopolitics: Lectures at the Collège de France, 1978–79*, ed. Michel Senellart, trans. Graham Burchell (Houndmills, Basingstoke, Hampshire: Palgrave Macmillan, 2008), 66.

173. A concise summary is Francis H. Dowley, "Thoughts on Poussin, Time, and Narrative: *The Israelites Gathering Manna in the Desert*," *Simiolus* 25, no. 4 (1997): 329–48.

174. Anthony Ashley Cooper, *An Essay on Painting: Being A Notion of the Historical Draught or Tablature of the Judgment of Hercules* (London: John Darby, 1714), 9.

175. Jonathan Richardson, *An Essay on the Theory of Painting*, 2nd ed. (London: Printed for A. C. and sold by A. Bettesworth in Pater-Noster-Row, 1725), 55.

176. For one such argument, see Michael Fried, *Absorption and Theatricality: Painting and Beholder in the Age of Diderot* (Chicago: University of Chicago Press, 1980), 90.

177. Samuel Johnson, *The Idler* 45 (February 24, 1759), in *The Works of Samuel Johnson* (New York: Lamb Publishing Company, 1903), 5:232.

178. Gotthold E. Lessing, *Laocöon* (1766), trans. E. A. McCormick (New York: Bobbs-Merrill, 1962), 77. For Reynolds's own endorsement of these temporal principles, see his annotations to William Mason's translation of C. A. Dufresnoy's *The Art of Painting*, in Malone, *Literary Works*, 3:104–5.

179. See John Galt, *The Life, Studies, and Works of Benjamin West, Esq.* (London: T. Cadell and W. Davies, 1820), 2:50.

180. See Edgar Wind, "The Revolution in History Painting" (1938), in *Hume and the Heroic Portrait*, ed. Jaynie Anderson (New York: Oxford University Press, 1986), 90.

181. David H. Solkin, *Art in Britain 1660–1815* (New Haven, CT: Yale University Press, 2015), 165.

182. Galt, *Life, Studies, and Works*, 2:48. An illuminating account of the Galt/West biography is Susan Rather, "Benjamin West, John Galt, and the Biography of 1816," *Art Bulletin* 86, no. 2 (2004): 324–45.

183. Benjamin West, *A Discourse, Delivered to the Students of the Royal Academy, on the Distribution of the Prizes, December 10, 1792, by the President. Humbly Inscribed by Permission to His Majesty: To Which Is Prefixed the Speech of the President to the Royal Academicians on the 24th of March 1792* (London: Printed by Thomas Cadell, printer to the Royal Academy, 1793), 22–23 (my emphasis).

184. Galt, *Life, Studies, and Works*, 2:47.

185. Compare *London Evening Post*, no. 8285 (May 4–6, 1775); "A Lover of Decency," *Public Advertiser*, no. 14238 (May 20, 1775).

186. For its art-historical references, see Robyn Asleson and Shelley M. Bennett, *British Paintings at the Huntington* (New Haven, CT: Yale University Press, 2001), 376–81.

187. On this painting, now at the Huntington Art Gallery, see Mannings and Postle, *Complete Catalogue*, 1:425.

188. Helen Brett to Shelley M. Bennett, February 2, 2002; Huntington Art Gallery, conservation file, acc. no. 23.62.

189. On the rendering of future ruling classes, see Patricia Crown, "Portraits and Fancy Pictures by Gainsborough and Reynolds: Contrasting Images of Childhood," *Journal for Eighteenth-Century Studies* 7, no. 2 (1984): 159–67, esp. 162.

190. See Martin Postle, *Sir Joshua Reynolds: The Subject Pictures* (New York: Cambridge University Press, 1995), 62. Cf. John Chu, "Joshua Reynolds and Fancy Painting in the 1770s," in Davis and Hallett, *Experiments in Paint*, 86–99.

191. "Reynolds, of all artists, painted children best," one Victorian commentator opined. "That childless man knew most of childhood, depicted its appearances in the truest and happiest spirit of comedy, entered into its changeful soil with the tenderest, heartiest sympathy, played with the playful, sighed with the sorrowful, and mastered all the craft of infancy"; Frederic G. Stephens, *English Children, as Painted by Sir Joshua Reynolds: An Essay on Some of the Characteristics of Reynolds as a Painter, with Especial Reference to His Portraiture of Children* (London: Seeley, Jackson, and Halliday, 1867), v. See also Marcia R. Pointon, *Hanging the Head: Portraiture and Social Formation in Eighteenth-Century England* (New Haven, CT: Yale University Press, 1993), esp. 186–88.

192. Walpole, *Anecdotes*, 4:vii.

193. Charles Dempsey, *Inventing the Renaissance Putto* (Chapel Hill: University of North Carolina Press, 2001), xiv. Thanks to Katie Scott for calling this connection to my attention.

194. Walpole, *Anecdotes*, 4:vii. For the pictures named, see Mannings and Postle, *Complete Catalogue*, 1:384 (#1484), 541–42 (#2098).

195. Edgar Wind, "Hume and the Heroic Portrait" (1930–1931), in Anderson, *Hume and the Heroic Portrait*, 23. For *Hannibal* (current location unknown), see Mannings and Postle, *Complete Catalogue*, 1:536 (#2088).

196. Wind, "Hume and the Heroic Portrait," 24.

197. Ronald Paulson, *Emblem and Expression: Meaning in English Art of the Eighteenth Century* (Cambridge, MA: Harvard University Press, 1975), 90. See further, Mannings and Postle, *Complete Catalogue*, 1:541 (#2097). See also Carey McIntosh, "Reynolds's Portrait of the Infant Johnson," in *Eighteenth-Century Studies in Honor of Donald F. Hyde*, ed. W. H. Bond (New York: Grolier Club, 1970), 279–96.

198. Paulson, *Emblem and Expression*, 90.

199. Paulson, *Emblem and Expression*, 91–92.

200. Seneca, "On Gathering Ideas," in *Seneca Ad Lucilium Epistulae Morales*, trans. R. H. Gummere (London: Heinemann, 1920), 2:281. A brilliant meditation on this theme is Carolina Mangone, "Like Father, Like Son: Bernini's Filial Imitation of Michelangelo," *Art History* 37, no. 4 (2014): 666–87.

201. [Unknown writer], "Extract of a letter from a Gentleman at Paris to his Friend in London, Jan. 6," *Morning Post and Daily Advertiser*, no. 1006 (January 16, 1776), n.p. For the identification, see "Extract of a letter from Paris to his Friend in London, dated January 21," *Morning Post and Daily Advertiser*, no. 1016 (January 27, 1776): n.p.

202. "Extract . . . Jan. 6," n.p.

203. Cf. Robert Rosenblum, "Reynolds in an International Milieu," in *Reynolds*, ed. Nicholas Penny (New York: Harry N. Abrams, 1986), 43–54, esp. 46–47; Julius Bryant,

Kenwood: Paintings in the Iveagh Bequest (New Haven, CT: Yale University Press, 2003), 334–37; Postle, *Subject Pictures*, 104–5. I thank Alexander Nemerov for pushing me on this point.

204. Postle, *Subject Pictures*, 60.

205. Conservation was undertaken in 1957, 1962, 1970, 1987, 1995, and 1997. As one cataloguer notes with laconic understatement: "Wrinkling of paint in localised areas and raised craquelure in white paint [can be] seen around infant with hat. Paint has tendency to lift and flake"; Bryant, *Kenwood*, 339. I thank Alexandra Gent for very kindly sharing her expertise and photographs with me.

206. Newman, *Promethean Ambitions*, 165.

207. Newman, *Promethean Ambitions*, 227.

208. The *Morning Post* claimed, "This Persic chymist is now in his forty-ninth year, the age at which Paracelsus died"; January 21, 1776, n.p. For articles on the life of Paracelsus in the London press, compare J. Cooke, "To the Morning Chronicle," *Morning Chronicle and London Advertiser*, no. 1297 (July 20, 1773), n.p. On Sterne's success, see Thomas Keymer, *Sterne, Moderns, and the Novel* (New York: Oxford University Press, 2002), esp. 49–82. For a reading of Sterne and Reynolds together, see Peter J. De Voogd, "Laurence Sterne, the Marbled Page, and the 'Use of Accidents,'" *Word & Image* 1, no. 3 (1985): 279–87.

209. Laurence Sterne, *The Life and Opinions of Tristram Shandy, Gentleman* (1759–1766), ed. Ian C. Ross (New York: Oxford University Press, 1983), book 1, chap. 2, p. 6. On the larger medical context, see Louis A. Landa, "The Shandean Homunculus: The Background of Sterne's 'Little Gentleman,'" in *Restoration and Eighteenth-Century Literature: Essays in Honor of Alan Dugald McKillop*, ed. Carroll Camden (Chicago: University of Chicago Press, 1963), 49–68.

210. See Jacques Fabien Gautier D'Agoty, "Découvertes sur la generation, contre les Oviparistes & les Vermiculistes," in *Observations sur l'histoire naturelle, sur la physique et sur la peinture* (Paris: Chez Delaguette, 1752), 1:9–16. On relations between Le Blon and Gautier D'Agoty, see Sarah Lowengard, *The Creation of Color in Eighteenth-Century Europe* (New York: Columbia University Press, 2008), 586–612.

211. On this point, see *The Clockmakers Outcry against the Author of the Life and Opinions of Tristram Shandy . . .* (London: Printed for J. Burd, near the Temple Gate, Fleet-Street, 1760).

212. I thank Ray Bates for informing me that spermaceti oil, which has been identified in Reynolds's *Self-Portrait as a Deaf Man* (fig. 2.3), was the period's standard lubricant for clockworks.

213. For the identification of the child as a figuration of fellow painter Angelica Kauffman, see Newman, "Conjuror Unmasked," 353–54.

214. Such a reading might complement in chemical terms the attention to "animation" ascribed to Reynolds's pictorial project in Mark Hallett's recent analysis; see Hallett, *Portraiture in Action*, esp. 18–19, 100–103.

215. See Paul Erickson et al., *How Reason Almost Lost Its Mind: The Strange Career of Cold War Rationality* (Chicago: University of Chicago Press, 2013), 39–41.

216. Immanuel Kant, *Critique of Judgment* (1790), trans. W. Pluhar (Indianapolis: Hackett, 1987), § 47; 177; for the German, see Immanuel Kant, *Kritik Der Urteilskraft*, 5th ed. (Leipzig: Felix Meiner, 1920), 163.

217. Kant, *Critique of Judgment*, §46, 175.

218. Kant, *Critique of Judgment*, §46; 175; German, 161.

219. Kant, *Critique of Judgment*, §46; 175; German, 161.

220. Kant, *Critique of Judgment*, §49; 186–87; German, 173.

221. On Kant's view of chemistry, see Joseph Agassi, "Kant's Program," *Synthese* 23, no. 1 (1971): 18–23.

Chapter 3

1. James Northcote, *The Life of Sir Joshua Reynolds* (London: Henry Colburn, 1819), 1:237. For Northcote's source, see Hester Lynch Piozzi, *Anecdotes of the Late Samuel Johnson, L.L.D. during the Last Twenty Years of His Life* (London: Printed for T. Cadell in the Strand, 1786), 98–99.

2. Northcote, *Life of Sir Joshua Reynolds*, 1:237. In fact, Reynolds did subsequently paint a portrait of a brewer (Samuel Whitbread) on a large copper plate; see David Mannings and Martin Postle, *Sir Joshua Reynolds: A Complete Catalogue of His Paintings* (New Haven, CT: Yale University Press, 2000), 1:470.

3. Samuel Johnson, "Life of Boerhaave" (1739), in *Early Biographical Writings of Dr. Johnson*, ed. J. D. Fleeman (Westmead, UK: Gregg International, 1973), 30.

4. For this exchange, see James Boswell, *The Life of Samuel Johnson, LL.D.* (London, 1791), 2:445. More broadly, see Richard B. Schwartz, *Samuel Johnson and the New Science* (Madison: University of Wisconsin Press, 1971).

5. Boswell, *Life of Johnson*, 1:236.

6. For more on Piozzi and Northcote, see Richard Wendorf, "'Well Said Mr. Northcote': Hester Thrale Piozzi's Annotated Copy of James Northcote's Biography of Sir Joshua Reynolds," *Harvard Library Bulletin* 9, no. 4 (1998): 29–40.

7. Northcote, *Life of Sir Joshua Reynolds*, 2:40.

8. For a period conception of patina as a "physico-chemico-economic" matter, see Gio Maironi da Ponte, *Sul Verderame: Memoria Fisico-Chimico-Economica* (Bergamo: Per Francesco Locatelli, 1784). On shifting attitudes toward patina more broadly, see Randolph Starn, "Three Ages of 'Patina' in Painting," *Representations* 78, no. 1 (2002): 86–115.

9. Northcote, *Life of Sir Joshua Reynolds*, 1:239.

10. Andrew McClellan, *Inventing the Louvre: Art, Politics, and the Origins of the Modern Museum in Eighteenth-Century Paris* (Cambridge: Cambridge University Press, 1994), 71.

11. A concise account is Alden R. Gordon, "Subverting the Secret of Herculaneum: Archaeological Espionage in the Kingdom of Naples," in *Antiquity Recovered: The Legacy of Pompeii and Herculaneum*, ed. Victoria C. Gardner Coates and Jon L. Seydl (Los Angeles: J. Paul Getty Museum, 2007), 37–58.

12. See Thomas W. Gaehtgens and Jacques Lugand, *Joseph-Marie Vien: Peintre du Roi (1716–1809)* (Paris: Arthena, 1988), 24–25, 74–76.

13. Comte de Caylus (Anne Claude Philippe de Pestels de Lévis de Tubières-Grimoard) and Michel-Joseph Majault, *Mémoire sur la peinture à l'encaustique et sur la peinture à la cire* (Geneva: Pissot libraire, à la Croix d'or, quai de Conti, 1755), 17–18. "Cette peinture ne s'écaille point. La chaleur du Soleil & celle des apartemens ne la peut altérer. Les années ne lui doivent causer aucun dérangement, c'est-à-dire, que les Ouvrages sont encore plus à l'abri de tous les dangers que les fresques; & que généralement ils doivent les surpasser en durée lorsqu'ils sont peints sur le bois." Here, as throughout, translations are my own unless otherwise noted.

14. Jean-Henri Müntz, *Encaustic; or, Count Caylus's Method of Painting in the Manner of the Ancients . . .* (London: Printed for the author, and A. Webley, at the Bible and Crown near Chancery Lane, Holborn, 1760), 18.

15. [Denis Diderot], *L'histoire et le secret de la peinture en cire* (Paris [?], 1755), 71. "La nature des couleurs qui produit avec les tems des effects si bizarres dans la peinture à l'huile, n'altérera point les *Encaustiques*."

16. Danielle Rice, "The Fire of the Ancients: The Encaustic Painting Revival, 1755 to 1812" (PhD diss., Yale University, 1979), 34.

17. Caylus and Majault, *Mémoire*, 8. "La Physique & la Chymie devoient également

concourir à cette découverte; car ces deux Sciences sont le flambeau de presque tous les Arts."

18. Caylus and Majault, *Mémoire*, 29. "Nous nous proposâmes de faire usage de dissolvans, tels que les huiles essentielles, qui par leur analogie avec la cire, la pénétrent & la mettent en état de pouvoir être étendue avec le pinceau."

19. Caylus and Majault, *Mémoire*, 69–70. "Tous les Chymistes, même les moins éclairés, savent que les fluides gras, auxquels on donne le nom d'huiles essentielles, ont la propriété de pénétrer & de dissourdre les corps qui leur sont analogues, quleque-fois à l'aide du feu, souvent sans le secours de cet agent, & que la cire est de l'espece de ceux qui peuvent être dissous de l'une & de l'autre façon." Widely influential among Parisian medical circles from the 1720s, Stahl explained reactions between materials in terms of analogy between their component parts; see Mi Gyung Kim, *Affinity, That Elusive Dream: A Genealogy of the Chemical Revolution* (Cambridge, MA: MIT Press, 2003), 168–74.

20. For more on Bachelier, see Hélène Mouradian and Xavier Salmon, *Jean-Jacques Bachelier, 1724–1806: Peintre du roi et de Madame de Pompadour* (Paris: Somogy, 1999). For recent work on Diderot's chemistry, see Oliver Wunsch, "Diderot and the Materiality of Posterity," *Early Modern French Studies* 40, no. 1 (2018): 63–78.

21. Diderot describes Bachelier's initial reasoning this way: "With some murky chemical tincture in hand, he said to himself, 'it seems to me that wax is an unctuous or resinous body; and substances of this kind become soluble in water when combined with alkalis.' . . . This is all too rich for a man with only blurred ideas of chemistry; it is ridiculous"; Diderot, *L'histoire et le secret*, 40. "Il avoit quelque teinture brouillée de Chimie, & il se disoit à lui-même 'il me semble que la cire est une substance huileuse ou résineuse; or les substances de cette nature combinées avec les alcalis deviennent solubles dans l'eau.' . . . Je le fais raisonner trop bien pour un homme qui n'a que des idées brouillées de chimie; cela est ridicule."

22. Diderot, *L'histoire et le secret*, 83–84. "Si nos yeux pouvoient se défaire de leurs préjudgés & s'accoutumer à des tableaux mattes, M. Bachelier auroit trouvé le moyen de faire valoir les chefs-d'oeuvres des tems passés, & de conserver éternellement ceux qui se font de nos jours, dans leur beauté premiére & sans que les couleurs s'en altérassent."

23. For Caylus and Diderot aligned in opposition to the Rococo, see Michael Fried, *Absorption and Theatricality: Painting and Beholder in the Age of Diderot* (Chicago: University of Chicago Press, 1980), esp. 71–105.

24. "No previous English artist took such pains to ensure that his paintings were documented by such a large body of fine prints"; Timothy Clayton, *The English Print 1688–1802* (New Haven, CT: Yale University Press, 1997), 184.

25. Northcote, *Life of Sir Joshua Reynolds*, 1:83.

26. Compare Valentine Green's *Beauties of the Present Age* (1780) and Samuel William Reynolds's project of publishing 357 mezzotints after Reynolds as a four-volume set (1821–1826). Timothy Clayton, "Green, Valentine (1739–1813)," in *Oxford Dictionary of National Biography* (Oxford University Press, 2004), http://www.oxforddnb.com/view/article/11405 (accessed June 6, 2016); Felicity Owen, "Reynolds, Samuel William (1773–1835)," in *Oxford Dictionary of National Biography* (Oxford University Press, 2004), http://www.oxforddnb.com/view/article/23438 (accessed June 6, 2016).

27. John T. Smith, *Nollekens and His Times* (London: Henry Colburn, 1829), 2:292.

28. For an interesting meditation on the resonances of Reynolds's death, see Richard Wendorf, *After Sir Joshua: Essays on British Art and Cultural History* (New Haven, CT: Yale University Press, 2005).

29. "The Arts," *Morning Chronicle*, no. 7553 (August 19, 1793), n.p. (my emphasis). On Pearson and Reynolds's collaborative work in stained glass, see Joanna Cobb, "From

Parrots to Princes: Exhibitions of Contemporary Stained Glass in Late Eighteenth-Century London," *Vidimus* 53 (July/August 2011), http://vidimus.org/issues/issue-53/feature/ (accessed April 2, 2017). Pearson's rendering of Reynolds's George III survives in Ely Cathedral's Stained Glass Museum.

30. See William R. Newman, *Promethean Ambitions: Alchemy and the Quest to Perfect Nature* (Chicago: University of Chicago Press, 2004).

31. Edmund Burke, *Reflections on the Revolution in France* (1790), ed. Frank M. Turner (New Haven, CT: Yale University Press, 2003), 7. Cf. Maurice Crosland, "The Image of Science as a Threat: Burke versus Priestley and the 'Philosophic Revolution,'" *British Journal for the History of Science* 20, no. 3 (1987): 277–307.

32. Walter Benjamin, "The Work of Art in the Age of Its Technological Reproducibility: Second Version" (1935–1936), in *The Work of Art in the Age of Its Technological Reproducibility, and Other Writings on Media*, ed. Michael W. Jennings, Brigid Doherty, and Thomas Y. Levin, trans. E. Jephcott et al. (Cambridge, MA: Harvard University Press, 2008), 23. While Benjamin makes the argument himself, a familiar account stressing the catalytic effect of lithography on the historical origins of photography is Helmut Gernsheim, *The Origins of Photography* (New York: Thames & Hudson, 1982), 27–40.

33. Geoffrey Batchen, *Burning with Desire: The Conception of Photography* (Cambridge, MA: MIT Press, 1997), 91.

34. Steven Shapin, "The Invisible Technician," *American Scientist* 77, no. 6 (1989): 554–63.

35. Alois Senefelder, *The Invention of Lithography* (1818), trans. J. W. Muller (New York: Fuchs & Lang Manufacturing Co., 1911), 26. "Dieser Stein hat ein sehr hestiges Bestreben sich mit Fett zu verbinden, welches sich oft so tief einsaught, daß, ihn ganz savon zu trennen in manchen Fällen selbst durch betrachtliches Abschleifen nicht moglich ist"; Senefelder, *Vollständiges Lehrbuch Der Steindruckerey* (Munich: Senefelder and E. A. Fleischmann, 1821), 37.

36. Senefelder, *Invention*, 179. For this basic process, see *Invention*, 168–71.

37. See Philipp André, *Specimens of Polyautography: Consisting of Impressions taken from Original Drawings made purposely for this Work* (London: P. André and J. Heath, 1803–1807), unnumbered. I thank Susan Dackerman for a particularly illuminating discussion of this print.

38. Benjamin, "Work of Art," 20.

39. Senefelder, *Invention*, 7. ". . . wozu seit 1799 der erste Anfang von mir gemacht wurde, rein chemisch zu kennen ist"; *Vollständiges Lehrbuch*, 11. Senefelder enlarges on the chemical nature of his art at great length later in the treatise; *Invention*, 95–98; *Vollständiges Lehrbuch*, 137–40.

40. Thomas Fisher, "The Process of Polyautographic Printing," *Gentleman's Magazine* 78, no. 1 (1808): 194.

41. John Landseer, *Lectures on the Art of Engraving: Delivered at the Royal Institution of Great Britain* (London: Longman, Hurst, Rees and Orme, 1807), 143. On Landseer, see Lucy Peltz, "Landseer, John George (1762/3–1852)," in *Oxford Dictionary of National Biography* (Oxford University Press, 2004), http://www.oxforddnb.com.proxy3.library.mcgill.ca/view/article/15986 (accessed July 30, 2016).

42. Landseer, *Lectures*, 143–44.

43. A classic account of Barry's position is John Gage, "Magilphs and Mysteries," *Apollo* 80 (1964): 38–41.

44. My account draws from *The Diary of Joseph Farington*, vol. 3, *September 1796–December 1798*, ed. Kenneth Garlick and Angus Macintyre (New Haven, CT: Yale University Press, 1979); Eric McCauley Lee, "'Titianus Redivivus': Titian in British Art Theory, Criticism, and Practice, 1768–1830" (PhD diss., Yale University, 1997), esp.

129–245; Mark N. Aronson and Angus Trumble, *Benjamin West and the Venetian Secret* (New Haven, CT: Yale Center for British Art, 2008); and Lance Mayer and Gay Myers, *American Painters on Technique: The Colonial Period to 1860* (Los Angeles: J. Paul Getty Museum, 2011), 11–24.

45. Cosway won the first drawing prize offered by the Society of Arts, exhibited in the inaugural show of 1760, and went on to win four other prizes offered by the Society; cf. Derek Hudson and Kenneth W. Luckhurst, *The Royal Society of Arts, 1754–1954* (London: John Murray, 1954), 10, 38; Celina Fox, *The Arts of Industry in the Age of Enlightenment* (New Haven, CT: Yale University Press, 2009), 209.

46. For these figures, see David H. Solkin, ed., *Art on the Line: The Royal Academy Exhibitions at Somerset House 1780–1836* (New Haven, CT: Yale University Press, 2001), 23, 37.

47. See D. L. Simms, "The Trail for Archimedes's Tomb," *Journal of the Warburg and Courtauld Institutes* 53 (1990): 281–86.

48. January 18, 1797; Farington, *Diary*, 751.

49. Stephen Francis Dutilh Rigaud, "Facts and Recollections of the XVIIIth Century in a Memoir of John Francis Rigaud Esq., R.A.," ed. William L. Pressly, *Walpole Society* 50 (1984): 100. For more on the importance of absorbent grounds to the Venetian secret, see Farington, *Diary*, esp. 745, 749–51, 757; and Mayer and Myers, *American Painters*, 16–18.

50. [Unknown writer], "Royal Academy," *General Evening Post*, no. 10,066 (April 29, 1797–May 2, 1797).

51. Anthony Pasquin [John Williams], *A Critical Guide to the Present Exhibition at the Royal Academy, for 1797: Containing Admonitions to the Artists on the Misconception of Theological Subjects, and a Complete Developement of the Venetian Art of Colouring, as is now so much the Rage of Imitation* (London: H. D. Symonds, 1797), 5. Cf. James Sambrook, "Williams, John (1754–1818)," in *Oxford Dictionary of National Biography* (Oxford University Press, 2004), http://www.oxforddnb.com.proxy3.library.mcgill.ca/view/article/29520 (accessed July 30, 2016).

52. Edmond Malone, *The Works of Sir Joshua Reynolds, Knt.* (London: T. Cadell and W. Davies, 1797), 1:xxii. For Malone's priority, see Lee, "Titianus Redivivus," 168–69.

53. Malone, *Works of Sir Joshua Reynolds*, 1:xxiii.

54. Charles Robert Leslie and Tom Taylor, *Life and Times of Sir Joshua Reynolds* (London: John Murray, 1865), 1:112. On Leslie and his posthumous publication, see Dorinda Evans, "Leslie, Charles Robert (1794–1859)," in *Oxford Dictionary of National Biography* (Oxford University Press, 2004), http://www.oxforddnb.com/view/article/16485 (accessed April 13, 2017).

55. Benjamin West, reproduced in John Galt, *The Life, Studies, and Works of Benjamin West, Esq.* (London: T. Cadell and W. Davies, 1820), 2:136–37. For the dating of this discourse to December 1797, see Lee, "Titianus Redivivus," 236.

56. Cf. David Bjelajac, *Washington Allston, Secret Societies, and the Alchemy of Anglo-American Painting* (New York: Cambridge University Press, 1997), esp. 32–65.

57. [Joseph Booth], *An Address to the Public on the Polygraphic Art* (London: J. Walter, 1788[?]), 4.

58. Joseph Booth, *A Treatise Explanatory of the Nature and Properties of Pollaplasiasmos; or, the Original Invention of Multiplying Pictures in Oil Colours . . .* (London: J. Rozea, 1784), 9.

59. Booth's enterprise was not alone, facing competition from a "Mimeographic Society"; see Martin Postle, *Sir Joshua Reynolds: The Subject Pictures* (New York: Cambridge University Press, 1995), 107.

60. Booth, *Treatise Explanatory*, 57.

61. Polygraphic Society, *A Catalogue of Pictures Copied or Multiplied (for Sale) by*

A Chymical and Mechanical Process . . . (London: A. Grant, 1792), 11, 9. On the politics of Fox images, see N. B. Penny, "The Whig Cult of Fox in Early Nineteenth-Century Sculpture," *Past & Present* 70 (1976): 94–105.

62. See Booth, *Treatise Explanatory*, 57, 61–62. The theme of Jupiter's rape of Europa answers to several possible pictures in West's oeuvre; see Galt, *Life, Studies, and Works*, 2:224, 231.

63. For Northcote, see items 31, 47–49, in Polygraphic Society, *Catalogue*, 10, 12. Kauffman's *Departure of Telemachus* was valued at more than twice the next highest price; see items 41, 60 in Polygraphic Society, *Catalogue*, 11, 14. For more on Kauffman's series on the Odysseus/Penelope/Telemachus family group, see Angela Rosenthal, "Angelica's Odyssey: Kauffman's Paintings of Penelope and the Weaving of Narrative," in *Women, Art and the Politics of Identity in Eighteenth-Century Europe*, ed. Melissa L. Hyde and Jennifer D. Milam (Burlington: Ashgate, 2003), 211–36.

64. Kauffman had departed London for Rome in 1781, three years prior to Booth's earliest publication. For Opie, see item 2, 43 in Polygraphic Society, *Catalogue*, 5, 12.

65. Pasquin, *Critical Guide*, 20–21 (emphasis added).

66. David Saunders and Antony Griffiths, "Two Mechanical Oil Paintings after de Loutherbourg: History and Technique," in *Studying Old Master Paintings: Technology and Practice*, ed. Marika Spring (London: Archetype, 2011), 191. Cf. David Saunders, "Joseph Booth's Chymical and Mechanical Paintings," in *European Paintings 15–18th Century: Copying, Replicating and Emulating*, ed. Erma Hermens (London/Copenhagen: Archetype/CATS, 2014), 113–21.

67. T. F——r, "To the Editor of the Monthly Magazine," *Monthly Magazine or British Register* (January 1, 1809). 517.

68. T. F——r, "To the Editor." A classic account of the importance to painting of grinding pigments is Thierry de Duve, "The Readymade and the Tube of Paint," in *Kant after Duchamp* (Cambridge, MA: MIT Press, 1996), 147–96.

69. For imperial benefits, see especially Booth's 1788 *Address to the Public*, 3, 16–17.

70. On comparison of London and Athens, see Booth, *Treatise Explanatory*, 51.

71. Booth, *Address to the Public*, 9.

72. Booth, *Address to the Public*, 5.

73. Carl L. Becker, *The Heavenly City of the Eighteenth-Century Philosophers* (New Haven, CT: Yale University Press, 1932), 130.

74. Booth, *Treatise Explanatory*, 60.

75. Jean-Etienne Liotard, *Traité des Principes et des Regles de La Peinture* (1781) (Geneva: Minkoff Reprint, 1973), 13. "La Peinture est le miroir immuable de tout ce que l'univers nous offre de plus beau. Par elle, le passé est toujours le présent. Victorieuse du temps, ses ouvrages sont immortels & invariables (a); ils enchaînent le présent, & reproduisent le passé. (a) Le peinture en émail ne s'altere jamais." Cf. Gudrun Swoboda, "Jean-Etienne Liotard (1702–1789), Painter of Extremes," *Points of View* (Vienna: Kunsthistorisches Museum, 2012), 4–18.

76. For Reynolds and Liotard in 1750s London, see David H. Solkin, *Art in Britain 1660–1815* (New Haven, CT: Yale University Press, 2015), 138–39.

77. See Richard Walker, "Henry Bone's Pencil Drawings in the National Portrait Gallery," *Walpole Society* 61 (1999): 306.

78. For more on Bone, see R. J. B. Walker, "Bone, Henry (1755–1834)," in *Oxford Dictionary of National Biography* (Oxford University Press, 2004), http://www.oxforddnb.com/view/article/2836 (accessed May 16, 2013).

79. For Bone's technique, see Sarah Coffin and Bodo Hofstetter, *Portrait Miniatures in Enamel* (London: Philip Wilson Publishers, 2000), 9–11.

80. *The Memoirs of Susan Sibbald (1783–1812)*, ed. Francis Paget Hett (London: John Lang, 1926), 143.

81. Smith, *Nollekens and His Times*, 2:292–93.

82. On Birch and his career, see Emily T. Cooperman and Lea Carson Sherk, *William Birch: Picturing the American Scene* (Philadelphia: University of Pennsylvania Press, 2011).

83. William Birch, "The Life and Anecdotes of William Russell Birch, Enamel Painter"; reproduced in Cooperman and Sherk, *William Birch*, 171.

84. December 3, 1784; RSA Archives, Committee Minutes 1784: ff87–89.

85. For a reproduction of Birch's pictorial palette (dated to 1815), see Cooperman and Sherk, *William Birch*, 170.

86. Birch, in Cooperman and Sherk, *William Birch*, 169–70, 172.

87. See Alois Riegl, "The Modern Cult of Monuments: Its Character and Its Origin" (1903), in *Oppositions Reader: Selected Readings from a Journal of Ideas and Criticism in Architecture 1973–1984*, ed. K. Michael Hays (New York: Princeton Architectural Press, 1998), 621–51.

88. Birch, in Cooperman and Sherk, *William Birch*, 170.

89. Adrian Forty, *Objects of Desire* (New York: Pantheon, 1986), 15, 17.

90. Cf. Michael Durey, *Transatlantic Radicals and the Early American Republic* (Lawrence: University of Kansas Press, 1997); Kenneth R. Johnston, *Unusual Suspects: Pitt's Reign of Alarm and the Lost Generation of the 1790s* (New York: Oxford University Press, 2013).

91. See Birch, in Cooperman and Sherk, *William Birch*, 184 (on Birch's politics, 27–30).

92. According to the influential argument of Reinhart Koselleck, the concept of revolution in politics shifted in the 1790s from a circular movement akin to natural cycles to signifying a break in the previous order accomplished by human action; see Reinhart Koselleck, "Historical Criteria of the Modern Concept of Revolution," in *Futures Past: On the Semantics of Historical Time* (1985), trans. Keith Tribe (New York: Columbia University Press, 2004), 43–58.

93. Joseph Priestley, *A Description of a Chart of Biography . . .*, 2nd ed. (Warrington: Printed for the Author, 1765), 7.

94. Daniel Rosenberg and Anthony Grafton, *Cartographies of Time* (New York: Princeton Architectural Press, 2010), 116. Cf. Jonathan Sachs, "Scales of Time and the Anticipation of the Future: Gibbon, Smith, Playfair," *Modern Intellectual History* 11, no. 3 (2014): 697–718.

95. Priestley, *Chart*, 8, 9.

96. Cf. Senefelder, *Invention*, 20, 34–37. On the chemistry of textile printing, see Linda Eaton, *Printed Textiles: British and American Cottons and Linens 1700–1850* (New York: Monacelli, 2014), 127–47.

97. Booth, *Address to the Public*, 9.

98. For this assessment, see Peter M. Jones, "'I had L[o]rds and Ladys to wait on yesterday . . .': Visitors to the Soho Manufactory," in *Matthew Boulton: Selling What All the World Desires*, ed. Shena Mason (Birmingham, England: Birmingham City Council, 2009), 74.

99. See Eric Robinson and Keith R. Thompson, "Matthew Boulton's Mechanical Paintings," *Burlington Magazine* 112, no. 809 (1970): 497–507; Barbara Fogarty, "Matthew Boulton and Francis Eginton's Mechanical Paintings: Production and Consumption 1777 to 1781" (MPhil thesis, University of Birmingham, 2010); Barbara Fogarty, "The Mechanical Paintings of Matthew Boulton and Francis Eginton," in *Matthew Boulton: Enterprising Industrialist of the Enlightenment*, ed. Kenneth Quickenden, Sally Baggott, and Malcolm Dick (Farnham: Ashgate, 2013), 111–26.

100. An inventory of these pictures, dated December 1780, lists fourteen items by Kauffman versus two by Reynolds; see Robinson and Thompson, "Boulton's Mechanical Paintings," 507.

101. For ongoing interest in replicating works of art, represented by James Watt's sculpturing machines, see Richard L. Hills, *James Watt*, vol. 3, *Triumph through Adversity, 1785–1819* (Ashbourne, UK: Landmark, 2006), 234–37.

102. For this assessment, see Richard L. Hills, *James Watt*, vol. 2, *The Years of Toil, 1775–1785* (Ashbourne, UK: Landmark, 2005), 211.

103. Watt to Darwin, May 12 [1779]; reproduced in *The Essential Writings of Erasmus Darwin*, ed. Desmond King-Hele (London: MacGibbon & Kee, 1968), 127.

104. Thomas Jefferson to Ebenezer Hazard, February 18, 1791, in *The Papers of Thomas Jefferson*, ed. Julian P. Boyd (Princeton, NJ: Princeton University Press, 1950), 19:287. For Jefferson's copying devices, see Silvio A. Bedini, *Thomas Jefferson and His Copying Machines* (Charlottesville: University Press of Virginia, 1984).

105. For engineer John Smeaton's use of a Watt & Co. copying press as early as 1783 in replicating designs for the piers of Harraton Bridge, see Royal Society Library, JS/4/92. I thank Rupert Baker for helping me to identify this drawing.

106. See James Watt & Co., *Directions for Using the Patent Portable Copying Machines . . .* (Birmingham, n.d.), 17.

107. Watt & Co., *Directions*, 11.

108. Advertisement for the Watt & Co. copying machine (Library of Birmingham, MS 3147/20/1), n.p.

109. James Watt to James Woodmason, September 2, 1780; James Watt & Co. Letter Book, April 7–December 12, 1780 (Library of Birmingham, MS 3147/19/1), f24.

110. On Watt's contacts to paper manufacturers, see Hills, *Years of Toil*, 201. On the Whatmans' "blueing" techniques, see Theresa Fairbanks Harris and Scott Wilcox, *Papermaking and the Art of Watercolor in Eighteenth-Century Britain: Paul Sandby and the Whatman Paper Mill* (New Haven, CT: Yale University Press, 2006), esp. 84–85.

111. Barbara Rhodes and William W. Streeter, *Before Photocopying: The Art & History of Mechanical Copying, 1780–1938* (New Castle, DE: Oak Knoll Press, 1999), 31.

112. See Watt to Woodmason, September 2, 1780; Library of Birmingham, MS 3147/19/1, f24.

113. James Watt to James Woodmason, September 4, 1780; Library of Birmingham, MS 3147/19/1, f25. Woodmason in turn looked to Soho when seeking to source chemical materials at industrial scale. In March 1783, he requested of Watt "a favour: a friend of mine wants some Tons of Sal Ence. or Caput Mortuum, as I learn you have some works near you where this is made. I shall be obliged to you to inquire whether you can procure a good quantity at 5 pounds sterling per Ton delivered in London." James Woodmason to James Watt, March 12, 1783; James Watt & Co. Correspondence, Library of Birmingham, MS 3147/19/4/43.

114. See Adrian Johns, "Ink," in *Materials and Expertise in Early Modern Europe: Between Market and Laboratory*, ed. Ursula Klein and E. C. Spary (Chicago: University of Chicago Press, 2010), 101–24.

115. For example, Reynolds wrote and then canceled the following in his juvenile commonplace book: "Receipts for making of Ink / Galls Pounded / a little Copperas / Gum Arabick Pounded / & Rain wat<er>"; Yale Center for British Art, MS Reynolds 33, f25.

116. Watt & Co., *Directions*, 15.

117. Watt & Co., *Directions*, 15.

118. Rhodes and Streeter, *Before Photocopying*, 31.

119. James Watt & Co. to James Woodmason, August 30, 1780; Library of Birmingham, MS 3147/19/1, f28.

120. James Watt & Co. to James Woodmason, October 27, 1780; Library of Birmingham, MS 3147/19/1, f52. For the charges levied on Woodmason for the ink and direc-

tions on how it should be priced, see James Watt & Co. to James Woodmason, n.d.; Library of Birmingham, MS 3147/19/1, f26.

121. On Watt's attempts to conceal the true composition of his inks, see Richard Hills, "James Watt and His Copying Machine," in *The Oxford Papers: Proceedings of the British Association of Paper Historians Fourth Annual Conference*, ed. Peter Bower, *Papers of the Bibliographical Society of America* 92, no. 1 (1998): 83.

122. "Directions for making Copying Ink & for Copying Drawings for which this ink is used"; Library of Birmingham, MS 3147/20/10.

123. "Directions for making Copying Ink"; Library of Birmingham, MS 3147/20/10.

124. In what follows, I draw on J. R. Partington, *A History of Chemistry* (London: Macmillan, 1962), 3:325–62; and David P. Miller, *Discovering Water: James Watt, Henry Cavendish and the Nineteenth-Century "Water Controversy"* (Aldershot, UK: Ashgate, 2004), esp. 27–57.

125. See Arthur L. Donovan, *Philosophical Chemistry in the Scottish Enlightenment: The Doctrines and Discoveries of William Cullen and Joseph Black* (Edinburgh: Edinburgh University Press, 1975).

126. Henry Guerlac, "Joseph Black's Work on Heat," in *Joseph Black 1728–1799: A Commemorative Symposium*, ed. A. D. C. Simpson (Edinburgh: Royal Scottish Museum, 1982), 16.

127. James Watt to Joseph Black, December 13, 1782; reproduced in *Partners in Science: Letters of James Watt & Joseph Black*, ed. Eric Robinson and Douglas McKie (Cambridge, MA: Harvard University Press, 1970), 118.

128. For extensive discussion of these debates, see Miller, *Discovering Water*.

129. William Godwin, *An Enquiry Concerning Political Justice, and Its Influence on General Virtue and Happiness* (London: G. G. J. and J. Robinson, 1793), 1:30.

130. Thomas Wedgwood to James Watt [Jr.], November 19, 1795; Library of Birmingham, MS 3147/19/4 (letter not numbered).

131. On Watt Jr's radical politics in the early 1790s, see Peter M. Jones, "Living the Enlightenment and the French Revolution: James Watt, Matthew Boulton, and Their Sons," *Historical Journal* 42, no. 1 (1999): 157–82; Eric Robinson, "An English Jacobin: James Watt, Junior, 1769–1848," *Cambridge Historical Journal* 11, no. 3 (1955): 349–55.

132. On Wedgwood's gift to Godwin (delivered in late November/early December 1795), see Pamela Clemit, "William Godwin and James Watt's Copying Machine: Wet-Transfer Copies in the Abinger Papers," *Bodleian Library Record* 18, no. 5 (2005): 532–60; Pamela Clemit, "Commerce of Luminaries: Eight Holograph Letters between William Godwin and Thomas Wedgwood," in *Godwinian Moments: From the Enlightenment to Romanticism*, ed. Robert M. Maniquis and Victoria Myers (Toronto: Toronto University Press, 2011), 261–82.

133. Thomas Wedgwood to James Watt, November 28, 1795; Library of Birmingham, MS 3147/19/4 (not numbered).

134. Thomas Wedgwood to James Watt, December 5, 1795; Library of Birmingham, MS 3147/19/4 (not numbered).

135. Thomas Wedgwood to James Watt Jr., December 9, 1795; Library of Birmingham, MS 3147/19/4 (not numbered).

136. Cf. Hilary Young, ed., *The Genius of Wedgwood* (London: Victoria & Albert Museum, 1995); Robin Reilley, *Josiah Wedgwood 1730–1795* (London: Macmillan, 1992).

137. In the first stages, areas of the print to be transferred would be coated in a metal or stone pigment suspended in a varnish solution; then, a "pounce" layer prepared from one portion of the same pigment to six parts of a borax flux would be dusted onto the sticky varnish and fired glossy; see Colin Wyman, "The Early Tech-

niques of Transfer Printing," *Transactions of the English Ceramic Circle* 10, nos. 4–5 (1980): 187–88, 193–94.

138. For discussion of which printmaker's mezzotint after Allan Ramsay's portrait of John Stuart, third earl of Bute, should be transfer-printed, see Guy Green to Josiah Wedgwood, August 2, 1763; Victoria & Albert Museum, Wedgwood Collection, MS 1431, f1r. For the claimed speed, see Wyman, "Early Techniques," 192.

139. Peter Perez Burdett to Thomas Bentley, May 26, 1771, reproduced in Martin Hopkinson, "Burdett, Wedgwood and Bentley," *Print Quarterly* 25, no. 2 (2008): 133.

140. See Elizabeth E. Barker and Alex Kidson, *Joseph Wright of Derby in Liverpool* (New Haven, CT: Yale University Press, 2007).

141. See Antony Griffiths, "Notes on Early Aquatint in England and France," *Print Quarterly* 4, no. 3 (1987): 255–70.

142. *A Catalogue of the Pictures, Sculptures, Models, Designs in Architecture, Drawings, Prints, &c. Exhibited by the Society of Artists of Great Britain, at the New Room, Near Exeter-Exchange, Strand, April the Twenty-Ninth, 1773, the Fourteenth Year of Exhibiting* (London: Printed by Harriot Bunce, printer to the Society, 1773), 3.

143. Compare Josiah Wedgwood to Thomas Bentley, November 15, 1771, and Wedgwood to Bentley, November 21, 1771, in Hopkinson, "Burdett, Wedgwood and Bentley," 135.

144. Wedgwood to Bentley, November 21, 1771, in Hopkinson, "Burdett, Wedgwood and Bentley," 136.

145. See entry for "Aqua-tint Engraving," in Josiah Wedgwood's commonplace book; Victoria & Albert Museum, Wedgwood Collection, MS 39 28408, ff173–74. For more on Burdett, see Colin Wyman, "P. P. Burdett," *Transactions of the English Ceramic Circle* 10, nos. 4–5 (1980): 305–13.

146. In his first log of experiments, Wedgwood explains his procedure this way: "In the following experiments I have expressed the materials by Number, which in this instance are a species of a short hand & save much writing. They have likewise an advantage, of not being intelligible without the key, to any person who might happened to taken up the book, which is often, in the course of making the experiments, unavoidably exposed to such accidents"; Victoria & Albert Museum, Wedgwood Collection, MS 26 19117.

147. See Ian Jenkins and Kim Sloan, *Vases and Volcanoes: Sir William Hamilton and His Collection* (London: British Museum, 1996).

148. *A Catalogue of Cameos: Intaglios, Medals, and Bas-Reliefs; with a General Account of Vases and Other Ornaments, After the Antique, Made by Wedgwood and Bentley; and Sold at Their Rooms in Great Newport-Street, London* (London: Printed in the year M. DCC. LXXIII. and sold by Cadel, 1773), 58.

149. The patent specification is reproduced in Llewellynn F. W. Jewitt, *The Wedgwoods: Being a Life of Josiah Wedgwood; with Notices of His Works and Their Productions . . .* (London: Virtue Brothers and Co., 1865), 200–203; quotation, 202.

150. Wedgwood to Bentley, November 21, 1771, reproduced in Hopkinson, "Burdett, Wedgwood and Bentley," 136.

151. For Wedgwood's credentials as a chemist, see Robert E. Schofield, "Josiah Wedgwood, Industrial Chemist," *Chymia* 5 (1959): 180–92.

152. Josiah Wedgwood, "Passing Colours," January 30, 1779; MS 39 28408, f290, f291.

153. For more on Chisholm, see Larry Stewart, "Assistants to Enlightenment: William Lewis, Alexander Chisholm and Invisible Technicians in the Industrial Revolution," *Notes and Records of the Royal Society* 62, no. 1 (2008): 17–29; F. W. Gibbs, "William Lewis, M.B., F.R.S. (1708–1781)," *Annals of Science* 8, no. 2 (1952): 122–51; F. W. Gibbs,

"A Notebook of William Lewis and Alexander Chisholm," *Annals of Science* 8, no. 3 (1952): 202–20.

154. Josiah Wedgwood &c., "Observations on Coloured Stones"; Victoria & Albert Museum, Wedgwood Collection, MS 26 19111, ff3, 15, 17.

155. Wedgwood, "Coloured Stones," MS 26 19111, f21.

156. On the chemistry of light in the late eighteenth century, see Victor D. Boantza, "Light in the Pneumatic Context: Aspects of Interplay between Theory and Practice in Early Photochemical Research," *Historia Scientiarum* 16 (2006): 105–27.

157. See R. B. Litchfield, *Tom Wedgwood: The First Photographer* (London: Duckworth, 1903), 5–9. Chemistry was equally key to Josiah Wedgwood's vision of his sons; in the spring of 1779, Wedgwood *père* rhapsodized to Bentley of commissioning pendant portraits of his family from either Wright of Derby or George Stubbs. The portrait of his sons would show "Jack standing at the table making fixable air with the glass apparatus &c; & his two brothers accompanying him. Tom jumping up and clapping his hands in joy & surprise at seeing the stream of bubbles rise up just as Jack has put a little chalk to the acid. Joss with the chemical dictionary before him in a thoughtful mood"; Josiah Wedgwood to Thomas Bentley, May 30, 1779, reproduced in *The Selected Letters of Josiah Wedgwood*, ed. Ann Finer and George Savage (London: Cory, Adams & Mackay, 1965), 234.

158. For this milieu, see Lynn A. Hunt and Margaret C. Jacob, "The Affective Revolution in 1790s Britain," *Eighteenth-Century Studies* 34, no. 4 (2001): 491–521; Mike Jay, *The Atmosphere of Heaven: The Unnatural Experiments of Dr. Beddoes and His Sons of Genius* (New Haven, CT: Yale University Press, 2009); Jan Golinski, *Science as Public Culture: Chemistry and Enlightenment in Britain, 1760–1820* (New York: Cambridge University Press, 1992), esp. 153–87.

159. "An Account of a Method of Copying Paintings upon Glass, and of Making Profiles, by the Agency of Light upon Nitrate of Silver Invented by T. Wedgwood, Esq. with Observations by H. Davy," *Journals of the Royal Institution of Great Britain* 1 (1802): 170–74. Emphasizing the fact that Davy then coedited the Royal Institution's journal (in addition to serving as its professor of chemistry and director of its chemical laboratory), R. B. Litchfield attributes this article to Davy; see Litchfield, *Tom Wedgwood*, 187–88. The paper was subsequently included in *The Collected Works of Sir Humphry Davy*, ed. John Davy (London: Smith, Elder and Co., 1839), 2:240–45.

160. Henry William Fox Talbot, *The Pencil of Nature* (London, Longman, Brown, Green, & Longmans, 1844), 11. For evocation of Wedgwood and Davy post-1839, see Douglas R. Nickel, "Notes towards New Accounts of Photography's Invention," in *Photography and Its Origins*, ed. Tanya Sheehan and Andrés Mario Zervigón (New York: Routledge, 2015), 82–93.

161. Talbot's techniques of the early 1840s are, strictly speaking, not fixed. As one photo conservator observes, Talbot's early photographs are now almost impossible to exhibit in museums since "even the most stringent conditions of gallery illumination are likely to cause perceptible changes in many of the images within a relatively short time"; Mike Ware, *Mechanisms of Image Deterioration in Early Photographs: The Sensitivity to Light of W. H. F. Talbot's Halide-Fixed Images 1834–1844* (London: London: Science Museum, 1994), 7.

162. Talbot, *Pencil of Nature*, 11. Five years earlier, Talbot had used similar language as he described experiments made "by Mr. Wedgwood . . . and Sir Humphry Davy, which however ended in failure"; H. F. Talbot, "Some Account of the Art of Photogenic Drawing," *London and Edinburgh Philosophical Magazine*, series 3, 14, 88 (February 1839): 197. See also Jordan Bear, "Self-Reflections: The Nature of Sir Humphry Davy's Photographic 'Failures,'" in Sheehan and Zervigón, *Photography and Its Origins*, 185–94.

163. Victoria & Albert Museum, Wedgwood Collection, MS E40 28515/23, f2_r.

164. Wedgwood Collection, MS E40 28515/23, f2_r.

165. Robert Waring Darwin's theory of "ocular spectra" first appeared in the *Philosophical Transactions of the Royal Society of London* in 1786; it was reprinted by Erasmus Darwin (Wedgwood's childhood tutor) in *Zoonomia; or The Laws of Organic Life* (London: J. Johnson, 1796), 1:538–70. For the relationship between Erasmus Darwin and Tom Wedgwood, see Alan Barnes, "Negative and Positive Images: Erasmus Darwin, Tom Wedgwood and the Origin of Photography," in *The Genius of Erasmus Darwin*, ed. C. U. M. Smith and Robert Arnott (Aldershot, UK: Ashgate, 2005), 237–53.

166. Darwin, *Zoonomia*, 1:549, 541–42.

167. See, for example, William Godwin to Thomas Wedgwood, April 19, 1797, as reproduced in *The Letters of William Godwin*, vol. 1, 1778–1797, ed. Pamela Clemit (New York: Oxford University Press, 2011), 197–200.

168. Tom Wedgwood, "Experiments and Observations on the Production of Light from Different Bodies, by Heat and by Attrition," *Philosophical Transactions of the Royal Society of London* 82 (1792): 36. On this paper, see E. Newton Harvey, *A History of Luminescence from the Earliest Times until 1900* (Philadelphia: American Philosophical Society, 1957), 370–71, 383–84.

169. Cf. Wedgwood, "Experiments and Observations," 34–35.

170. Wedgwood, "Experiments and Observations," 33–34, 37.

171. Time was the subject of a much-anticipated publication by Wedgwood, which was to be edited and seen through the press by his brother-in-law, James Mackintosh, who possibly lost it at sea when returning from India in 1811; see Alan Barnes, "Coleridge, Tom Wedgwood and the Relationship between Time and Space in Midlands Enlightenment Thought," *Journal for Eighteenth-Century Studies* 30, no. 2 (2007): 243–60. On Wedgwood's thought more broadly, compare Geoffrey Batchen, "Tom Wedgwood and Humphry Davy: 'An Account of a Method,'" *History of Photography* 17, no. 2 (1993): 172–83; John Beer, "Coleridge, Mackintosh and the Wedgwoods: A Reassessment, Including Some Unpublished Records," *Romanticism* 7, no. 1 (2001): 16–40; Nicholas J. Wade, "Accentuating the Negative: Tom Wedgwood (1771–1805), Photography and Perception," *Perception* 34, no. 5 (2005): 513–20.

172. Thomas Wedgwood, "Essay on Time"; Thomas Wedgwood Papers, Victoria & Albert Museum, Wedgwood Collection, MS E40 28459, f3 + r.

173. Wedgwood, "Essay on Time," MS E40 28459, ff4 + r–5 + r.

174. Barnes dates this letter, following its postmark, to December 18, 1801. However, evidence in its text leads me to think it was composed nearly a year earlier and sent in December. Specifically, on the recto of the second page, Wedgwood asserts, "Today, <January 10>, an event happened 5 days ago . . . this event happened <at 12 o'clock, Jany. 5> A.D. 1801." The "1" signaling the last digit of the year has been strengthened in pen over what appears to be a zero—as though the writer had forgotten that a new year had just begun and retained 1800; compare Tom Wedgwood to James Mackintosh, January 1801 (Victoria & Albert Museum, Wedgwood Collection, MS E40 28489 a); Barnes, "Coleridge, Tom Wedgwood," 248. A concise reading of Mackintosh's politics is Johnston, *Unusual Suspects*, 205–23.

175. Wedgwood to Mackintosh, January 1801; Victoria & Albert Museum, Wedgwood Collection, MS E40 28489 a f2_r.

176. See Immanuel Kant, *Critique of Pure Reason* (1781; 1787), trans. Norman Kemp Smith (Basingstoke: Palgrave Macmillan, 2007), esp. 68–78. Wedgwood and his brother had awarded an annuity of one hundred fifty pounds sterling to Coleridge in 1798; see Trevor H. Levere, *Poetry Realized in Nature: Samuel Taylor Coleridge and Early Nineteenth-Century Science* (New York: Cambridge University Press, 1981).

177. Thomas Wedgwood to James Mackintosh, April 1, 1801; Victoria & Albert Museum, Wedgwood Collection, MS E40 28492, f1$_r$.

178. Wedgwood to Mackintosh, April 1, 1801, MS E40 28492, f1$_v$.

179. Wedgwood to Mackintosh, April 1, 1801, MS E40 28492, f1$_v$.

180. Nicholas Salmon, *Stemmata Latinitatis; or, An Etymological Latin Dictionary: Wherein the Whole Mechanism of the Latin Tongue Is Methodically and Conspicuously Exhibited* (London: Printed for the author, by W. and C. Spilsbury, 1796), vol. 2, n.p.

181. Wedgwood to Mackintosh, April 1, 1801; Victoria & Albert Museum, Wedgwood Collection, MS E40 28492, ff1$_v$–2$_r$. The last phrase, "always be," is canceled in the MS.

182. Neil McKendrick, "Josiah Wedgwood and Factory Discipline," *Historical Journal* 4, no. 1 (1961): 51. Cf. E. P. Thompson, "Time, Work-Discipline, and Industrial Capitalism," *Past and Present* 38, no. 1 (1967): 56–97. On Blake's prints, see Robert N. Essick, *William Blake's Commercial Book Illustrations* (Oxford: Clarendon Press, 1991), 96–100.

183. Victoria & Albert Museum, Wedgwood Collection, MS E40 28515/14, f1$_{r-v}$.

184. Victoria & Albert Museum, Wedgwood Collection , MS E40 28496 C, f15.

185. Wedgwood to Mackintosh, January 1801; Victoria & Albert Museum, Wedgwood Collection, MS E40 28489 a, f2$_r$.

186. Davy, "Account of a Method," 170.

187. Davy, "Account of a Method," 172.

188. For Davy's volte-face at the Royal Institution, see Golinski, *Science as Public Culture*, esp. 188–203.

189. Cf. "An Account of Some Experiments and Observations on the Constituent Parts of Some Astringent Vegetables and on their Operation in Tanning" (1803) and "Observations on the Process of Tanning" (1803) in Davy, *Collected Works*, 2:247–96.

190. Davy, "Account of a Method," 171.

191. Aristotle, *Generation of Animals*, trans. A. L. Peck (Cambridge, MA: Harvard University Press, 1943), 729.a.11–12 (p. 108–9).

192. On Aristotle's persistence, compare Angus McLaren, "The Pleasures of Procreation: Traditional and Biomedical Theories of Conception," in *William Hunter and the Eighteenth-Century Medical World*, ed. W. F. Bynum and Roy Porter (New York: Cambridge University Press, 1985), 323–41; Nancy Tuana, "The Weaker Seed: The Sexist Bias of Reproductive Theory," *Hypatia* 3, no. 1 (1988): 35–59.

193. For Darwin's adoption of Lavoisier, see David Philip Miller, *James Watt, Chemist: Understanding the Origins of the Steam Age* (London: Pickering & Chatto, 2009), 136–39.

194. Darwin, *Zoonomia*, 1:484.

195. Darwin, *Zoonomia*, 1:484.

196. Darwin, *Zoonomia*, 1:493.

197. Mary Shelley, *Frankenstein; or, The Modern Prometheus: The 1818 Text*, intro. and notes by Marilyn Butler (New York: Oxford University Press, 1993), vol. 1, chap. 1, pp. 22–23.

198. Shelley, *Frankenstein*, 23. For a recent treatment of Shelley's chemistry emphasizing connections to Davy, see Mary Fairclough, *Literature, Electricity and Politics 1740–1840: "Electrick Communication Every Where"* (London: Palgrave Macmillan, 2017), esp. 192–98.

199. Shelley, *Frankenstein*, vol. 1, chap. 3, pp. 32, 36.

200. Shelley, *Frankenstein*, 191.

Chapter 4

1. E. V. Rippingille, "Personal Recollections of Great Artists," *Art Journal* (1860): 100. See also Katherine Solender, *Dreadful Fire! Burning of the Houses of Parliament* (Bloomington, IN: Cleveland Museum of Art/Indiana University Press, 1984).

2. Rippingille, "Personal Recollections," 99, 100.

3. See Joyce Townsend, *Turner's Painting Techniques* (London: Tate Publishing, 2005).

4. For an overview, see Philippa Levine, *The Amateur and the Professional: Antiquarians, Historians and Archaeologists in Victorian England, 1838–1886* (New York: Cambridge University Press, 1986).

5. Charles Lock Eastlake, *Materials for a History of Oil Painting* (London: Longman, Brown, Green and Longmans, 1847), 538.

6. Cf. Eastlake, *Materials*, 539–44; 444.

7. Eastlake, *Materials*, 539. For Eastlake's Plymouth roots, see David Robertson, *Sir Charles Eastlake and the Victorian Art World* (Princeton, NJ: Princeton University Press, 1978), 2–4.

8. [Unknown writer], "Materials for a History of Oil Painting," *Blackwood's Edinburgh Magazine* 62 (September 1847): 311 (301–11).

9. For this accusation, see Wilhelm de Winterton, "The Present State, Prospects, and Theory of Painting, no. II," *Civil Engineer and Architect's Journal*, October 1844, 378. On Laxton and his journal, see G. C. Boase, "Laxton, William (1802–1854)," rev. Elizabeth Baigent, in *Oxford Dictionary of National Biography* (Oxford University Press, 2004), http://www.oxforddnb.com.huntington.idm.oclc.org/view/article/16215 (accessed November 9, 2016).

10. For the provenance of the Fitzwilliam Museum Ledger Books, see Malcolm Cormack, "The Ledgers of Sir Joshua Reynolds," *Walpole Society* 42 (1968–1970): 105–6.

11. On Haydon's view of Reynolds as "the great master of colour in modern times," see Benjamin Robert Haydon, "Lecture VI," in *Lectures on Painting and Design* (London: Longman, Brown, Green and Longmans, 1844), 249.

12. For more on Beechey's technical interests, see his endorsement of color-maker George Field's varnish in Field, "Colourless Lac Varnish," *Transactions of the Society, Instituted at London, for the Encouragement of Arts, Manufactures, and Commerce* 45 (1827): 59–67, esp. 62.

13. See for example William Cotton and John Burnet, eds., *Sir Joshua Reynolds, and His Works: Gleanings from His Diary, Unpublished Manuscripts, and from Other Sources* (London: Longman, Brown, Green, Longmans, and Roberts, 1856); William Cotton, *A Catalogue of the Portraits Painted by Sir Joshua Reynolds, Knt., P.R.A.* (London: Longman, Brown, Green, Longmans, and Roberts, 1857); William Cotton, ed., *Sir Joshua Reynolds' Notes and Observations on Pictures: Chiefly of the Venetian School . . .* (London: J. R. Smith, 1859).

14. For a concise summary, see Alison Smith, "Medium and Method in Pre-Raphaelite Painting," in *Pre-Raphaelites: Victorian Art and Design*, ed. Tim Barringer, Jason Rosenfeld, and Alison Smith (New Haven, CT: Yale University Press, 2012), 18–23.

15. [Lady Elizabeth Eastlake], "Photography," *Quarterly Review* 101 (1857): 443.

16. William Lake Price, "On Composition and Chiar'oscuro. I," *Photographic News* 3, no. 76 (1860): 281.

17. See Diane Waggoner and T. J. Barringer, *The Pre-Raphaelite Lens: British Photography and Painting, 1848–1875* (Washington, DC: National Gallery of Art, 2010).

18. Steve Edwards, *The Making of English Photography: Allegories* (University Park: Pennsylvania State University Press, 2006), 136.

19. Peter Galassi, *Before Photography: Painting and the Invention of Photography* (New York: Museum of Modern Art, 1981), 25.

20. Geoffrey Batchen, *Burning with Desire: The Conception of Photography* (Cambridge, MA: MIT Press, 1997), 38.

21. Important recent work has begun to dismantle this intuition; e.g., Stephen Pinson, *Speculating Daguerre: Art and Enterprise in the Work of L. J. M. Daguerre* (Chicago: University of Chicago Press, 2012). Yet the ongoing need to defend and begin such

dismantling against what contributor Stephen Bann calls the myth of "photographic exceptionalism" persists through recent collections such as Tanya Sheehan and Andrés Mario Zervigón, eds., *Photography and Its Origins* (New York: Routledge, 2015). See also Nicoletta Leonardi and Simone Natale, eds., *Photography and Other Media in the Nineteenth Century* (University Park: Pennsylvania State University Press, 2018).

22. Roland Barthes, *Camera Lucida: Reflections on Photography* (1980), trans. Richard Howard (New York: Hill and Wang, 1981), chap. 32, p. 76.

23. Rosalind Krauss, "Notes on the Index: Seventies Art in America, Part 2," *October*, no. 4 (1977): 59.

24. Walter Benn Michaels, "Photographs and Fossils," in *Photography Theory*, ed. James Elkins (New York: Routledge, 2007), 432.

25. André Bazin, "The Ontology of the Photographic Image" (1945), *Film Quarterly* 13, no. 4 (1960): 8. See also Kendall Walton, "Transparent Pictures: On the Nature of Photographic Realism," *Critical Inquiry* 11, no. 2 (1984): 246–77.

26. Kaja Silverman, *The Miracle of Analogy; or, The History of Photography, Part 1* (Stanford, CA: Stanford University Press, 2015), 10–11.

27. Alfred Gell, "The Technology of Enchantment and the Enchantment of Technology," in *The Art of Anthropology: Essays and Diagrams*, ed. Eric Hirch (New York: Berg, 1999), 50.

28. See Jordan Bear, *Disillusioned: Victorian Photography and the Discerning Subject* (State College: Penn State University Press, 2015), 119–30. I thank Jordan Bear for generously sharing his work with me prior to its publication, and for his astute comments on my own interpretation.

29. Gaston Bachelard, *Earth and Reveries of Will: An Essay on the Imagination of Matter* (1943), trans. K. Haltmann (Dallas: Dallas Institute Publications, 2002), 7.

30. Gaston Bachelard, *Air and Dreams: An Essay on the Imagination of Movement* (1943), trans. E. R. Farrell and C. F. Farrell (Dallas: Dallas Institute Publications, 1988), 8.

31. For this chronology, see George Wallis, "The Ghost of an Art-Process practiced at Soho, near Birmingham, about 1777 to 1780, Erroneously Supposed to have been Photography," *Art Journal*, August 1, 1866, 251.

32. See David K. Brown, "Smith, Sir Francis Pettit (1808–1874)," in *Oxford Dictionary of National Biography* (Oxford University Press, 2004), http://www.oxforddnb.com.huntington.idm.oclc.org/view/article/25798 (accessed November 14, 2016).

33. See D. K. Brown, *Before the Ironclad: Development of Ship Design, Propulsion, and Armament in the Royal Navy, 1815–60* (Annapolis, MD: Naval Institute Press, 1990), 102. A fuller account of Smith's experiments is John Fincham, *A History of Naval Architecture* (London: Whittaker and Co., 1851), 342–58.

34. On Perkins and his gallery, see Iwan R. Morus, *Frankenstein's Children: Electricity, Exhibition, and Experiment in Early-Nineteenth-Century London* (Princeton, NJ: Princeton University Press, 1998), esp. 75–83.

35. See Don Leggett, *Shaping the Royal Navy: Technology, Authority and Naval Architecture, ca. 1830–1906* (Manchester: Manchester University Press, 2015), 71–72.

36. Edward Price to Francis Pettit Smith, November 12, 1862; Science Museum Library & Archives, Wroughton (hereafter SMLA), MS 2117 E 1.

37. Matthew Piers Watt Boulton to C. J. Chubb, February 21, 1849; Library of Birmingham, Chubb Papers 1849; MS 3782/11/12, f1_{r-v}.

38. For Price's claim, see "Who Discovered Photography?" *Illustrated London News* 43 (November 21, 1863): 530; for dispute of this, see M. P. W. Boulton, *Remarks Concerning Certain Photographs Supposed to Be of Early Date* (London: Bradbury & Evans, 1864), 14. Edward Price to Francis Pettit Smith, August 7, 1863; SMLA, MS 2117 F 24 ii_v.

39. Price to Smith, December 1, 1862 (SMLA, MS 2117 E 3_v); Price to Smith, August 7, 1863 (SMLA, MS 2117 F 24 ii_v).

40. According to one reviewer in 1775, the subject was an exemplary demonstration of painting's temporal limits: "Every picture is confined to a single point of time, and every fact requires a succession of time for its being fully completed, so as many pictures may be founded upon it, as there are points in the time, in which it was transacting: Of this there cannot be a more striking proof than this very story of Antiochus and Stratonice. For not to mention the great number of pictures that have been built upon it formerly, there are no less than four or five in the present Exhibitions [i.e., the spring shows of the Royal Academy and the Society of Artists of Great Britain], and yet all of them distinguished by some peculiar circumstance"; [Unknown writer], "Remarks on the Pictures in the Exhibition of the Royal Academy, concluded," *London Chronicle*, no. 2875 (May 11–13, 1775): 452.

41. David's *Antiochus and Stratonice* (oil on canvas, 1774) is now in the collection of the École des Beaux Arts, Paris.

42. See Helmut Von Erffa and Allen Staley, *The Paintings of Benjamin West* (New Haven, CT: Yale University Press, 1986), 169–70.

43. For the sourcing of this silver plate, see Price to Smith, November 26, 1862; SMLA, MS 2117 E 2 ii_{r}. See also Dugald Campbell to Smith, October 30, 1863; SMLA, MS 2117 F 32.

44. Compare Eric Robinson and Keith R. Thompson, "Matthew Boulton's Mechanical Paintings," *Burlington Magazine* 112, no. 809 (1970): 497–507; David Saunders and Antony Griffiths, "Two Mechanical Oil Paintings after de Loutherbourg: History and Technique," in *Studying Old Master Paintings: Technology and Practice*, ed. Marika Spring (London: Archetype, 2011), 186–93; Barbara Fogarty, "Matthew Boulton and Francis Eginton's Mechanical Paintings: Production and Consumption 1777 to 1781" (MPhil thesis, University of Birmingham, 2010); Barbara Fogarty, "The Mechanical Paintings of Matthew Boulton and Francis Eginton," in *Matthew Boulton: Enterprising Industrialist of the Enlightenment*, ed. Kenneth Quickenden, Sally Baggott, and Malcolm Dick (Farnham: Ashgate, 2013), 111–26; David Saunders, "Joseph Booth's Chymical and Mechanical Paintings," in *European Paintings 15–18th Century: Copying, Replicating and Emulating*, ed. Erma Hermens (London/Copenhagen: Archetype/CATS, 2014), 113–21; and Bear, *Disillusioned*, esp. 121–23.

45. On this transition, see Anthony Burton, *Vision and Accident: The Story of the Victoria and Albert Museum* (London: V&A Publications, 1999), 120.

46. Price to Smith, November 26, 1862; SMLA, MS 2117 E 2 iv_{r}.

47. Samuel Lewis, *A Topographical Dictionary of England*, 5th ed. (London: S. Lewis & Co., 1844): 2:395. This note is part of a significant expansion of the entry on Handsworth apparent between Lewis's fourth edition (1840) and the fifth edition (1844). Since the 1840s were an active moment of James Watt Jr.'s promotion of his father's inventive activities, it is tempting to assign authorship of the 1844 entry to him. On Watt Jr.'s activities, see David P. Miller, *Discovering Water: James Watt, Henry Cavendish and the Nineteenth-Century "Water Controversy"* (Aldershot, UK: Ashgate, 2004), esp. 222–43.

48. Price to Smith, January 3, 1863; SMLA, MS 2117 F $1_{\text{r–v}}$.

49. For a concise account of Daguerre's process, see Laurent Mannoni, *The Great Art of Light and Shadow: Archaeology of the Cinema*, trans. Richard Crangle (Exeter: University of Exeter Press, 2000), 195–96.

50. Dr. [Hugh] Diamond, "Photography and Photographic Apparatus," in *The Record of the International Exhibition* (Glasgow: William Mackenzie, 1862), 565. For the controversial installation of photography as a subset of mechanics at this exhibition, see Edwards, *Making of English Photography*, 165–203.

51. See Diamond, "Photographic Apparatus," 565.

52. Sir David Brewster, "Photogenic Drawing; or, Drawing by the Agency of Light," *Edinburgh Review* 86, no. 154 (1843): 309–44. Issued in discrete fascicles, *The Pencil of*

Nature included a page advising the reader that the work's plates "are the sun pictures themselves and not, as some persons have imagined, engravings in imitation"; see Joel Snyder, "Inventing Photography, 1839–1879," in *On the Art of Fixing a Shadow: One Hundred Fifty Years of Photography*, ed. Sarah Greenough et al. (Washington, DC: National Gallery of Art, 1989), 4. For the larger conversation in which such language moves, compare Joel Snyder, "Res Ipsa Loquitur," in *Things That Talk: Object Lessons from Art and Science*, ed. Lorraine Daston (New York: Zone Books, 2004), 195–221; Ted Underwood, *The Work of the Sun: Literature, Science, and Political Economy, 1760–1860* (New York: Palgrave Macmillan, 2005).

53. Price to Smith, December 16, 1862; SMLA, MS 2117 E 8_{r-v}.

54. Price to Smith, December 2, 1862; SMLA, MS 2117 E 4 ii_{r}.

55. Price to Smith, December 26, 1862; SMLA, MS 2117 E 17_{r-v}.

56. Price to Smith, December 22, 1862; SMLA, MS 2117 E 15_{r-v}.

57. Price to Smith, February 12, 1863; SMLA, MS 2117 F 8_{v}.

58. On the immediate descendants of Boulton and Watt more broadly, see Peter M. Jones, "Living the Enlightenment and the French Revolution: James Watt, Matthew Boulton, and Their Sons," *Historical Journal* 42, no. 1 (1999): 157–82; Eric Robinson, "An English Jacobin: James Watt, Junior, 1769–1848," *Cambridge Historical Journal* 11, no. 3 (1955): 349–55.

59. Price to Smith, November 11, 1863; SMLA, MS 2117 F 42_{v}–ii_{r}.

60. See Peter Gaskell, *The Manufacturing Population of England: Its Moral, Social, and Physical Conditions . . .* (London: Baldwin and Cradock, 1833), 6–9.

61. Frederick Engels, *The Condition of the Working-Class in England* (1844; 1892), trans. Florence Kelley Wischnewetzky (London: Swan Sonnenschein & Co., 1892), 15.

62. Compare Dennis Smith, *Conflict and Compromise: Class Formation in English Society, 1830–1914: A Comparative Study of Birmingham and Sheffield* (London: Routledge & Kegan Paul, 1982), esp. 28–38, 53, 71–72; Michael Durey, *Transatlantic Radicals and the Early American Republic* (Lawrence: University of Kansas Press, 1997), esp. 28–31.

63. See Anne B. Rodrick, *Self-Help and Civic Culture: Citizenship in Victorian Birmingham* (Aldershot, UK: Ashgate, 2004), esp. 93–132. For this concluding quote, see "A Strange Discovery"; unidentified newspaper clipping, SMLA, MS 2117 F 12. Birmingham was also thinkable in the period as home of what August Welby Northmore Pugin called "Brummagem Gothic"; see the anonymous review of Pugin's *True Principals of Christian Architecture* (1841), *Gentleman's Magazine*, January 1842, 59–62. I thank Tim Barringer for this reference.

64. Price to Smith, October 15, 1863; SMLA, MS 2117 F 30, $f1_{r-v}$.

65. Price to Smith, December 26, 1862; SMLA, MS 2117 E 17 ii_{v}. "J. W. G. Watt" was J. W. Gibson, a twenty-five-year-old nephew who took over Watt Jr.'s estate at his death in 1848; see Miller, *Discovering Water*, 243.

66. Price to Smith, December 16, 1862; SMLA, MS 2117 E 8 ii_{v}.

67. For this claim, see Christine MacLeod, *Heroes of Invention: Technology, Liberalism and British Identity, 1750–1914* (New York: Cambridge University Press, 2007), 10.

68. For more on Meteyard, see Fred Hunter, "Meteyard, Eliza (1816–1879)," in *Oxford Dictionary of National Biography* (Oxford University Press, 2004), http://www.oxforddnb.com.huntington.idm.oclc.org/view/article/18624 (accessed November 21, 2016).

69. Eliza Meteyard, *The Life of Josiah Wedgwood: From His Private Correspondence and Family Papers . . .*, 2 vols. (London: Hurst and Blackett, 1865, 1866), 2:557–58.

70. Meteyard, *Life of Josiah Wedgwood*, 1:245.

71. "Be good enough to use my name in connexion with the controversy as little as possible. It may make enemies when I could wish to have friends"; Eliza Meteyard to F. P. Smith, November 11, 1863; SMLA, MS 2117 F 39 f ii_{r}.

72. Eliza Meteyard to Bennet Woodcroft, August 26, 1863; SLMA, MS 2117 F 25 i_{r-v}.

73. Eliza Meteyard to F. P. Smith, November 2, 1863; SLMA, MS 2117 F 35 i_v–ii_r.

74. Meteyard credits this invention to Wedgwood's collaboration with some familiar figures. He worked with his father, Josiah Wedgwood, who had used a camera obscura for producing transfer prints used in the famous "Frog Service" he executed for Russian empress Catherine the Great. He also collaborated with Liverpool-based physician Matthew Turner, who, as noted in chapter 2, delivered chemical directions for Wright of Derby's *Alchymist*. Turner, Meteyard claims, had mastered "the art of copying prints upon glass by striking off impressions with a coloured solution of silver, and fixing them on the glass by baking on an iron plate in a heat sufficient to incorporate the solution with the glass"; see Meteyard, *Life of Wedgwood*, 2:586. Of course, claims that this plate had any relation to the Wedgwoods had been dashed by early 1864 when correspondence through the Photographic Society of London identified it instead as an early work by Talbot; for this recognition, see *Photographic Journal* 141 (January 15, 1864): 438.

75. Eliza Meteyard, *A Group of Englishmen (1795 to 1815) Being Records of the Younger Wedgwoods and Their Friends* . . . (London: Longmans, Green, and Co, 1871), frontispiece, 155–56.

76. Meteyard, *Group of Englishmen*, 156. For an instructively scathing critique of Meteyard's claims (made amid counterarguments assigning photography's invention to Joseph Nicéphore Niépce), see Helmut Gernsheim and Alison Gernsheim, *The History of Photography from the Earliest Use of the Camera Obscura in the Eleventh Century up to 1954* (New York: Oxford University Press, 1955), esp. 34. On the Gersheims, see Jessica S. McDonald, "A Sensational Story: Helmut Gernsheim and 'the World's First Photograph,'" in Sheehan and Zervigón, *Photography and Its Origins*, 15–28.

77. Meteyard, *Group of Englishmen*, 155.

78. Compare Peter Jones, *Industrial Enlightenment: Science, Technology and Culture in Birmingham and the West Midlands, 1760–1820* (Manchester: Manchester University Press, 2008); Kenneth Quickenden, Sally Baggott, and Malcolm Dick, eds., *Matthew Boulton: Enterprising Industrialist of the Enlightenment* (Farnham: Ashgate, 2013).

79. Price to Smith, November 26, 1862; SMLA, MS 2117 E 2 ii_{r-v}.

80. According to Boulton's diary, Beechey along with his wife and son were resident at Soho between September 26 and October 29, 1798. Additional meetings and sittings with Beechey in London followed on November 14 and 17, 1798. See Boulton's diary, Library of Birmingham, MS 3782/12/107/26, n.p. See John Wilson, "Beechey, Sir William (1753–1839)," in *Oxford Dictionary of National Biography* (Oxford University Press, 2004), http://www.oxforddnb.com.huntington.idm.oclc.org/view/article/1949 (accessed November 14, 2016).

81. Price to Smith, November 26, 1862; SMLA, MS 2117 E 2 iii_{r-v}.

82. For this very intriguing argument, see Eric McCauley Lee, "'Titianus Redivivus': Titian in British Art Theory, Criticism, and Practice, 1768–1830" (PhD diss., Yale University, 1997), 215–20.

83. Price to Smith, May 23, 1863; SMLA, MS 2117 F 19 ii_{r-v}.

84. Price to Smith, August 7, 1863; SMLA, MS 2117 F 24 ii_r. See also Price to Smith, October 15, 1863; SMLA, MS 2117 F 30.

85. Price to Smith, July 9, 1863; SMLA, MS 2117 F 22 ii_v.

86. *The Photographic Journal, being the Journal of the Photographic Society* 139 (November 16, 1863): 395–96.

87. *Photographic Journal* 139 (1863): 397. On Shadbolt, see Ian Sumner, "Shadbolt, George (1819–1901)," in *Encyclopedia of Nineteenth-Century Photography*, ed. John Hannavy (New York: Routledge, 2008), 2:1265.

88. F. W. Fairholt to Dr. Hugh Diamond, as reproduced in *Photographic Journal* 141 (January 15, 1864): 431. See Joanna Selborne, "Fairholt, Frederick William (*bap.* 1813, *d.* 1866)," in *Oxford Dictionary of National Biography* (Oxford University Press, 2004), http://www.oxforddnb.com.huntington.idm.oclc.org/view/article/9098 (accessed November 27, 2016).

89. John J. Cole to Dr. Diamond, as reproduced in *Photographic Journal* 141 (1864): 434–35.

90. Eastlake, "Photography," 458.

91. See Alan Pearson, "Hunt, Robert (1807–1887)," in *Oxford Dictionary of National Biography* (Oxford University Press, 2004), http://www.oxforddnb.com.huntington.idm.oclc.org/view/article/14203 (accessed November 27, 2016).

92. Robert Hunt to Dr. Diamond, November 23, 1863, as reproduced in *Photographic Journal* 141 (1864): 431. On the photographic reproduction of painting and other arts in the period, see Anthony J. Hamber, *"A Higher Branch of the Arts": Photographing the Fine Arts in England, 1839–1880* (Amsterdam: Gordon and Breach, 1996).

93. On Malone, see W. H. Brock, *William Crookes (1832–1919) and the Commercialization of Science* (Aldershot, UK: Ashgate, 2008), 39–41.

94. For Malone's claims, see *Photographic Journal* 139 (1863): 397.

95. David K. Brown, "Lloyd, Thomas (1803–1875)," in *Oxford Dictionary of National Biography* (Oxford University Press, 2004), http://www.oxforddnb.com.huntington.idm.oclc.org/view/article/56934 (accessed November 27, 2016).

96. Thomas Lloyd to F. P. Smith, January 15, 1863; SMLA, MS 2117 F 3 i_{v}–ii_{v}. While it is doubtful that the "Landseer" in question actually was Sir Edwin Landseer (1802–1873), a fascinating meditation on that painter's use of photographs is Diana Donald, "'A Mind and Conscience Akin to Our Own': Darwin's Theory of Expression and the Depiction of Animals in Nineteenth-Century Britain," in *Endless Forms: Charles Darwin, Natural Science and the Visual Arts*, ed. Diana Donald and Jane Munro (Cambridge: Fitzwilliam Museum, 2009), 195–213.

97. "James Watt and Photography," *Birmingham Journal*, April 18, 1863.

98. "Who Discovered Photography?," *Saturday Review of Politics, Literature, Science and Art* 16, no. 419 (November 7, 1863): 613–14.

99. For a concise biography, see Bruce Kinzer, "Flying Under the Radar: The Strange Case of Matthew Piers Watt Boulton," *Times Literary Supplement*, May 1, 2009, 14–15.

100. M. P. W. Boulton to Smith, November 19, 1863; SMLA, MS 2117 F 49_{v}–ii_{r}.

101. Boulton to Smith, November 1863; SMLA, MS 2117 F 52.

102. M. P. W. Boulton, *Remarks on Some Evidence Recently Communicated to the Photographic Society* (London: Bradbury & Evans, 1863), 4, 5–6.

103. Boulton to Smith, December 12, 1863 (SMLA, MS 2117 F 62 ii_{v}); Boulton to Smith, December 20, 1863 (SMLA, MS 2117 F 67 ii_{v}).

104. Price to Smith, December 20, 1863; SMLA, MS 2117 F 66 I_{v}.

105. Price to Smith, December 20, 1863; SMLA, MS 2117 F 66 ii_{r}. See Julie Hankey, *A Passion for Egypt: Arthur Weigall, Tutankhamun, and the "Curse of the Pharaohs"* (London: I. B. Tauris, 2001).

106. Price to Smith, December 20, 1863; SMLA, MS 2117 F 66 ii_{r}.

107. Boulton, *Remarks Concerning Certain Photographs*, 3.

108. Boulton, *Remarks Concerning Certain Photographs*, 13.

109. Boulton, *Remarks Concerning Certain Photographs*, 9. Boulton quotes the dimensions of paintings in his possession made "by the secret process of copying pictures formerly practiced at Soho" after Reynolds's *Faith, Hope* and *Charity*, his paintings of the Christian virtues prepared in part for use by glass painter Thomas Jervais, who installed the designs in New College Chapel, Oxford. For these Soho replicas, see M. P. W. Boul-

ton to Dr. Diamond, in *Photographic Journal* 141 (1864): 433. For Reynolds's paintings, see David Mannings and Martin Postle, *Sir Joshua Reynolds: A Complete Catalogue of His Paintings* (New Haven, CT: Yale University Press, 2000), 1:545–52.

110. Boulton, *Remarks Concerning Certain Photographs*, 46. Later in the same pamphlet, Boulton changes his view on period usage of the phrase (67).

111. Boulton, *Remarks Concerning Certain Photographs*, 10–12.

112. M. P. W. Boulton, *Remarks Concerning Certain Pictures Supposed to Be Photographs of Early Date* (London: Bradbury & Evans, 1865), 9–10, 55–56.

113. Smith to Boulton, February 10, 1864; SMLA, MS 2117 G 11, f1v.

114. Smith to Boulton, February 10, 1864; SMLA, MS 2117 G 11, f2r.

115. Jonathan Sperber, *The European Revolutions, 1848–1851* (New York: Cambridge University Press, 1994), 19.

116. Smith to Boulton, February 10, 1864; SMLA, MS 2117 G 11, f2$_{r-v}$ (emphasis added). For Smith's source letter, see extract from Jas. Brown to Mr. Brown Sr., May 3, 1863; SMLA, MS 2117 F 14.

117. Smith to Boulton, February 10, 1864; SMLA, MS 2117 G 11, f3$_{r}$.

118. See Edwards, *Making of English Photography*, 165–203.

119. For M. P. W. Boulton's role in the closure of the Soho Manufactory and associated facilities, see George Demidowicz, *The Soho Industrial Buildings: Manufactory, Mint and Foundry* (forthcoming). I thank the author for sharing his unpublished work with me.

120. Edwards, *Making of English Photography*, 31.

121. For Boulton's recognition that mechanical painting devices replace "up to a certain point for the intelligent labour of an artist" and thus might well draw painters' ire, see Boulton, *Remarks Concerning Certain Pictures*, 55.

122. *Photographic Journal* 139 (1863): 397.

123. Boulton, *Remarks Concerning Certain Photographs*, 22.

124. Boulton, *Remarks Concerning Certain Photographs*, 4. For a similar statement, see the cutting of an article entitled "The Watt and Wedgwood Photographs," *Manchester Guardian*, February 26, 1864; SMLA, MS 2117 G 19, f1$_{r}$.

125. Boulton, *Remarks Concerning Certain Photographs*, 60.

126. For the claim that "the circumstance that photography was not invented earlier remains the greatest mystery in its history," see Gernsheim and Gernsheim, *History of Photography*, xvii. For positive endorsement of this mystery, see Batchen, *Burning with Desire*, esp. 24–53.

127. Snyder, "Inventing Photography," 8.

128. To be clear, none of the images discussed in the 1860s as sun pictures were replications after Reynolds, but the Soho partnership had been replicating Reynolds's designs in the years around 1780; see Robinson and Thompson, "Boulton's Mechanical Paintings," 505, 507.

129. On claims for the Poor Law, see Solender, *Dreadful Fire!*, 58. A more recent and expansive account is Leo Costello, "Power, Creativity, and Destruction in Turner's Fires," *19: Interdisciplinary Studies in the Long Nineteenth Century*, 25 (2017), https://doi.org/10.16995/ntn.791 (accessed September 3, 2018).

130. Michel Serres, "Turner Translates Carnot," in *Calligram: Essays in New Art History from France*, ed. Norman Bryson (New York: Cambridge University Press, 1988), 158. An expanded argument in this direction is Michel Serres, "Science and the Humanities: The Case of Turner," trans. Catherine Brown and William Paulson, *SubStance* 26, no. 2 (1997): 6–21.

131. See William S. Rodner, *J. M. W. Turner: Romantic Painter of the Industrial Revolution* (Berkeley: University of California Press, 1997).

132. Serres, "Turner Translates," 158.

133. D. S. L. Cardwell, *From Watt to Clausius: The Rise of Thermodynamics in the Early Industrial Age* (London: Heinemann, 1971), 52. For a more recent account that frames Watt's contribution in slightly different terms, see Ingo Müller, *A History of Thermodynamics: The Doctrine of Energy and Entropy* (New York: Springer, 2007), 49–51.

134. Compare the deflationary view of Black-Watt relations in Cardwell, *Watt to Clausius*, 40–57.

135. David Philip Miller, *James Watt, Chemist: Understanding the Origins of the Steam Age* (London: Pickering & Chatto, 2009), 96. See also Miller, "Seeing the Chemical Steam through the Historical Fog: Watt's Steam Engine as Chemistry," *Annals of Science* 65, no. 1 (2008): 47–72.

136. Miller, *Watt, Chemist*, 146.

137. Miller, *Watt, Chemist*, 36.

138. Miller, *Watt, Chemist*, 35, 43–45.

139. Miller, *Watt, Chemist*, 43–57; Miller, *Discovering Water*, esp. 215–51.

140. See MacLeod, *Heroes of Invention*, esp. 91–180.

141. For the influential argument, made in 1819 by *Edinburgh Review* founder Francis Jeffrey, that Watt's steam engine constituted a military advantage crucial to the defeat of Napoleon, see MacLeod, *Heroes of Invention*, 94–97, 123.

142. *Biographical Sketch of Mr. F. P. Smith, first practical introducer of the Screw Propeller* (Greenwich, 1856), 22.

143. *Biographical Sketch*, 50.

144. See *Biographical Sketch*, 64.

145. Boulton, *Remarks Concerning Certain Pictures*, 18.

146. See for example James P. Harrison, *Mastering the Sky: A History of Aviation from Ancient Times to the Present* (New York: Sarpedon, 1996), 48.

147. M. P. W. Boulton, "Improvements in Propulsion and in Aerial Locomotion, and in Apparatus Connected Therewith" (patent no. 392; 1868), as reproduced in Charles H. Gibbs-Smith, *The Aeroplane: An Historical Survey of Its Origins and Development* (London: HM Stationery Office, 1960), 314.

148. Gibbs-Smith, *Aeroplane*, 314.

149. Erasmus Darwin, *The Botanic Garden, A Poem in Two Parts*, 2nd American ed. (New York: T. & J. Swords, 1807), canto I, lines 289–90, p. 22.

150. Matthew Piers Watt Boulton, *On Aërial Locomotion* (London: Bradbury & Evans, 1864), 3.

151. Boulton, *Aërial Locomotion*, 4 (my emphasis).

152. See M. P. W. Boulton, patent no. 915 (1867), *Patents for Inventions: Abridgments of Specifications. Class 4, Aeronautics. Period.—A.D. 1855–1900* (London: HMSO, 1905), 2.

153. Boulton, *Aërial Locomotion*, 4.

154. Boulton, *Aërial Locomotion*, 6.

155. Boulton, *Aërial Locomotion*, n.p.

156. See Sir Dugald Clerk, *The Gas and Oil Engine: Seventh Edition* (New York: J. Wiley & Sons, 1896), 484–85.

157. Boulton, *Aërial Locomotion*, 6.

158. For analysis of the composition produced in the era of Boulton's publication by a leading chemist and manufacturer, see F. A. Abel, "Researches on Gun-Cotton: On the Manufacture and Composition of Gun-Cotton," *Philosophical Transactions of the Royal Society of London* 156 (1866): 269–308.

159. "Accidents and Offences," *Illustrated London News*, no. 272 (July 17, 1847), 39. Cf. Arthur Percival, *The Faversham Gunpowder Industry*, 3rd ed. (Faversham: Faversham Society, 1986).

160. See Gilbert T. Morgan and David Doig Pratt, *British Chemical Industry: Its Rise and Development* (New York: Longmans, Green & Co., 1938), 243.

161. See H. T. Wood, "Archer, Frederick Scott (1813–1857)," rev. John Ward, in *Oxford Dictionary of National Biography* (Oxford University Press, 2004), http://www.oxforddnb.com.huntington.idm.oclc.org/view/article/620 (accessed December 2, 2016).

162. Robert Hunt, *Researches on Light in Its Chemical Relations . . .*, 2nd ed. (London: Longman, Brown, Green, and Longmans, 1854), 143.

163. For a concise statement of the explosive dangers to photographers, see Lynn Ann Davis, "Wet-Collodion Positive Process," in Hannavy, *Encyclopedia of Nineteenth-Century Photography*, 2:1486–88.

164. Eastlake, "Photography," 453.

165. For this summary, see M. P. W. Bolton, *Inquisitio Philosophica: An Examination of the Principles of Kant and Hamilton* (London: Chapman and Hall, 1866), esp. 91–96. For the attribution of this work by "Bolton" to M. P. W. Boulton, I follow Kinzer, "Flying Under the Radar," 14–15.

166. Bolton, *Inquisitio*, 96.

167. Boulton, *Aërial Locomotion*, 7.

168. Sir Walter Scott, prefatory letter to *The Monastery: A Romance* (Philadelphia: M'Carty and Davis, 1825), 32.

169. *Biographical Sketch*, 22.

170. Comparing the language and concepts of Boulton's specifications to period models of the Lenoir engine, it seems likely that Boulton would have understood his own engines' operations at least in thermodynamic terms. For comparison, see E. F. C. Somerscales, "Internal Combustion Engines," in *An Encyclopaedia of the History of Technology*, ed. Ian McNeil (London: Routledge, 1990), 303–29. I thank Leslie Tomory for his guidance on these points.

171. By this I mean that, by 1864, Boulton was interested in the use of chemical ingredients for picture-making as potential fuels for his airplane engines. By the time of his *Remarks Concerning Certain Pictures* of 1865–1866, he was also researching the chemical composition of the sun pictures. I don't mean to suggest that the sun pictures Smith found had been made with gun-cotton, which of course postdated their likely creation by half a century. But I take the concluding lines of *Aërial Locomotion* (1864) to suggest that Boulton was keen to find suitable chemicals beyond gun-cotton, chemicals he might plausibly have believed to be in use in the sun pictures, even if he had come to see those artifacts as nonphotographic. I thank Omar Nasim for pressing me on this point.

172. This conjunction has gone strangely unexamined in histories of photography. A recent edited collection aiming to move research on Talbot "beyond photography" gives no freestanding and only passing mention of his engines (discussed below); Mirjam Brusius, Katrina Dean, and Chitra Ramalingam, eds., *William Henry Fox Talbot: Beyond Photography* (New Haven, CT: Yale Center for British Art/Paul Mellon Centre for Studies in British Art, 2013). Brief mention is made in Patrick Maynard, *The Engine of Visualization: Thinking Through Photography* (Ithaca, NY: Cornell University Press, 1997), 64–67.

173. See Gernsheim and Gernsheim, *History of Photography*, 36.

174. Their 1807 patent specification stipulates, "Nous imaginâmes que s'il [the gas] se trouvait pénétré brusquement dans un vase clos par la flamme d'une substance éminemment combustible, réduite en poussière très-fine, et disséminée dans toute la capacité de ce vase, il déploierait alors une énergie beaucoup plus grande" (French patent no. 409; 1807), as reproduced in Horst O. Hardenberg, *The Niepce Brothers' Boat Engines* (Warrendale, PA: Society of Automotive Engineers, 1993), 83. For subsequent use of essential oils, see Hardenberg, *Boat Engines*, 61.

175. Cf. Joseph Maria Eder, *History of Photography*, trans. Edward Epstean (New

York: Columbia University Press, 1945), 195–97; Gernsheim and Gernsheim, *History of Photography*, 36–47.

176. The Pyréolophore was "le principal objet des sérieuses préoccupations des deux frères Niépce. Aussi, leur lettres sont remplies des descriptions de nouveaux appareils, des essais, des expériences relatifs au *Pyréolophore*, et des perfectionnements, des ameliorations apportées à cette machine; des résultats de leurs expérimentations sur des matières combustibles propres à imprimer le mouvement à leurs moteurs. Claude et Nicéphore employaient tour leur temps et toute leur intelligence, non seulement à ces travaux, mais encore à faire les désmarches nécessaires, afin de placer avantageusement leur appareil de prédeliction. Aussi *la lithographie et surtout l'héliographie ne tiennent qu'une très-petite place dans leur volumineuse correspondance*"; Victor Fouque, *La vérité sur l'invention de la Photographie: Nicéphore Niépce, sa vie . . .* (Paris: Librairie des auteurs et de l'Académie des bibliophiles, 1867), 82 (my emphasis)

177. See Emily Doucet, "Anticipating Machines Heavier than Air: Nadar, Photography and the Objects of Technology" (MA thesis, University College London, 2013).

178. The most detailed accounts of Talbot's engines I have encountered are Horst O. Hardenberg, *The Middle Ages of the Internal-Combustion Engine, 1794–1886* (Warrendale, PA: Society of Automotive Engineers, 1999), 144–51, 173–80; and H. J. P. Arnold, *William Henry Fox Talbot: Pioneer of Photography and Man of Science* (London: Hutchinson, 1977), esp. 217–31. See also Larry J. Schaaf, *Records of the Dawn of Photography: Talbot's Notebooks P & Q* (Cambridge University Press/National Museum of Photography, Film & Television, 1996), P 101, Q 23, Q 33, Q 53–55; Brian Bowers, *A History of Electric Light & Power* (Stevenage, Hertfordshire: P. Peregrinus in association with the Science Museum, 1982), 262.

179. William Henry Fox Talbot, "Obtaining and Applying Motive Power," patent A.D. 1846 (signed by Talbot June 3, 1847), no. 11,475, p. 2.

180. On the publication of Talbot's project, see Larry J. Schaaf, *H. Fox Talbot's Pencil of Nature: Anniversary Facsimile* (New York: Hans P. Kraus, Jr, 1989). For an overview of Talbot's patent litigation and the importance of wet-plate collodion to it, see Arnold, *William Henry Fox Talbot*, esp. 195–216.

181. See Elizabeth Bacon Eger, "Creative Combustion: Image, Imagination and the Work of Robert Fulton," *Panorama: Journal of the Association of Historians of American Art* 2, no. 1 (Summer 2016), http://journalpanorama.org/creative-combustion-image-imagination-and-the-work-of-robert-fulton/ (accessed August 2, 2017).

182. See Samuel F. B. Morse, *Lectures on the Affinity of Painting with the Other Fine Arts*, ed. Nicolai Cikovsky Jr. (Columbia: University of Missouri Press, 1983), 59. More broadly on Morse, see Jennifer L. Roberts, *Transporting Visions: The Movement of Images in Early America* (Berkeley: University of California Press, 2014), 140–60.

183. Milton Friedman, *Essays in Positive Economics* (Chicago: University of Chicago Press, 1953), 35.

184. See Arnold, *Fox Talbot*, 175–216.

185. See Bennet Woodcroft, *A Sketch of the Origin and Progress of Steam Navigation from Authentic Documents* (London: Taylor, Walton, and Maberly, 1848).

186. See Burton, *Vision and Accident*, esp. 45–56; Anita McConnell, "Woodcroft, Bennet (1803–1879)," in *Oxford Dictionary of National Biography* (Oxford University Press, 2004), http://www.oxforddnb.com.huntington.idm.oclc.org/view/article/29908 (accessed December 7, 2016).

187. Bear, *Disillusioned*, 130.

188. For a brilliant reading of these debates in *ancien régime* France, see Katie Scott, *Becoming Property: Art, Theory and Law in Early Modern France* (New Haven, CT: Yale University Press, 2018), esp. chap. 2.

189. A classic foray in this direction is Jennifer Mnookin, "The Image of Truth: Photographic Evidence and the Power of Analogy," *Yale Journal of Law and the Humanities* 10, no. 1 (1998): 1–74.

190. Joel Snyder, "Review of Geoffrey Batchen's *Burning with Desire*," *Art Bulletin* 81, no. 3 (1999): 541.

Conclusion

1. Paul J. Crutzen, "Geology of Mankind," *Nature* 415 (2002): 23.

2. Instructive accounts in a humanist direction include Dipesh Chakrabarty, "The Climate of History: Four Theses," *Critical Inquiry* 35, no. 2 (2009): 197–222; McKenzie Wark, *Molecular Red: Theory for the Anthropocene* (London: Verso, 2015); Adam M. Romero et al., "Chemical Geographies," *GeoHumanities* 3, no. 1 (2017): 158–77.

3. Maja-Lena Bränvall, Richard Bindler, and Ingemar Renberg, "The Medieval Metal Industry Was the Cradle of Modern Large-Scale Atmospheric Lead Pollution in Northern Europe," *Environmental Science and Technology* 33, no. 24 (1999): 4391–95.

4. William F. Ruddiman, "The Anthropogenic Greenhouse Era Began Thousands of Years Ago," *Climate Change* 61 (2003): 261–93.

5. Jan Zalasiewicz et al., "When Did the Anthropocene Begin? A Mid-Twentieth-Century Boundary Level Is Stratigraphically Optimal," *Quaternary International* 383 (2015): 196–203. Cf. Colin N. Waters et al., *A Stratigraphical Basis for the Anthropocene* (London: Geological Society, 2014).

6. Will Steffen, Paul J. Crutzen, and John R. McNeill, "The Anthropocene: Are Humans Now Overwhelming the Great Forces of Nature?" *Ambio* 36, no. 8 (December 2007): 616 (614–21).

7. Among the instructive critiques of the Anthropocene narrative's Eurocentric/colonialist assumptions and scientistic erasures of politics, see Kathleen D. Morrison, "Provincializing the Anthropocene," *Seminar*, no. 673, "Nature and History" (2015), 75–80; Andreas Malm and Alf Hornborg, "The Geology of Mankind? A Critique of the Anthropocene Narrative," *Anthropocene Review* 1, no. 1 (2014): 62–69; Andreas Malm, *Fossil Capital: The Rise of Steam-Power and the Roots of Global Warming* (London: Verso, 2016); and T. J. Demos, *Against the Anthropocene: Visual Culture and the Environment Today* (Berlin: Sternberg Press, 2017).

8. For an insightful analysis of the confrontation with anthropogenic climate change in the work of John Ruskin, a figure with far greater art-historical consequence than Marsh, see Vicky Albritton and Fredrik Albritton Jonsson, *Green Victorians: The Simple Life in John Ruskin's Lake District* (Chicago: University of Chicago Press, 2016).

9. For insightful assessments of Marsh's work, see Marcus Hall, ed., "The Nature of G. P. Marsh: Tradition and Historical Judgment," special issue, *Environment and History* 10, no. 2 (May 2004): 123–235. For Marsh's place in narratives of the Anthropocene, cf. Crutzen, "Geology," 23; Paul J. Crutzen and Will Steffen, "How Long Have We Been in the Anthropocene Era? An Editorial Comment," *Climate Change* 61 (2003): 253; Will Steffen et al., "The Anthropocene: Conceptual and Historical Perspectives," *Philosophical Transactions A* 369, no. 1938 (2011): 844.

10. George Perkins Marsh, *Man and Nature; or, Physical Geography as Modified by Human Action* (New York: Charles Scribner, 1865), 13.

11. Marsh, *Man and Nature*, 35.

12. Marsh, *Man and Nature*, 5. On Marsh's ownership of Winckelmann's works and his art-collecting more broadly, see Helena Wright, *The First Smithsonian Collection: The European Engravings of George Perkins Marsh and the Role of Prints in the U.S. National Museum* (Washington, DC: Smithsonian Institution Scholarly Press, 2015), 23. For claims such as "the independence of Greece is to be regarded as the most promi-

nent of the causes, originating in its constitution and government, of its superiority in art," in period translation, see J. J. Winckelmann, *The History of Ancient Art* (1764), trans. G. Henry Lodge (Boston: Little, Brown, and Company, 1856), 2:9–10. On efforts to reconcile histories of American art and ecological awareness, see Alan C. Braddock and Christoph Irmscher, eds., *A Keener Perception: Ecological Studies in American Art History* (Tuscaloosa: University of Alabama Press, 2009).

13. Marshall Sahlins, "The Sadness of Sweetness: The Native Anthropology of Western Cosmology," *Current Anthropology* 27, no. 3 (1996): 395.

14. See Esther Leslie, *Synthetic Worlds: Nature, Art and the Chemical Industry* (London: Reaktion, 2005).

15. See Gilbert T. Morgan and David Doig Pratt, *British Chemical Industry: Its Rise and Development* (New York: Longmans, Green, & Co., 1938), 325–62. On the modernity enacted by chemical warfare, see Peter Sloterdijk, *Terror from the Air* (2002), trans. Amy Patton and Steve Corcoran (Los Angeles: Semiotext(e), 2009).

16. A concise account is *Paint & Painting: An Exhibition and Working Studio Sponsored by Winsor & Newton to celebrate their 150th Anniversary* (London: Tate Gallery, 1982).

17. Laura Anne Kalba, *Color in the Age of Impressionism: Commerce, Technology and Art* (University Park: Pennsylvania State University Press, 2017), 108; Thierry de Duve, "The Readymade and the Tube of Paint," in *Kant after Duchamp* (Cambridge, MA: MIT Press, 1996), 147–96.

18. Jan Zalasiewicz et al., "The Technofossil Record of Humans," *Anthropocene Review* 1, no. 1 (2014): 34–43.

19. Ross Barrett and Daniel Worden, "Introduction," in *Oil Culture*, ed. Ross Barrett and Daniel Worden (Minneapolis: University of Minnesota Press, 2014), xix.

20. Alois Riegl, *Problems of Style: Foundations for a History of Ornament* (1893), trans. Evelyn Kain (Princeton, NJ: Princeton University Press, 1992), 4.

21. Riegl, *Problems of Style*, 5, 229. On the history of academic art history, cf. Craig H. Smyth and Peter M. Lukehart, eds., *The Early Years of Art History in the United States* (Princeton, NJ: Department of Art and Archaeology, Princeton University, 1993); Elizabeth Mansfield, ed., *Making Art History: A Changing Discipline and Its Institutions* (New York: Routledge, 2007).

22. Alois Riegl, *Late Roman Art Industry* (1901), trans. Rolf Winkes (Rome: G. Bretschneider, 1985), 8.

23. For different ways of counting the elements, see Jeffrey Jerome Cohen and Lowell Duckert, eds., *Elemental Ecocriticism: Thinking with Earth, Air, Water, and Fire* (Minneapolis: University of Minnesota Press, 2015).

24. Jeff Wall, "Photography and Liquid Intelligence," in *Jeff Wall*, ed. Thierry de Duve et al. (London: Phaidon, 1996), 90.

25. For an exemplary *Begriffsgeschichte*, see Eric Schatzberg, "*Technik* Comes to America: Changing Meanings of *Technology* before 1930," *Technology and Culture* 47, no. 3 (2006): 486–512. I am delighted to acknowledge the following participants in my seminar at McGill University in Fall 2017 for their brilliant contributions to the possibilities and prospects of elemental art history: Nina Chabelnik, Brianne Chapelle, Laurence Charlebois, Chloë Frantani, Samantha Holmes, Anya Kowalchuk, Sarah MacRae-Korobkov, and Gabby Herie Marcuzzi.

BIBLIOGRAPHY

Archival Sources

British Library: "Mayerne manuscript" (Sir Theodore de Mayerne, "Pictoria, sculptoria et quae subalternarum artium"; MS 2052)
Fitzwilliam Museum, Cambridge: Joshua Reynolds's ledger books (MS 1–2.1916)
Huntington Art Gallery: conservation/object files for collection items 10.09, 11.16, 21.02, 23.13, 23.62, 25.20, 26.85, 26.107, 44.100, 44.101, 44.108, 58.16, 78.20.34
Library of Birmingham: advertisement for the James Watt & Co. copying machine (MS 3147/20/1); Chubb Papers (MS 3782/11/12); "Directions for making Copying Ink & for Copying Drawings for which this ink is used" (MS 3147/20/10); James Watt & Co. Correspondence (MS 3147/19/4); James Watt & Co. Letter Book, April 7–December 12, 1780 (MS 3147/19/1)
Royal Academy of Arts: James Northcote Papers (MS NOR); Sir Joshua Reynolds Papers (MS REY); object file for 03/576
Scottish National Gallery: conservation/object file for NG 338
Royal Society of Arts: Guard Books (RSA/PR/GE/110/7/47; RSA/PR/GE/110/8/145; RSA/PR/GE/110/25/1); Minutes of the Society, 1754–1762 (RSA/AD/MA/100/12/01/01–RSA/AD/MA/100/12/01/07); Minutes of Various Committees, 1758–1761 (RSA/PR/GE/112/12/1–2)
Royal Society Library: election certificates (EC/1739/02, EC/1760/14, EC/1777/08)
Science Museum Library & Archives, Wroughton (SMLA): "Sun Pictures" Papers (MS 2117)
Tate Britain: conservation/object files for collection items 79, 106, 162, 305, 885, 889, 890, 891, 1840, 3343, 4505, 4694, 5546, 5635, 5640, 5750, 6244

Victoria & Albert Museum, Wedgwood Collection: Alexander Chisholm's notebook (MS 39 28405); Josiah Wedgwood's commonplace books (MS 39 28408, MS 39 28410); Josiah Wedgwood's correspondence (MS 1431); Josiah Wedgwood's experiment books (MS 26 19117, MS 36 19121); Notes and Experiments by Josiah Wedgwood and Alexander Chisholm (MS 26 19109, MS 26 19111); Thomas Wedgwood Papers (MS E40 28419, MS E40 28459, MS E40 28479, MSS E40 28483–28515)

Yale Center for British Art: Joshua Reynolds's commonplace book (MS Reynolds 33); "A Painters Journal through Flanders, Brabant, Holland and part of Germany June 24th 1781" (MS Reynolds 38)

Selected Bibliography

Abel, F. A. "Researches on Gun-Cotton: On the Manufacture and Composition of Gun-Cotton." *Philosophical Transactions of the Royal Society of London* 156 (1866): 269–308.

Agassi, Joseph. "Kant's Program." *Synthese: An International Journal for Epistemology, Methodology and Philosophy of Science* 23, no. 1 (1971): 18–23.

Albritton, Vicky, and Fredrik Albritton Jonsson. *Green Victorians: The Simple Life in John Ruskin's Lake District*. Chicago: University of Chicago Press, 2016.

André, Philipp. *Specimens of Polyautography: Consisting of Impressions taken from Original Drawings made purposely for this Work*. London: P. André and J. Heath, 1803–1807.

Archer, Mary D., and Christopher D. Haley, eds. *The 1702 Chair of Chemistry at Cambridge: Transformation and Change*. New York: Cambridge University Press, 2005.

Aristotle. *Generation of Animals*. Translated by A. L. Peck. Cambridge: Harvard University Press, 1943.

Arnold, H. J. P. *William Henry Fox Talbot: Pioneer of Photography and Man of Science*. London: Hutchinson, 1977.

Aronson, Mark N., and Angus Trumble. *Benjamin West and the Venetian Secret*. New Haven, CT: Yale Center for British Art, 2008.

Asleson, Robyn, and Shelley M Bennett. *British Paintings at the Huntington*. New Haven, CT: Huntington Library, Art Collections, and Botanical Gardens, in association with Yale University Press, 2001.

A. Z. *Remarks on Mr. Robert Dossie's Institutes of Experimental Chemistry*. London: S. Hooper, 1760.

Bachelard, Gaston. *Air and Dreams: An Essay on the Imagination of Movement*. Translated by E. R. Farrell and C. F. Farrell. Dallas: Dallas Institute Publications, 1988.

———. *Earth and Reveries of Will: An Essay on the Imagination of Matter*. Translated by K. Haltmann. Dallas: Dallas Institute Publications, 2002.

Baker, Robert. *Observations on the Pictures Now in Exhibition at the Royal Academy, Spring Gardens, and Mr. Christie's by the Author of the Remarks on the English Language*. London: Printed for John Bell, 1771.

Barker, Elizabeth E., and Alex Kidson. *Joseph Wright of Derby in Liverpool*. New Haven, CT: Yale University Press, 2007.

Barnes, Alan. "Coleridge, Tom Wedgwood and the Relationship between Time and Space in Midlands Enlightenment Thought." *Journal for Eighteenth-Century Studies* 30, no. 2 (2007): 243–60.

———. "Negative and Positive Images: Erasmus Darwin, Tom Wedgwood and the Origin of Photography." In *The Genius of Erasmus Darwin*, edited by C. U. M. Smith and Robert Arnott, 237–53. Aldershot, UK: Ashgate, 2005.

Barrell, John. *The Political Theory of Painting from Reynolds to Hazlitt: "The Body of the Public."* New Haven, CT: Yale University Press, 1986.

Barrett, Ross, and Daniel Worden, eds. *Oil Culture*. Minneapolis: University of Minnesota Press, 2014.
Barry, Fabio. "Painting in Stone: The Symbolism of Colored Marbles in the Visual Arts and Literature from Antiquity until the Enlightenment." PhD dissertation, Columbia University, 2011.
Barthes, Roland. *Camera Lucida: Reflections on Photography*. Translated by Richard Howard. New York: Hill and Wang, 1981.
———. "The World as Object." In *Calligram: Essays in New Art History from France*, edited by Norman Bryson, 106–15. New York: Cambridge University Press, 1988.
Batchen, Geoffrey. *Burning with Desire: The Conception of Photography*. Cambridge, MA: MIT Press, 1997.
———. "Tom Wedgwood and Humphry Davy: 'An Account of a Method.'" *History of Photography* 17, no. 2 (1993): 172–83.
Baudot, Laura. "An Air of History: Joseph Wright's and Robert Boyle's Air Pump Narratives." *Eighteenth-Century Studies* 46, no. 1 (2012): 1–28.
Baxandall, Michael. *The Limewood Sculptors of Renaissance Germany*. New Haven, CT: Yale University Press, 1980.
———. *Patterns of Intention: On the Historical Explanation of Pictures*. New Haven, CT: Yale University Press, 1985.
Bazin, André. "The Ontology of the Photographic Image." *Film Quarterly* 13, no. 4 (Summer 1960): 4–9.
[Beale, John]. "Two Instances of Something Remarkable in Shining Flesh, from Dr. J. Beal of Yeavel in Somersetshire." *Philosophical Transactions of the Royal Society of London* 125 (1676), 599–603.
Bear, Jordan. *Disillusioned: Victorian Photography and the Discerning Subject*. State College: Penn State University Press, 2015.
Becker, Carl L. *The Heavenly City of the Eighteenth-Century Philosophers*. New Haven, CT: Yale University Press, 1932.
Bedini, Silvio A. *Thomas Jefferson and His Copying Machines*. Charlottesville: University Press of Virginia, 1984.
Beer, John. "Coleridge, Mackintosh and the Wedgwoods: A Reassessment, Including Some Unpublished Records." *Romanticism* 7, no. 1 (2001): 16–40.
Benjamin, Walter. *The Work of Art in the Age of Its Technological Reproducibility, and Other Writings on Media*, edited by Michael W. Jennings, Brigid Doherty, and Thomas Y. Levin. Cambridge, Massachusetts: Harvard University Press, 2008.
Bennett, Jim. "Hooke's Instruments." In *London's Leonardo: The Life and Work of Robert Hooke*, edited by Jim Bennett et al., 63–104. Oxford: Oxford University Press, 2003.
———. "The Mechanics' Philosophy and the Mechanical Philosophy." *History of Science* 24 (1986): 1–28.
Beretta, Marco. *Imaging a Career in Science: The Iconography of Antoine Laurent Lavoisier*. Canton, MA: Science History Publications, 2001.
Berg, Maxine. "New Commodities, Luxuries and their Consumers in Eighteenth-Century England." In *Consumers and Luxury: Consumer Culture in Europe 1650–1850*, edited by Maxine Berg and Helen Clifford, 63–85. New York: Manchester University Press, 1999.
Bermingham, Ann. *Learning to Draw: Studies in the Cultural History of a Polite and Useful Art*. New Haven, CT: Paul Mellon Centre for Studies in British Art/Yale University Press, 2000.
———. *Sensation and Sensibility: Viewing Gainsborough's Cottage Door*. New Haven, CT: Yale Center for British Art, 2005.
———, ed. "Technologies of Illusion: The Art of Special Effects in Eighteenth-Century Britain." Special issue. *Huntington Library Quarterly* 70, no. 2 (2007): 203–310.

Bignamini, Ilaria. "George Vertue, Art Historian and Art Institutions in London, 1689–1768: A Study of Clubs and Academies." *Volume of the Walpole Society* 54 (1988): 1–148.

Birch, Thomas. *History of the Royal-Society of London, for Improving of Natural Knowledge*. 4 vols. London: A. Millar, 1756.

Bjelajac, David. *Washington Allston, Secret Societies and the Alchemy of Anglo-American Painting*. Cambridge University Press: Cambridge, 1997.

Boantza, Victor D. "Light in the Pneumatic Context: Aspects of Interplay between Theory and Practice in Early Photochemical Research." *Historia Scientiarum* 16 (2006): 105–27.

———. *Matter and Method in the Long Chemical Revolution: Laws of Another Order*. Farnham, Surrey: Ashgate, 2013.

Bolton, M. P. W. *Inquisitio Philosophica: An Examination of the Principles of Kant and Hamilton*. London: Chapman and Hall, 1866.

[Booth, Joseph]. *An Address to the Public on the Polygraphic Art*. London: J. Walter, 1788 (?).

———. *A Treatise Explanatory of the Nature and Properties of Pollaplasiasmos; or, the Original Invention of Multiplying Pictures in Oil Colours*. London: J. Rozea, 1784.

Borg, Alan. *The History of the Worshipful Company of Painters, Otherwise Painter-Stainers*. Huddersfield: Jeremy Mills, 2005.

Boswell, James. *The Life of Samuel Johnson, LL.D.* 2 vols. London, 1791.

Boulton, Matthew P. W. *On Aërial Locomotion*. London: Bradbury & Evans, 1864.

———. *Remarks Concerning Certain Photographs Supposed to Be of Early Date*. London: Bradbury & Evans, 1864.

———. *Remarks Concerning Certain Pictures Supposed to Be Photographs of Early Date*. London: Bradbury & Evans, 1865.

———. *Remarks on Some Evidence Recently Communicated to the Photographic Society*. London: Bradbury & Evans, 1863.

Bowers, Brian. *A History of Electric Light & Power*. Stevenage, Hertfordshire: P. Peregrinus in association with the Science Museum, 1982.

Boyle, Robert. *The Aerial Noctiluca*. In *The Works of Robert Boyle*. Vol. 9, *Publications of 1678–83*. Edited by Michael Hunter and Edward B. Davis, 266–305. London: Pickering & Chatto, 2000.

———. *Experiments and Considerations Touching Colours*. In *The Works of Robert Boyle*. Vol. 4, *Colours and Cold, 1664–5*. Edited by Michael Hunter and Edward B. Davis. London: Pickering & Chatto, 1999.

———. "A Paper of the Honourable Robert Boyl's despoited with the Secretaries of the Royal Society, Octob. 14, 1680. and opened since his Death; being an Account of his making the Phosphorous, &c." *Philosophical Transactions* 196 (1693): 583–84.

———. "A Physico-Chymical Essay, Containing An Experiment with some Considerations touching the differing Parts and Redintegration of Salt-Petre." In *The Works of Robert Boyle*. Vol. 2, *The Sceptical Chymist and Other Publications of 1661*. Edited by Michael Hunter and Edward B. Davis, 93–114. London: Pickering & Chatto, 1999.

———. "A Short Memorial of some Observations made upon an Artificial Substance, that shines without any precedent Illustration." In *Lectures and Collections made by Robert Hooke, Secretary of the Royal Society*. London: J. Martyn, 1678.

———. "Some Observations about Shining Flesh, made by the Honourable Robert Boyle . . ." *Philosophical Transactions*, no. 89 (1672), 5108–16.

Braddock, Alan C., and Christoph Irmscher, eds. *A Keener Perception: Ecological Studies in American Art History*. Tuscaloosa: University of Alabama Press, 2009.

Brewster, David. "Photogenic Drawing; or, Drawing by the Agency of Light." *Edinburgh Review* 86, no. 154 (1843): 309–44.

Brotton, Jerry. *The Sale of the Late King's Goods: Charles I and His Art Collection*. London: Macmillan, 2006.
Brown, Bill, ed. *Things*. Chicago: University of Chicago Press, 2004.
Brusius, Mirjam, Katrina Dean, and Chitra Ramalingam, eds. *William Henry Fox Talbot: Beyond Photography*. New Haven, CT: Yale Center for British Art and Paul Mellon Centre for Studies in British Art, 2013.
Bucklow, Spike. *The Alchemy of Paint: Art, Science, and Secrets from the Middle Ages*. London: Marion Boyars, 2009.
Burke, Edmund. *Reflections on the Revolution in France*. Edited by Frank M. Turner. New Haven, CT: Yale University Press, 2003.
Burton, Anthony. *Vision and Accident: The Story of the Victoria and Albert Museum*. London: V&A Publications, 1999.
Butlin, Martin. "An Eighteenth-Century Art Scandal: Nathaniel Hone's 'The Conjuror.'" *Connoisseur* 174, no. 699 (1970): 1–9.
Cardwell, D. S. L. *From Watt to Clausius: The Rise of Thermodynamics in the Early Industrial Age*. London: Heinemann, 1971.
Carlyle, Leslie. "The Artist's Anticipation of Change as Discussed in British Nineteenth Century Instruction Books on Oil Painting." In *Appearance, Opinion, Change: Evaluating the Look of Paintings*, 62–67. United Kingdom Institute for Conservation, 1990.
A Catalogue of the Pictures, Sculptures, Models, Designs in Architecture, Drawings, Prints, &c. Exhibited at the Great Room, in Spring-Gardens, Charing-Cross, April the Twenty-Sixth, 1771, by the Royal Incorporated Society of Artists of Great-Britain. London: Printed for the Society, 1771.
A Catalogue of the Pictures, Sculptures, Models, Designs in Architecture, Drawings, Prints, &c. Exhibited by the Society of Artists of Great Britain, at the New Room, Near Exeter-Exchange, Strand, April the Twenty-Ninth, 1773, the Fourteenth Year of Exhibiting. London: Printed by Harriot Bunce, printer to the Society, 1773.
A Catalogue of the Valuable Collection of Paintings by Ancient and Modern Masters of Sir Thomas Lawrence. London: Catalogues had at Mr. Christie's Office, 1830.
Caylus, Comte de (Anne Claude Philippe de Tubières-Grimoard de Pestels de Lévis), and Michel-Joseph Majault. *Mémoire sur la peinture à l'encaustique et sur la peinture à la cire*. Geneva: Pissot libraire, à la Croix d'or, quai de Conti, 1755.
Chakrabarty, Dipesh. "The Climate of History: Four Theses." *Critical Inquiry* 35, no. 2 (2009): 197–222.
Chang, Ku-Ming. "Fermentation, Phlogiston and Matter Theory: Chemistry and Natural Philosophy in Georg Ernst Stahl's *Zymotechnia Fundamentalis*." *Early Science and Medicine* 7, no. 1 (2002): 31–64.
Chu, John. "High Art and High Stakes: The 3rd Duke of Dorset's Gamble on Reynolds." *British Art Studies* 2 (2016). http://dx.doi.org/10.17658/issn.2058-5462/issue-02/jchu.
Clayton, Timothy. *The English Print 1688–1802*. New Haven, CT: Paul Mellon Centre for Studies in British Art/Yale University Press, 1997.
Clemit, Pamela. "Commerce of Luminaries: Eight Holograph Letters between William Godwin and Thomas Wedgwood." In *Godwinian Moments: From the Enlightenment to Romanticism*, edited by Robert M. Maniquis and Victoria Myers, 261–82. Toronto: Toronto University Press, 2011.
———. "William Godwin and James Watt's Copying Machine: Wet-Transfer Copies in the Abinger Papers." *Bodleian Library Record* 18, no. 5 (April 2005): 532–60.
Clericuzio, Antonio. "Carneades and the Chemists: A Study of *The Sceptical Chymist* and its Impact on Seventeenth-Century Chemistry." In *Robert Boyle Reconsidered*, edited by Michael C. W. Hunter, 79–90. Cambridge: Cambridge University Press, 1994.
The Clockmakers Outcry against the Author of the Life and Opinions of Tristram Shandy:

Dedicated to the Most Humble of Christian Prelates. London: Printed for J. Burd, near the Temple Gate, Fleet-Street, 1760.
Clow, Archibald, and Nan L. Clow. "Vitriol in the Industrial Revolution." *Economic History Review* 15, nos. 1–2 (1945): 44–55.
Cohen, Jeffrey Jerome, and Lowell Duckert, eds. *Elemental Ecocriticism: Thinking with Earth, Air, Water, and Fire*. Minneapolis: University of Minnesota Press, 2015.
Cole, Michael. *Cellini and the Principles of Sculpture*. Cambridge: Cambridge University Press, 2002.
Cole, William. "A Letter from Mr. William Cole of Bristol, to the Philosophical Society of Oxford, containing his Observations on the Purple Fish." *Philosophical Transactions* 15, no. 178 (December 1685): 1278–86.
[Combe, William]. *A Poetical Epistle to Sir Joshua Reynolds, Knt*. London: Fielding and Walker, 1777.
Cooper, Anthony Ashley. *An Essay on Painting: Being A Notion of the Historical Draught or Tablature of the Judgment of Hercules*. London: John Darby, 1714.
Cooper, M. A. R. "Robert Hooke (1635–1703): Proto-Photogrammetrist." *Photogrammetric Record* 15, no. 87 (1996): 403–17.
Cooperman, Emily T., and Lea Carson Sherk. *William Birch: Picturing the American Scene*. Philadelphia: University of Pennsylvania Press, 2011.
Cormack, Malcolm. "The Ledgers of Sir Joshua Reynolds." *Walpole Society* 42 (1968–1970): 105–69.
Costelloe, Timothy M. *The British Aesthetic Tradition: From Shaftesbury to Wittgenstein*. New York: Cambridge University Press, 2013.
Cotton, William, and John Burnet, eds. *Sir Joshua Reynolds, and His Works: Gleanings from His Diary, Unpublished Manuscripts, and from Other Sources*. London: Longman, Brown, Green, Longmans, and Roberts, 1856.
Crosland, Maurice. "The Image of Science as a Threat: Burke versus Priestley and the 'Philosophic Revolution'." *British Journal for the History of Science* 20, no. 3 (1987): 277–307.
Crow, Thomas E. *Emulation: Making Artists for Revolutionary France*. New Haven, CT: Yale University Press, 1995.
Crutzen, Paul J. "Geology of Mankind." *Nature* 415 (January 3, 2002): 23.
Crutzen, Paul J., and Will Steffen. "How Long Have We Been in the Anthropocene Era? An Editorial Comment." *Climate Change* 61 (2003): 251–57.
Cunningham, Allan. *The Lives of the Most Eminent British Painters, Sculptors, and Architects*. London: J. Murray, 1830.
Darwin, Erasmus. *The Botanic Garden, a Poem in Two Parts*. 2nd ed. New York: T. & J. Swords, 1807.
———. *The Essential Writings of Erasmus Darwin*. Edited by Desmond King-Hele. London: MacGibbon & Kee, 1968.
———. *Zoonomia; or The Laws of Organic Life*. 2 vols. London: J. Johnson, 1796.
Daston, Lorraine, and Peter Galison. *Objectivity*. New York: Zone Books, 2007.
Davis, Lucy, and Mark Hallett, eds. *Joshua Reynolds: Experiments in Paint*. London: Wallace Collection/Paul Holberton Publishing, 2015.
Davy, Humphry. "An Account of a Method of Copying Paintings upon Glass, and of Making Profiles, by the Agency of Light upon Nitrate of Silver. Invented by T. Wedgwood, Esq. with Observations by H. Davy." *Journals of the Royal Institution of Great Britain* 1 (1802): 170–74.
Davy, John, ed. *The Collected Works of Sir Humphry Davy*. 9 vols. London: Smith, Elder and Co., 1839.
De Duve, Thierry. "The Readymade and the Tube of Paint." In *Kant after Duchamp*, 147–96. Cambridge, MA: MIT Press, 1996.

Degalan, Aimee Marcereau. "Dangerous Beauty: Painted Canvases and Painted Faces in Eighteenth-Century Britain." PhD dissertation, Case Western Reserve University, 2007.

De Marchi, Neil, and Hans J. Van Miegroet. "Ingenuity, Preference, and the Pricing of Pictures: The Smith-Reynolds Connection." In *Economic Engagements with Art*, edited by Neil De Marchi and Craufurd Goodwin, 379–412. Durham, NC: Duke University Press, 1999.

Demos, T. J. *Against the Anthropocene: Visual Culture and the Environment Today*. Berlin: Sternberg, 2017.

Dempsey, Charles. *Inventing the Renaissance Putto*. Chapel Hill: University of North Carolina Press, 2001.

De Voogd, Peter J. "Laurence Sterne, the Marbled Page, and the 'Use of Accidents.'" *Word & Image* 1, no. 3 (1985): 279–87.

Diamond, Hugh. "Photography and Photographic Apparatus." In *The Record of the International Exhibition*. Glasgow: William Mackenzie, 1862.

[Diderot, Denis]. *L'histoire et le secret de la peinture en cire*. Paris: 1755.

Didi-Huberman, Georges. *Devant le temps: Histoire de l'art et anachronism des images*. Paris: Editions de Minuit, 2005.

Dobbs, Betty J. T. *The Foundations of Newton's Alchemy; or, "The Hunting of the Greene Lyon."* Cambridge: Cambridge University Press, 1975.

Doherty, Meghan C. "Discovering the 'True Form': Hooke's *Micrographia* and the Visual Vocabulary of Engraved Portraits." *Notes and Records of the Royal Society* 66, no. 3 (2012): 211–34.

Donald, Diana, and Jane Munro, eds. *Endless Forms: Charles Darwin, Natural Science and the Visual Arts*. Cambridge: Fitzwilliam Museum, 2009.

Donovan, Arthur L. *Philosophical Chemistry in the Scottish Enlightenment: The Doctrines and Discoveries of William Cullen and Joseph Black*. Edinburgh: Edinburgh University Press, 1975.

Dossie, Robert. *The Handmaid to the Arts*. London: Printed for J. Nourse, 1764.

Doucet, Emily. "Anticipating Machines Heavier than Air: Nadar, Photography and the Objects of Technology." MA thesis, University College London, 2013.

Draaisma, Douwe. *Metaphors of Memory: A History of Ideas about the Mind*. Translated by Paul Vincent. New York: Cambridge University Press, 2000.

Drake, Ellen Tan. *Restless Genius: Robert Hooke and His Earthly Thoughts*. New York: Oxford University Press, 1996.

Dubin, Nina. *Futures and Ruins: Eighteenth-Century Paris and the Art of Hubert Robert*. Los Angeles: Getty Research Institute, 2010.

Dubois, Hélène. "'Use a little wax with your colours, but don't tell anybody': Joshua Reynolds's Painting Experiments with Wax and His Sources." *Hamilton Kerr Institute Bulletin* 3 (2000): 97–106.

Durey, Michael. *Transatlantic Radicals and the Early American Republic*. Lawrence: University of Kansas Press, 1997.

Eamon, William. "Robert Boyle and the Discovery of Colour Indicators." *Ambix* 27, no. 3 (1980): 204–9.

Eastlake, Charles Lock. *Materials for a History of Oil Painting*. London: Longman, Brown, Green & Longmans, 1847.

[Eastlake, Elizabeth]. "Photography." *Quarterly Review* 101 (April 1857): 443–68.

Eaton, Linda. *Printed Textiles: British and American Cottons and Linens 1700–1850*. New York: Monacelli, 2014.

Eder, Josef Maria. *History of Photography*. Translated by Edward Epstean. New York: Columbia University Press, 1945.

Edwards, Steve. *The Making of English Photography: Allegories*. University Park: Pennsylvania State University Press, 2006.
Egerton, Judy, ed. *Wright of Derby*. New York: Metropolitan Museum of Art, 1990.
Eigen, Edward. "On Purple and the Genesis of Photography or the Natural History of an Exposure." In *Ocean Flowers: Impressions from Nature*, edited by Carol Armstrong and M. Catherine Zegher, 271–87. New York: Drawing Center, 2004.
Elkins, James. *What Painting Is: How to Think about Oil Painting Using the Language of Alchemy*. New York: Routledge, 1999.
———, ed. *Photography Theory*. New York: Routledge, 2007.
Ellenbogen, Josh. *Reasoned and Unreasoned Images: The Photography of Bertillon, Galton, and Marey*. University Park: Pennsylvania State University Press, 2012.
Elsner, Jas'. "From Empirical Evidence to the Big Picture: Some Reflections on Riegl's Concept of *Kunstwollen*." *Critical Inquiry* 32, no. 4 (2006): 741–66.
Engels, Frederick. *The Condition of the Working-Class in England*. Translated by Florence Kelley Wischnewetzky. London: Swan Sonnenschein & Co., 1892.
Erffa, Helmut Von, and Allen Staley. *The Paintings of Benjamin West*. New Haven, CT: Yale University Press, 1986.
Erickson, Paul, Judy L. Klein, Lorraine Daston, Rebecca Lemov, Thomas Sturm, and Michael D. Gordin. *How Reason Almost Lost Its Mind: The Strange Career of Cold War Rationality*. Chicago: University of Chicago Press, 2013.
Evelyn, John, *Sculptura; or, The History, and Art of Chalcography and Engraving in Copper*. London: J. C. for T. Beedle, 1662.
Fairclough, Mary. *Literature, Electricity and Politics 1740–1840: "Electrick Communication Every Where."* Palgrave Studies in Literature, Science, and Medicine. London: Palgrave Macmillan, 2017.
Farington, Joseph. *The Diary of Joseph Farington*. Vol. 3, *September 1796–December 1798*. Edited by Kenneth Garlick and Angus Macintyre. New Haven, CT: Paul Mellon Centre for Studies in British Art/Yale University Press, 1979.
———. *Memoirs of the Life of Sir Joshua Reynolds with Observations on his Talents and Character*. London: Strahan and Spottiswoode, for T. Cadell & W. Davies, 1819.
Fend, Mechthild. *Fleshing Out Surfaces: Skin in French Art and Medicine, 1650–1850*. Manchester: Manchester University Press, 2017.
Field, George. "Colourless Lac Varnish." *Transactions of the Society, Instituted at London, for the Encouragement of Arts, Manufactures, and Commerce* 45 (1827): 59–67.
Fincham, John. *A History of Naval Architecture*. London: Whittaker and Co., 1851.
Fisher, Thomas. "The Process of Polyautographic Printing." *Gentleman's Magazine* 78, no. 1 (1808): 193–96.
Flint, Stamford R. *Mudge Memoirs: Being a Record of Zachariah Mudge, and Some Members of His Family*. Truro: Netheron and Worth, 1883.
Fogarty, Barbara. "Matthew Boulton and Francis Eginton's Mechanical Paintings: Production and Consumption 1777 to 1781." MPhil thesis, University of Birmingham, 2010.
———. "The Mechanical Paintings of Matthew Boulton and Francis Eginton." In *Matthew Boulton: Enterprising Industrialist of the Enlightenment*, edited by K. Quickenden, S. Baggott, and M. Dick, 111–26. Farnham: Ashgate, 2013.
Fordham, Douglas. *British Art and the Seven Years' War: Allegiance and Autonomy*. Philadelphia: University of Pennsylvania Press, 2010.
Forty, Adrian. *Objects of Desire*. New York: Pantheon, 1986.
Foucault, Michel. *The Birth of Biopolitics: Lectures at the Collège de France, 1978–79*, edited by Michel Senellart; translated by Graham Burchell. Basingstoke: Palgrave Macmillan, 2008.

Fouque, Victor. *La vérité sur l'invention de la Photographie: Nicéphore Niépce*. Paris: Librairie des auteurs et de l'Académie des bibliophiles, 1867.

Fowles, John. *A Maggot*. London: Jonathan Cape, 1985.

Fox, Celina. *The Arts of Industry in the Age of Enlightenment*. New Haven, CT: Paul Mellon Centre for Studies in British Art/Yale University Press, 2009.

Foy, Anna M. "The Convention of Georgic Circumlocution and the Proper Use of Human Dung in Samuel Martin's *Essay upon Plantership*." *Eighteenth-Century Studies* 49, no. 4 (2016): 475–506.

Frank, Robert G. *Harvey and the Oxford Physiologists: Scientific Ideas and Social Interaction*. Berkeley: University of California Press, 1980.

Fried, Michael. *Absorption and Theatricality: Painting and Beholder in the Age of Diderot*. Chicago: University of Chicago Press, 1980.

———. "Toward a Supreme Fiction: Genre and Beholder in the Art Criticism of Diderot and His Contemporaries." *New Literary History* 6, no. 3 (1975): 543–85.

Gaehtgens, Thomas W., and Jacques Lugand. *Joseph-Marie Vien: Peintre du Roi (1716–1809)*. Paris: Arthena, 1988.

Gage, John. "Magilphs and Mysteries." *Apollo* 80 (1964): 38–41.

Galassi, Peter. *Before Photography: Painting and the Invention of Photography*. New York: Museum of Modern Art, 1981.

Galt, John. *The Life, Studies, and Works of Benjamin West, Esq.* 2 vols. London: T. Cadell and W. Davies, 1820.

Gaskell, Peter. *The Manufacturing Population of England: Its Moral, Social, and Physical Conditions, and the Changes Which Have Arisen from the Use of Steam Machinery; with an Examination of Infant Labour*. London: Baldwin and Cradock, 1833.

Gautier D'Agoty, Jacques Fabien. "Découvertes sur la generation, contre les Oviparistes & les Vermiculistes." In *Observations sur l'histoire naturelle, sur la physique et sur la peinture*, 1:9–29. Paris: Chez Delaguette, 1752.

Gell, Alfred. *The Art of Anthropology: Essays and Diagrams*. Edited by Eric Hirch. New York: Berg, 1999.

Gerard, Alexander. *An Essay on Genius*. London: W. Strahan, T. Cadell, 1774.

———. *An Essay on Taste: By Alexander Gerard, D.D. Professor of Divinity in the Marischal College of Aberdeen. The Second Edition, with Corrections and Additions. to Which Are Annexed, Three Dissertations on the Same Subject, by Mr De Voltaire, Mr D'alembert, and Mr De Montesquieu*. Edinburgh: Printed for A. Millar, London, 1764.

Gernsheim, Helmut. *The Origins of Photography*. New York: Thames & Hudson, 1982.

Gernsheim, Helmut, and Alison Gernsheim. *The History of Photography from the Earliest Use of the Camera Obscura in the Eleventh Century up to 1954*. New York: Oxford University Press, 1955.

G. G. L. "Historia Inventionis Phosphori." In *Miscellanea Berolinensia ad Incrementum Scientarum*, 91–98. Berlin: Johan. Christ. Papenii, 1710.

Gibbs, F. W. "A Notebook of William Lewis and Alexander Chisholm." *Annals of Science* 8, no. 3 (1952): 202–20.

———. "Robert Dossie (1717–1777) and the Society of Arts." *Annals of Science* 7, no. 2 (1951): 149–72.

———. "William Lewis, M.B., F.R.S. (1708–1781)." *Annals of Science* 8, no. 2 (1952): 122–51.

Gibbs-Smith, Charles H. *The Aeroplane: An Historical Survey of Its Origins and Development*. London: HM Stationery Office, 1960.

Gibson-Wood, Carol. *Jonathan Richardson: Art Theorist of the English Enlightenment*. New Haven, CT: Paul Mellon Centre for Studies in British Art/Yale University Press, 2000.

———. "Picture Consumption in London at the End of the Seventeenth Century." *Art Bulletin* 84, no. 3 (2002): 491–500.

Glennie, Paul, and Nigel J. Thrift. *Shaping the Day: A History of Timekeeping in England and Wales 1300–1800*. Oxford: Oxford University Press, 2009.

Godwin, William. *An Enquiry Concerning Political Justice, and Its Influence on General Virtue and Happiness*. 2 vols. London: G. G. J. and J. Robinson, 1793.

———. *The Letters of William Godwin*. Vol. 1, *1778–1797*. Edited by Pamela Clemit. New York: Oxford University Press, 2011.

Golinski, Jan. "A Noble Spectacle: Phosphorus and the Public Culture of Science in the Early Royal Society." *Isis* 80 (1989): 11–39.

———. *Science as Public Culture: Chemistry and Enlightenment in Britain, 1760–1820*. New York: Cambridge University Press, 1992.

Gombrich, E. H. "Reynolds's Theory and Practice of Imitation." *Burlington Magazine* 80, no. 467 (1942): 40–45.

Grew, Nehemiah. *The Anatomy of Plants; With an Idea of a Philosophical History of Plants, and Several Other Lectures, Read before the Royal Society*. London: Printed by W. Rawlins, 1682.

———. *Musaeum Regalis Societatis; or, A Catalogue & Description of the Natural and Artificial Rarities belonging to the Royal Society and preserved at Gresham Colledge*. London: W. Rawlins, 1681.

Griffiths, Antony. "Notes on Early Aquatint in England and France." *Print Quarterly* 4, no. 3 (1987): 255–70.

Gruman, Gerald J. "A History of Ideas about the Prolongation of Life: The Evolution of Prolongevity Hypotheses to 1800." *Transactions of the American Philosophical Society* 56, no. 9 (1966): 1–102.

Guerlac, Henry. "Joseph Black's Work on Heat." In *Joseph Black 1728–1799: A Commemorative Symposium*, edited by A. D. C. Simpson, 13–22. Edinburgh: Royal Scottish Museum, 1982.

Gunther, R. T., ed. *Early Science at Oxford*. Vol. 12, *Dr. Plot and the Correspondence of the Philosophical Society of Oxford*. Oxford, 1939.

Hackney, Stephen, Rica Jones, and Joyce Townsend, eds. *Paint and Purpose: A Study of Technique in British Art*. London: Tate Gallery, 1999.

Hallett, Mark. "Reynolds, Celebrity, and the Exhibition Space." In *Joshua Reynolds: The Creation of Celebrity*, edited by Martin Postle, 34–47. London: Tate, 2005.

———. *Reynolds: Portraiture in Action*. New Haven, CT: Paul Mellon Centre for Studies in British Art/Yale University Press, 2014.

Hamber, Anthony J. *"A Higher Branch of the Arts": Photographing the Fine Arts in England, 1839–1880*. Amsterdam: Gordon and Breach, 1996.

Hanson, Craig Ashley. *The English Virtuoso: Art, Medicine, and Antiquarianism in the Age of Empiricism*. Chicago: University of Chicago Press, 2009.

Hardenberg, Horst O. *The Middle Ages of the Internal-Combustion Engine, 1794–1886*. Warrendale, PA: Society of Automotive Engineers, 1999.

———. *The Niepce Brothers' Boat Engines*. Warrendale: Society of Automotive Engineers, 1993.

Hargraves, Matthew. *"Candidates for Fame": The Society of Artists of Great Britain, 1760–1791*. New Haven, CT: Yale University Press, 2005.

Harris, Theresa Fairbanks, and Scott Wilcox. *Papermaking and the Art of Watercolor in Eighteenth-Century Britain: Paul Sandby and the Whatman Paper Mill*. New Haven, CT: Yale University Press, 2006.

Harrison, James. "Erasmus Darwin's View of Evolution." *Journal of the History of Ideas* 32, no. 2 (1971): 247–64.

Harvey, E. Newton. *A History of Luminescence from the Earliest Times until 1900*. Philadelphia: American Philosophical Society, 1957.
Haydon, Benjamin Robert. *The Diary of Benjamin Robert Haydon*. Edited by Willard Bissell Pope. 5 vols. Cambridge: Harvard University Press, 1963.
———. *Lectures on Painting and Design*. 2 vols. London: Longman, Brown, Green and Longmans, 1844.
Henderson, Felicity, Sachiko Kusukawa, and Alexander Marr, eds. "Curiously Drawn: Early Modern Science as a Visual Pursuit." Special issue. *Huntington Library Quarterly* 78, no. 2 (2015): 141–415.
Hilles, Frederick W. *The Literary Career of Sir Joshua Reynolds*. Cambridge: Cambridge University Press, 1936.
Hills, Richard L. *James Watt*. Vol. 2, *The Years of Toil, 1775–1785*. Ashbourne: Landmark, 2005.
———. *James Watt*. Vol. 3, *Triumph through Adversity, 1785–1819*. Ashbourne: Landmark, 2006.
———. "James Watt and His Copying Machine." *The Oxford Papers: Proceedings of the British Association of Paper Historians Fourth Annual Conference*, ed. Peter Bower, *The Papers of the Bibliographical Society of America*, 92, no. 1 (1998): 81–88.
Hogarth, William. *The Analysis of Beauty*. Edited by Ronald Paulson. New Haven, CT: Yale University Press, 1997.
Hoock, Holger. *The King's Artists: The Royal Academy of Arts and the Politics of British Culture 1760–1840*. New York: Oxford University Press, 2003.
Hooke, Robert. *Lectures and Collections made by Robert Hooke, Secretary of the Royal Society*. London: J. Martyn, 1678.
———. *Micrographia; or, Some Physiological Descriptions of Minute Bodies made by Magnifying Glasses*. London: John Martyn and James Allestry, 1665.
———. *Philosophical Experiments and Observations of the Late Eminent Dr. Robert Hooke, S.R.S. and Geom. Prof. Gresh., and Other Eminent Virtuoso's in His Time*. Edited by William Derham. London: Printed by W. and J. Innys, 1726.
———. *The Posthumous Works of Robert Hooke*. Edited by Richard Waller. London: Sam Smith and Benjamin Walford, 1705.
[Hooke, Robert]. "Review of Marc Antonio Cellio's *Il Fosforo overo la Pietra Bolognese* . . . (Roma: 1680)." In *Philosophical Collections*, no. 3 (London: Printed for Richard Chiswel, Printer to the RS, at the Rose and Crown in St. Paul's Church yard, December 10, 1681): 77–79.
Hopkinson, Martin. "Burdett, Wedgwood and Bentley." *Print Quarterly* 25, no. 2 (June 2008): 132–46.
Hudson, Derek. *Sir Joshua Reynolds: A Personal Study*. London: G. Bles, 1958.
Hudson, Derek, and Kenneth W. Luckhurst. *The Royal Society of Arts, 1754–1954*. London: Murray, 1954.
Hunt, Lynn A., and Margaret C. Jacob. "The Affective Revolution in 1790s Britain." *Eighteenth-Century Studies* 34, no. 4 (2001): 491–521.
Hunt, Robert. *Researches on Light in Its Chemical Relations Embracing a Consideration of all the Photographic Processes*. London: Longman, Brown, Green, and Longmans, 1854.
Hunter, Matthew C. "The Cunning of Sir Sloshua: Reynolds, the Sea, and Risk." *Grey Room* 69 (2018): 80–107.
———. "Mr. Hooke's Reflecting Box: Modeling the Camera Obscura in the Early Royal Society of London." *Huntington Library Quarterly* 78, no. 2 (2015): 301–28.
———. *Wicked Intelligence: Visual Art and the Science of Experiment in Restoration London*. Chicago: University of Chicago Press, 2013.

Hunter, Matthew C., and Francesco Lucchini, eds. *The Clever Object*. Hoboken: John Wiley & Sons, 2013.

Hunter, Michael. *Boyle: Between God and Science*. New Haven, CT: Yale University Press, 2009.

———, ed. *Printed Images in Early Modern Britain: Essays in Interpretation*. Farnham: Ashgate, 2010.

Huxham, John. *An Essay on Fevers to Which Is Now Added, A Dissertation on the Malignant, Ulcerous Sore-Throat*. 5th ed. London: J. Hinton, 1767.

Jacobson, David M. "Corinthian Bronze and the Gold of the Alchemists," *Gold Bulletin* 33, no. 2 (2000): 60–66.

Jacobson, David M., and M. P. Weitzman. "What Was Corinthian Bronze?" *American Journal of Archaeology* 96, no. 2 (1992): 237–47.

James Watt & Co. *Directions for Using the Patent Portable Copying Machines. Invented & Made by James Watt and Company of Soho, near Birmingham*. Birmingham, n.d.

Janson, H. W. "The 'Image Made by Chance' in Renaissance Thought." In *16 Studies*, 55–69. New York: Harry N. Abrams, 1974.

Jay, Mike. *The Atmosphere of Heaven: The Unnatural Experiments of Dr. Beddoes and His Sons of Genius*. New Haven, CT: Yale University Press, 2009.

Jefferson, Thomas. *The Papers of Thomas Jefferson*. Edited by Julian P. Boyd, 27 vols. Princeton, NJ: Princeton University Press, 1950.

Jenkins, Ian, and Kim Sloan. *Vases and Volcanoes: Sir William Hamilton and His Collection*. London: British Museum Press, 1996.

Jewitt, Llewellynn F. W. *The Wedgwoods: Being a Life of Josiah Wedgwood; with Notices of His Works and Their Productions, Memoirs of the Wedgewood and Other Families, and a History of the Early Potteries of Staffordshire*. London: Virtue Brothers and Co., 1865.

Johns, Adrian. "Ink." In *Materials and Expertise in Early Modern Europe: Between Market and Laboratory*, edited by Ursula Klein and E. C. Spary, 101–24. Chicago: University of Chicago Press, 2010.

[Johnson, Elizabeth]. *The Geography and Astronomy of the Created World, and of Course the Longitude: Being the Fourth Book by the Author of the Explanation of the Vision to Ezekiel*. Oxford: Printed for the author, 1785.

Johnson, Samuel. *A Dictionary of the English Language . . .* 2nd ed. 2 vols. London: J. and P. Knapton et al., 1755.

———. "Life of Boerhaave." *Gentleman's Magazine* 9 (1739): 37–38, 72–73, 114–16, 172–76.

———. *The Works of Samuel Johnson*. 16 vols. New York: Lamb Publishing Company, 1903.

Johnston, Kenneth R. *Unusual Suspects: Pitt's Reign of Alarm and the Lost Generation of the 1790s*. Oxford: Oxford University Press, 2013.

Jones, Peter M. "'I had L[o]rds and Ladys to wait on yesterday . . .': Visitors to the Soho Manufactory." In *Matthew Boulton: Selling What All the World Desires*, edited by Shena Mason, 71–79. Birmingham: Birmingham City Council, 2009.

———. *Industrial Enlightenment: Science, Technology and Culture in Birmingham and the West Midlands, 1760–1820*. Manchester: Manchester University Press, 2008.

———. "Living the Enlightenment and the French Revolution: James Watt, Matthew Boulton, and Their Sons." *Historical Journal* 42, no. 1 (1999): 157–82.

Jones, Rica. "Painting a Face All Red at the First Sitting: Ramsay's Technique for Portraits, 1735–1760." In *Allan Ramsey: Portraits of the Enlightenment*, edited by Mungo Campbell et al., 111–15. New York: Prestel, 2003.

———. "William Hogarth, *Mrs. Salter*." In *Paint and Purpose: A Study of Technique in British Art*, edited by Stephen Hackney, Rica Jones, and Joyce Townsend, 42–47. London: Tate Gallery, 1999.

Kant, Immanuel. *Critique of Judgment*. Translated by W. Pluhar. Indianapolis: Hackett, 1987.
———. *Critique of Pure Reason*. Translated by Norman Kemp Smith. Basingstoke: Palgrave Macmillan, 2007.
———. *Kritik Der Urteilskraft*. 5th ed. Leipzig: Felix Meiner, 1920.
Keller, Vera. "Scarlet Letters: Sir Theodore de Mayerne and the Early Stuart Color World in the Royal Society." In *Archival Afterlives: Life, Death, and Knowledge-Making in Early Modern British Scientific and Medical Archives*, edited by Vera Keller, Anna Marie Roos, and Elizabeth Yale, 72–119. Scientific and Learned Cultures and Their Institutions, vol. 23. Leiden: Brill, 2018.
Kelsey, Robin. *Photography and the Art of Chance*. Cambridge, MA: Belknap Press/Harvard University Press, 2015.
Kemp, Martin. "True to Their Natures: Sir Joshua Reynolds and Dr. William Hunter at the Royal Academy of Arts." *Notes and Records of the Royal Society of London* 46, no. 1 (1992): 77–88.
Keymer, Thomas. *Sterne, Moderns, and the Novel*. New York: Oxford University Press, 2002.
Kim, Mi Gyung. *Affinity, That Elusive Dream: A Genealogy of the Chemical Revolution*. Cambridge, MA: MIT Press, 2003.
Kinzer, Bruce. "Flying Under the Radar: The Strange Case of Matthew Piers Watt Boulton." *Times Literary Supplement*, May 1, 2009, 14–15.
Kitson, Michael. "Hogarth's 'Apology for Painters.'" *Walpole Society* 41 (1966–1968): 46–111.
Klein, Ursula, and Wolfgang Lefebvre. *Materials in Eighteenth-Century Science: A Historical Ontology*. Cambridge, MA: MIT Press, 2007.
Koselleck, Reinhart. *Futures Past: On the Semantics of Historical Time*. New York: Columbia University Press, 2004.
Krauss, Rosalind. "Notes on the Index: Seventies Art in America." *October*, no. 3 (1977), 68–81.
———. "Notes on the Index: Seventies Art in America, Part 2." *October*, no. 4 (1977), 58–67.
Kubler, George. *The Shape of Time: Remarks on the History of Things*. New Haven, CT: Yale University Press, 1962.
Kusukawa, Sachiko. "Richard Waller's Table of Colours (1686)." In *Colour Histories: Science, Art, and Technology in the 17th and 18th Centuries*, edited by Magdalena Bushart and Friedrich Steinle, 3–21. Boston: De Gruyter, 2015.
Landa, Louis A. "The Shandean Homunculus: The Background of Sterne's 'Little Gentleman'." In *Restoration and Eighteenth-Century Literature: Essays in Honor of Alan Dugald McKillop*, edited by Carroll Camden, 49–68. Chicago: University of Chicago Press, 1963.
Landseer, John. *Lectures on the Art of Engraving: Delivered at the Royal Institution of Great Britain*. London: Longman, Hurst, Rees and Orme, 1807.
Le Blon, J. C. *Coloritto; or, The Harmony of Colouring in Painting*. London, 1725.
Ledbury, Mark. *James Northcote, History Painting, and the Fables*. New Haven, CT: Yale University Press, 2014.
Lee, Eric McCauley. "'Titianus Redivivus': Titian in British Art Theory, Criticism, and Practice, 1768–1830." PhD dissertation, Yale University, 1997.
Lemire, Beverly. *Fashion's Favourite: The Cotton Trade and the Consumer in Britain, 1660–1800*. Oxford: Pasold Research Fund, 1991.
Leslie, Charles Robert, and Tom Taylor. *Life and Times of Sir Joshua Reynolds*. 2 vols. London: William Clowes and Sons, 1865.

Leslie, Esther. *Synthetic Worlds: Nature, Art and the Chemical Industry*. London: Reaktion, 2005.
Lessing, Gotthold E. *Laocöon*. Translated by E. A. McCormick. New York: Bobbs-Merrill, 1962.
Levere, Trevor H. *Poetry Realized in Nature: Samuel Taylor Coleridge and Early Nineteenth-Century Science*. New York: Cambridge University Press, 1981.
Levine, Philippa. *The Amateur and the Professional: Antiquarians, Historians and Archaeologists in Victorian England, 1838–1886*. New York: Cambridge University Press, 1986.
Lewis, Samuel. *A Topographical Dictionary of England*. 5th ed. 4 vols. London: S. Lewis & Co., 1844.
Lewis, William. *Commercium Philosophico-Technicum; or, The Philosophical Commerce of Arts: Designed as an Attempt to Improve Arts, Trades, and Manufactures*. London: H. Baldwin, 1763.
Liotard, Jean-Etienne. *Traité des Principes et des Regles de La Peinture*. Geneva: Minkoff, 1973.
Lippincott, Louise. *Selling Art in Georgian London: The Rise of Arthur Pond*. New Haven, CT: Paul Mellon Centre for Studies in British Art/Yale University Press, 1983.
Litchfield, R. B. *Tom Wedgwood: The First Photographer*. London: Duckworth, 1903.
Loh, Maria H. "New and Improved: Repetition as Originality in Italian Baroque Practice and Theory." *Art Bulletin* 86, no. 3 (September 2004): 477–504.
Lowengard, Sarah. *The Creation of Color in Eighteenth-Century Europe*. New York: Columbia University Press, 2008.
Lysacht, D. J. "Hooke's Theory of Combustion." *Ambix* 1, no. 2 (December 1937): 93–105.
MacLeod, Christine. *Heroes of Invention: Technology, Liberalism and British Identity, 1750–1914*. New York: Cambridge University Press, 2007.
Malm, Andreas. *Fossil Capital: The Rise of Steam-Power and the Roots of Global Warming*. London: Verso, 2016.
Malm, Andreas, and Alf Hornborg. "The Geology of Mankind? A Critique of the Anthropocene Narrative." *Anthropocene Review* 1, no. 1 (2014): 62–69.
Malone, Edmond. *The Works of Sir Joshua Reynolds*. 2 vols. London: T. Cadell and W. Davies, 1797.
Mangone, Carolina. "Like Father, Like Son: Bernini's Filial Imitation of Michelangelo." *Art History* 37, no. 4 (2014): 666–87.
Mannings, David, and Martin Postle. *Sir Joshua Reynolds: A Complete Catalogue of His Paintings*. 2 vols. New Haven, CT: Paul Mellon Centre for Studies in British Art/Yale University Press, 2000.
Mannoni, Laurent. *The Great Art of Light and Shadow: Archaeology of the Cinema*. Translated by Richard Crangle. Exeter: University of Exeter Press, 2000.
Mansfield, Elizabeth, ed. *Making Art History: A Changing Discipline and Its Institutions*. New York: Routledge, 2007.
Margócsy, Dániel. *Commercial Visions: Science, Trade, and Visual Culture in the Dutch Golden Age*. Chicago: University of Chicago Press, 2014.
Marsh, George Perkins. *Man and Nature; or, Physical Geography as Modified by Human Action*. New York: Charles Scribner, 1865.
Martin, Benjamin. *Typographia Naturalis; or, The Art of Printing, or Taking Impressions from Natural Subjects, As Leaves, Shells, Fossils, &c. As Also from Medals, Intaglios, &c. by Means of Isinglass*. London: Printed for the author, at No 171, in Fleet-Street, 1772.
Martin, Craig. *Renaissance Meteorology: Pomponazzi to Descartes*. Baltimore: Johns Hopkins University Press, 2011.

Mayer, Lance, and Gay Myers. *American Painters on Technique: The Colonial Period to 1860*. Los Angeles: J. Paul Getty Museum, 2011.
Maynard, Patrick. *The Engine of Visualization: Thinking Through Photography*. Ithaca, New York: Cornell University Press, 1997.
McClellan, Andrew. *Inventing the Louvre: Art, Politics, and the Origins of the Modern Museum in Eighteenth-Century Paris*. New York: Cambridge University Press, 1994.
McIntosh, Carey. "Reynolds's Portrait of the Infant Johnson." In *Eighteenth-Century Studies in Honor of Donald F. Hyde*, edited by W. H. Bond, 279–96. New York: Grolier Club, 1970.
McKendrick, Neil. "Josiah Wedgwood and Factory Discipline." *Historical Journal* 4, no. 1 (1961): 30–55.
McKendrick, Neil, John Brewer, and J. H. Plumb. *The Birth of a Consumer Society: The Commercialization of Eighteenth-Century England*. Bloomington: Indiana University Press, 1982.
McKie, Douglas. "Fire and the Flamma Vitalis: Boyle, Hooke, and Mayow." In *Science, Medicine and History: Essays on the Evolution of Scientific Thought and Medical Practice Written in Honour of Charles Singer*, edited by E. Ashworth Underwood, 1:469–88. 2 vols. New York: Geoffrey Cumberlege/Oxford University Press, 1953.
McLaren, Angus. "The Pleasures of Procreation: Traditional and Biomedical Theories of Conception." In *William Hunter and the Eighteenth-Century Medical World*, edited by W. F. Bynum and Roy Porter, 323–41. New York: Cambridge University Press, 1985.
Meteyard, Eliza. *A Group of Englishmen (1795 to 1815) Being Records of the Younger Wedgwoods and Their Friends: Embracing the History of the Discovery of Photography and a Facsimile of the First Photograph*. London: Longmans, Green, and Co, 1871.
———. *The Life of Josiah Wedgwood: From His Private Correspondence and Family Papers . . . with an Introductory Sketch of the Art of Pottery in England*. 2 vols. London: Hurst and Blackett, 1866.
Michaels, Walter Benn. *The Shape of the Signifier: 1967 to the End of History*. Princeton, NJ: Princeton University Press, 2004.
Miles, Ellen G. "Thomas Hudson (1701–1779): Portraitist to the British Establishment." PhD dissertation, Yale University, 1976.
Miller, David Philip *Discovering Water: James Watt, Henry Cavendish and the Nineteenth-Century "Water Controversy."* Aldershot, UK: Ashgate, 2004.
———. *James Watt, Chemist: Understanding the Origins of the Steam Age*. London: Pickering & Chatto, 2009.
———. "Seeing the Chemical Steam through the Historical Fog: Watt's Steam Engine as Chemistry." *Annals of Science* 65, no. 1 (2008): 47–72.
Mitchell, Charles. "Three Phases of Reynolds's Method." *Burlington Magazine* 80, no. 467 (1942): 35–40.
Mnookin, Jennifer. "The Image of Truth: Photographic Evidence and the Power of Analogy." *Yale Journal of Law and the Humanities* 10, no. 1 (1998): 1–74.
Morgan, Gilbert T., and David Doig Pratt. *British Chemical Industry: Its Rise and Development*. New York: Longmans, Green & Co., 1938.
Morrison, Kathleen D. "Provincializing the Anthropocene." *Seminar*, no. 673: "Nature and History" (September 2015), 75–80.
Morus, Iwan R. *Frankenstein's Children: Electricity, Exhibition, and Experiment in Early-Nineteenth-Century London*. Princeton, NJ: Princeton University Press, 1998.
Mouradian, Hélène, and Xavier Salmon. *Jean-Jacques Bachelier, 1724–1806: Peintre du roi et de Madame de Pompadour*. Paris: Somogy, 1999.
Moxey, Keith P. F. *Visual Time: The Image in History*. Durham, NC: Duke University Press, 2013.

Mudge, John. *Directions for Making the Best Composition for the Metals of Reflecting Telescopes*. London: W. Bowyer and J. Nichols, 1777.

Müntz, Jean-Henri. *Encaustic; or, Count Caylus's Method of Painting in the Manner of the Ancients: To Which Is Added a Sure and Easy Method for Fixing of Crayons*. London: Printed for the author, and A. Webley, at the Bible and Crown near Chancery Lane, Holborn, 1760.

Nagel, Alexander, and Christopher S. Wood. *Anachronic Renaissance*. New York: Zone Books, 2010.

Nead, Lynda. "Velocities of the Image *c.* 1900." *Art History* 27, no. 5 (2004): 745–69.

Neer, Richard T. "Poussin, Titian, and Tradition: *The Birth of Bacchus* and the Genealogy of Images." *Word & Image* 18, no. 3 (2002): 267–81.

Newman, John. "Reynolds and Hone: 'The Conjuror' Unmasked." In *Reynolds*, edited by Nicholas Penny, 344–54. New York: Harry N. Abrams, 1986.

Newman, William R. *Atoms and Alchemy: Chymistry and the Experimental Origins of the Scientific Revolution*. Chicago: University of Chicago Press, 2006.

———. *Promethean Ambitions: Alchemy and the Quest to Perfect Nature*. Chicago: University of Chicago Press, 2004.

Newman, William R., and Lawrence M. Principe. "Alchemy vs. Chemistry: The Etymological Origins of a Historiographic Mistake." *Early Science and Medicine* 3, no. 1 (1998): 32–65.

Nicolson, Benedict. *Joseph Wright of Derby: Painter of Light*. 2 vols. New York: Pantheon, 1968.

Northcote, James. "A Disappointed Genius." *Artist* 19 (July 18, 1807).

———. *The Life of Sir Joshua Reynolds: Comprising Original Anecdotes of Many Distinguished Persons, His Contemporaries; and a Brief Analysis of His Discourses*. 2nd ed. 2 vols. London: Henry Colburn, 1819.

Ochs, Kathleen H. "The Royal Society of London's History of Trades Programme: An Early Episode in Applied Science." *Notes and Records of the Royal Society of London* 39, no. 2 (1985): 129–58.

Ogden, Henry V. S., and Margaret S. Ogden. *English Taste in Landscape in the Seventeenth Century*. Ann Arbor: University of Michigan Press, 1955.

[Oldenburg, Henry]. "An Experiment of a Way of Preparing a Liquor, That Shall Sink into, and Colour the Whole Body of Marble, Causing a Picture, Drawn on a Surface, to Appear Also in the Inmost Parts of the Stone." *Philosophical Transactions of the Royal Society of London*, no. 7 (December 4, 1665), 126.

O'Quinn, Daniel. "Facing Past and Future Empires: Joshua Reynolds's Portraits of Augustus Keppel." In *The Culture of the Seven Years' War: Empire, Identity, and the Arts in the Eighteenth-Century Atlantic World*, edited by Frans De Bruyn and Shaun Regan, 307–38. Toronto: University of Toronto Press, 2014.

Partington, J. R. *A History of Chemistry*. 4 vols. London: Macmillan, 1962–1964.

Pasquin, Anthony [John Williams]. *A Critical Guide to the Present Exhibition at the Royal Academy, for 1797: Containing Admonitions to the Artists on the Misconception of Theological Subjects, and a Complete Developement of the Venetian Art of Colouring, as is now so much the Rage of Imitation*. London: H. D. Symonds, 1797.

Paulson, Ronald. *Emblem and Expression: Meaning in English Art of the Eighteenth Century*. Cambridge: Harvard University Press, 1975.

———. *Hogarth: His Life, Art, and Times*. 2 vols. New Haven, CT: Paul Mellon Centre for Studies in British Art/Yale University Press, 1971.

Penney, David. "Thomas Mudge and the Longitude: A Reason to Excel." In *The Quest for Longitude*, edited by William J. H. Andrewes, 293–310. Cambridge, MA: Collection of Historical Scientific Instruments, Harvard University, 1996.

Penny, N. B. "The Whig Cult of Fox in Early Nineteenth-Century Sculpture." *Past & Present* 70 (February 1976): 94–105.

Percival, Arthur. *The Faversham Gunpowder Industry*. 3rd ed. Faversham: The Faversham Society, 1986.

Phillips, H. *Unique Collection of Drawings & Prints: A Catalogue of All the Great and Valuable Collection of Ancient Drawings, Scarce Prints, and Books of Prints, Which Belonged to Sir Joshua Reynolds, Deceased, Late President of the Royal Academy* . . . London: 1798.

Pinson, Stephen. *Speculating Daguerre: Art and Enterprise in the Work of L. J. M. Daguerre*. Chicago: University of Chicago Press, 2012.

Piozzi, Hester Lynch. *Anecdotes of the Late Samuel Johnson, L.L.D. during the Last Twenty Years of His Life*. London: Printed for T. Cadell in the Strand, 1786.

Pointon, Marcia R. *Hanging the Head: Portraiture and Social Formation in Eighteenth-Century England*. New Haven, CT: Paul Mellon Centre for Studies in British Art/Yale University Press, 1993.

Polygraphic Society. *A Catalogue of Pictures Copied or Multiplied (for Sale) by A Chymical and Mechanical Process; Exhibited with the Capital Originals from which they have been taken, by The Polygraphic Society, at their Rooms, in Pall-Mall, Being their Eight Exhibition: Opened the 4th of April, 1792*. London: A. Grant, 1792.

Porter, Roy. *Doctor of Society: Thomas Beddoes and the Sick Trade in Late-Enlightenment England*. London: Routledge, 1992.

———. "Erasmus Darwin, Doctor of Evolution?" In *History, Humanity and Evolution: Essays for John C. Greene*, edited by James Moore, 39–70. New York: Cambridge University Press, 1989.

Postle, Martin. *Sir Joshua Reynolds: The Subject Pictures*. New York: Cambridge University Press, 1995.

Price, William Lake. "On Composition and Chiar'oscuro. I." *Photographic News* 3, no. 76 (February 17, 1860): 281.

Priestley, Joseph. *A Description of a Chart of Biography* . . . 2nd ed. Warrington: Printed for the Author, 1765.

———. *The History and Present State of Discoveries relating to Vision, Light and Colour*. London: J. Johnson, 1772.

Principe, Lawrence M. *The Aspiring Adept: Robert Boyle and His Alchemical Quest*. Princeton, NJ: Princeton University Press, 1998.

———. "Chymical Exotica in the Seventeenth Century; or, How to Make the Bologna Stone." *Ambix* 63, no. 2 (2016): 118–144.

———. "A Revolution Nobody Noticed? Changes in Early Eighteenth-Century Chymistry." In *New Narratives in Eighteenth-Century Chemistry*, edited by Lawrence M. Principe, 1–22. Dordrecht: Springer, 2007.

Principe, Lawrence M., and Lloyd DeWitt. *Transmutations—Alchemy in Art: Selected Works from the Eddleman and Fisher Collections at the Chemical Heritage Foundation*. Philadelphia: Chemical Heritage Foundation, 2002.

Prodger, Phillip. *Darwin's Camera: Art and Photography in the Theory of Evolution*. Oxford: Oxford University Press, 2009.

Pocock, J. G. A. *Virtue, Commerce, and History: Essays on Political Thought and History, Chiefly in the Eighteenth Century*. Cambridge: Cambridge University Press, 1995.

Quickenden, Kenneth, Sally Baggott, and Malcolm Dick, eds. *Matthew Boulton: Enterprising Industrialist of the Enlightenment*. Farnham: Ashgate, 2013.

Raspe, R. E. *A Critical Essay on Oil-Painting; Proving that the Art of Painting in Oil was Known before the Pretended Discovery of John and Hubert Van Eyck*. London: H. Goldney, 1781.

Rather, Susan. "Benjamin West, John Galt, and the Biography of 1816." *Art Bulletin* 86, no. 2 (2004): 324–45.

Reynolds, Joshua. *A discourse delivered to the students of the Royal Academy, on the distribution of the prizes, Dec. the 10th, 1774. By the president.* London: Thomas Davies, 1775.
———. *Discourses on Art*. Edited by Robert R. Wark. New Haven, CT: Paul Mellon Centre for Studies in British Art/Yale University Press, 1997.
Reynolds, Joshua. *The Letters of Sir Joshua Reynolds*. Edited by John Ingamells and John Edgcumbe. New Haven, CT: Yale University Press, 2000.
Rhodes, Barbara, and William W. Streeter. *Before Photocopying: The Art & History of Mechanical Copying, 1780–1938*. New Castle: Oak Knoll Press, 1999.
Rice, Danielle. "The Fire of the Ancients: The Encaustic Painting Revival, 1755 to 1812." PhD dissertation, Yale University, 1979.
Richardson, Jonathan. *An Essay on the Theory of Painting*. 2nd ed. London: Printed for A. C. and sold by A. Bettesworth in Pater-Noster-Row, 1725.
Riegl, Alois. *Late Roman Art Industry*. Translated by Rolf Winkes. Rome: G. Bretschneider, 1985.
———. "The Modern Cult of Monuments: Its Character and Its Origin." In *Oppositions Reader: Selected Readings from a Journal of Ideas and Criticism in Architecture 1973–1984*, edited by K. Michael Hays, 621–51. New York: Princeton Architectural Press, 1998.
———. *Problems of Style: Foundations for a History of Ornament*. Translated by Evelyn M. Kain. Princeton, NJ: Princeton University Press, 1992.
Rigaud, Stephen Francis Dutilh. "Facts and Recollections of the XVIIIth Century in a Memoir of John Francis Rigaud Esq., R.A.," edited by William L. Pressly, *Walpole Society* 50 (1984): 1–164.
Rippingille, E. V. "Personal Recollections of Great Artists. No. 8 - Sir Augustus W. Callcott, R.A." *Art Journal* 22 (1860): 99–100.
Roberts, Jennifer L. *Transporting Visions: The Movement of Images in Early America*. Berkeley: University of California Press, 2014.
———. "The Veins of Pennsylvania: Benjamin Franklin's Nature-Print Currency." *Grey Room* 69 (2017): 50–79.
Robertson, David. *Sir Charles Eastlake and the Victorian Art World*. Princeton, NJ: Princeton University Press, 1978.
Robinson, Eric. "An English Jacobin: James Watt, Junior, 1769–1848." *Cambridge Historical Journal* 11, no. 3 (1955): 349–55.
Robinson, Eric, and Douglas McKie, eds. *Partners in Science: Letters of James Watt & Joseph Black*. Cambridge: Harvard University Press, 1970.
Robinson, Eric, and Keith R. Thompson. "Matthew Boulton's Mechanical Paintings." *Burlington Magazine* 112, no. 809 (1970): 497–507.
Robinson, Henry Peach. *Pictorial Effect in Photography: Being Hints on Composition and Chiaroscuro for Photographers*. London: Piper & Carter, 1869.
Rodrick, Anne B. *Self-Help and Civic Culture: Citizenship in Victorian Birmingham*. Aldershot, UK: Ashgate, 2004.
Romero, Adam M., Julie Guthman, Ryan E. Galt, Matt Huber, Becky Mansfield, and Susan Sawyer. "Chemical Geographies." *GeoHumanities* 3, no. 1 (2017): 158–77.
Roos, Anna Marie. "Nehemiah Grew (1641–1712) and the Saline Chymistry of Plants." *Ambix* 54, no. 1 (2007): 51–68.
———. "The Saline Chymistry of Color in Seventeenth-Century English Natural History." *Early Science and Medicine* 20 (2015): 562–88.
Rosenberg, Daniel, and Anthony Grafton. *Cartographies of Time*. New York: Princeton Architectural Press, 2010.
Rosenblum, Robert. "Reynolds in an International Milieu." In *Reynolds*, edited by Nicholas Penny, 43–54. New York: Harry N. Abrams, 1986.
Rossi, Paolo. *The Dark Abyss of Time: The History of the Earth and the History of Nations*

from Hooke to Vico. Translated by Lydia G. Cochrane. Chicago: University of Chicago Press, 1984.
Roy, Ashok, et al. "Joshua Reynolds in the National Gallery and Wallace Collection." Special issue. *National Gallery Technical Bulletin* 35 (2015): 1–111.
Ruddiman, William F. "The Anthropogenic Greenhouse Era Began Thousands of Years Ago." *Climate Change* 61 (2003): 261–93.
Rudwick, Martin J. S. *Bursting the Limits of Time: The Reconstruction of Geohistory in the Age of Revolution*. Chicago: University of Chicago Press, 2005.
———. *The Meaning of Fossils: Episodes in the History of Paleontology*. New York: American Elsevier, 1972.
Sachs, Jonathan. "Scales of Time and the Anticipation of the Future: Gibbon, Smith, Playfair." *Modern Intellectual History* 11, no. 3 (2014): 697–718.
Sahlins, Marshall. "The Sadness of Sweetness: The Native Anthropology of Western Cosmology." *Current Anthropology* 27, no. 3 (June 1996): 395–415.
Saunders, David. "Joseph Booth's Chymical and Mechanical Paintings." In *European Paintings 15–18th Century: Copying, Replicating and Emulating*, edited by Erma Hermens, 113–21. London and Copenhagen: Archetype and CATS, 2014.
Saunders, David, and Antony Griffiths. "Two Mechanical Oil Paintings after de Loutherbourg: History and Technique." In *Studying Old Master Paintings: Technology and Practice*, edited by Marika Spring, 186–93. London: Archetype, 2011.
Schaaf, Larry J. *Records of the Dawn of Photography: Talbot's Notebooks P & Q*. Cambridge: Cambridge University Press/National Museum of Photography, Film & Television, 1996.
Schatzberg, Eric. "*Technik* Comes to America: Changing Meanings of *Technology* before 1930." *Technology and Culture* 47, no. 3 (July 2006): 486–512.
Schofield, Robert E. "Josiah Wedgwood, Industrial Chemist." *Chymia* 5 (1959): 180–92.
Schwartz, Richard B. *Samuel Johnson and the New Science*. Madison: University of Wisconsin Press, 1971.
Scott, Katie. *Becoming Property: Art, Theory, and Law in Early Modern France*. New Haven, CT: Yale University Press, 2018.
Schupbach, William. "The Fame and Notoriety of Dr. John Huxham." *Medical History* 25 (1981): 415–21.
Seneca. *Seneca Ad Lucilium Epistulae Morales*. Translated by R. H. Gummere. 3 vols. London: Heinemann, 1920.
Senefelder, Alois. *The Invention of Lithography*. Translated by J W. Muller. New York: Fuchs & Lang Manufacturing Co., 1911.
———. *Vollständiges Lehrbuch Der Steindruckerey*. Munich: Senefelder and E. A. Fleischmann, 1821.
Serres, Michel. "Turner Translates Carnot." In *Calligram: Essays in New Art History from France*, edited by Norman Bryson, 154–65. New York: Cambridge University Press, 1988.
Shapin, Steven. "The Invisible Technician." *American Scientist* 77, no. 6 (November–December 1989): 554–63.
Shapin, Steven, and Simon Schaffer. *The Leviathan and the Air-Pump: Hobbes, Boyle, and the Experimental Life*. Princeton, NJ: Princeton University Press, 1985.
Shee, Martin Archer. *The Commemoration of Reynolds, in Two Parts, with Notes, and Other Poems*. London: J. Murray, 1814.
Sheehan, Tanya, and Andrés Mario Zervigón, eds. *Photography and Its Origins*. New York: Routledge, 2015.
Shelley, Mary Wollstonecraft. *Frankenstein; or, The Modern Prometheus: The 1818 Text*. Edited by Marilyn Butler. New York: Oxford University Press, 1993.

Shurlock, F. W. "The Scientific Pictures of Joseph Wright." *Scientific Progress in the Twentieth Century* 17, no. 67 (January 1923): 432–38.

Siegert, Bernhard. *Relays: Literature as an Epoch of the Postal System*. Translated by Kevin Repp. Stanford, CA: Stanford University Press, 1999.

Silver, Sean. *The Mind Is a Collection: Case Studies in Eighteenth-Century Thought*. Philadelphia: University of Pennsylvania Press, 2015.

———. "The Prehistory of Serendipity, from Bacon to Walpole." *Isis* 106, no. 2 (June 2015): 235–56.

Silverman, Kaja. *The Miracle of Analogy; or, The History of Photography, Part 1*. Stanford, CA: Stanford University Press, 2015.

Simms, D. L. "The Trail for Archimedes's Tomb." *Journal of the Warburg and Courtauld Institutes* 53 (1990): 281–86.

Simond, Louis. *Journal of a Tour and Residence in Great Britain During the Years 1810 and 1811*. 2nd ed. 2 vols. Edinburgh: J. Ballantyne, 1817.

[Slare, Frederick]. "An account of several Experiments made with the shining substance of the liquid and of the solid *Phosphorous* . . ." *Philosophical Collections*, no. 3 (December 10, 1681), 48–50.

Sloterdijk, Peter. *Terror from the Air*. Translated by Amy Patton and Steve Corcoran. Los Angeles: Semiotext(e), 2009.

Smiles, Sam, ed. *Sir Joshua Reynolds: The Acquisition of Genius*. Bristol: Sansom & Company, 2009.

Smith, Adam. "Of the Nature of that Imitation which takes Place in What are called the Imitative Arts." In *Essays on Philosophical Subjects*, edited by Dugald Stewart, 133–84. Dublin: Printed for Messrs. Wogan, Byrne, J. Moore, Colbert, Rice W. Jones, Porter, and Folingsby, 1795.

Smith, Alison. "Medium and Method in Pre-Raphaelite Painting." In *Pre-Raphaelites: Victorian Art and Design*, edited by Tim Barringer, Jason Rosenfeld, and Alison Smith, 18–23. New Haven, CT: Yale University Press, 2012.

Smith, Dennis. *Conflict and Compromise: Class Formation in English Society, 1830–1914: A Comparative Study of Birmingham and Sheffield*. London: Routledge & Kegan Paul, 1982.

Smith, John T. *Nollekens and His Times: A Life of That Celebrated Sculptor and Memoirs of Several Contemporary Artists, from the Time of Roubilliac, Hogarth and Reynolds to That of Fuseli, Flaxman and Blake*. 2 vols. London: Henry Colburn, 1829.

Smith, Pamela H. *The Body of the Artisan: Art and Experience in the Scientific Revolution*. Chicago: University of Chicago Press, 2004.

———. *The Business of Alchemy: Science and Culture in the Holy Roman Empire*. Princeton, NJ: Princeton University Press, 1994.

———. "In the Workshop of History: Making, Writing, and Meaning." *West 86th* 19, no. 1 (2012): 4–31.

Smith, Robert. *A Compleat System of Opticks in Four Books; viz. A Popular, a Mathematical, a Mechanical, and a Philosophical Treatise; to Which Are Added Remarks upon the Whole. by Robert Smith LL.D. Professor of Astronomy and Experimental Philosophy at Cambridge, and Master of Mechanicks to His Majesty*. 2 vols. Cambridge: Printed for the author, 1738.

Smyth, Craig H., and Peter M. Lukehart, eds. *The Early Years of Art History in the United States*. Princeton, NJ: Department of Art and Archaeology, Princeton University, 1993.

Snyder, Joel. "Inventing Photography, 1839–1879." In *On the Art of Fixing a Shadow: One Hundred Fifty Years of Photography*, edited by Sarah Greenough et al., 3–38. Washington, DC: National Gallery of Art, 1989.

———. “Res Ipsa Loquitur.” In *Things That Talk: Object Lessons from Art and Science*, edited by Lorraine Daston, 195–221. New York: Zone Books, 2004.
———. “Review of Geoffrey Batchen’s *Burning with Desire*.” *Art Bulletin* 81, no. 3 (September 1999): 540–42.
Solender, Katherine. *Dreadful Fire! Burning of the Houses of Parliament*. Bloomington, IN: Cleveland Museum of Art/Indiana University Press, 1984.
Solkin, David H. *Art in Britain 1660–1815*. Pelican History of Art. New Haven, CT: Paul Mellon Centre for Studies in British Art/Yale University Press, 2015.
———. “Great Pictures or Great Men? Reynolds, Male Portraiture, and the Power of Art.” *Oxford Art Journal* 9, no. 2 (1986): 42–49.
———. “Joseph Wright of Derby and the Sublime Art of Labor.” *Representations* 83, no. 1 (2003): 167–94.
———. *Painting for Money: The Visual Arts and the Public Sphere in Eighteenth-Century England*. New Haven, CT: Paul Mellon Centre for Studies in British Art/Yale University Press, 1992.
———, ed. *Art on the Line: The Royal Academy Exhibitions at Somerset House 1780–1836*. New Haven, CT: Paul Mellon Centre for Studies in British Art/Yale University Press, 2001.
Sprat, Thomas. *The History of the Royal Society of London*. London: T. R. for J. Martyn, 1667.
Starn, Randolph. “Three Ages of ‘Patina’ in Painting.” *Representations* 78, no. 1 (2002): 86–115.
Sterne, Laurence. *The Life and Opinions of Tristram Shandy, Gentleman*. Edited by Ian C. Ross. New York: Oxford University Press, 1983.
Stefanis, Konstantinos. “Nathaniel Hone’s 1775 Exhibition: The First Single-Artist Retrospective.” *Visual Culture in Britain* 14, no. 2 (2013): 131–53.
Steffen, Will, Jacques Grinevald, Paul Crutzen, and John McNeill. “The Anthropocene: Conceptual and Historical Perspectives.” *Philosophical Transactions: Series A, Mathematical, Physical, and Engineering Sciences* 369, no. 1938 (2011): 842–67.
Stewart, Larry. “Assistants to Enlightenment: William Lewis, Alexander Chisholm, and Invisible Technicians in the Industrial Revolution.” *Notes and Records of the Royal Society of London* 62, no. 1 (March 2008): 17–29.
Styles, John. *The Dress of the People: Everyday Fashion in Eighteenth-Century England*. New Haven, CT: Yale University Press, 2007.
Sugimoto, Hiroshi. *Hiroshi Sugimoto: History of History*. New York: Japan Society, 2005.
Swoboda, Gudrun. “Jean-Etienne Liotard (1702–1789), Painter of Extremes.” *Points of View* (Vienna: Kunsthistorisches Museum, 2012): 4–18.
Talbot, [W.] H. F. “Some Account of the Art of Photogenic Drawing.” *The London and Edinburgh Philosophical Magazine*, series 3, vol. 14, no. 86 (February 1839): 196–211.
Talbot, William Henry Fox. *The Pencil of Nature*. London: Longman, Brown, Green, & Longmans, 1844–1846.
Talley, M. Kirby. “‘All Good Pictures Crack’: Sir Joshua Reynolds’s Practice and Studio.” In *Reynolds*, edited by Nicholas Penny, 55–70. New York: Harry N. Abrams, 1986.
Taws, Richard. *The Politics of the Provisional: Art and Ephemera in Revolutionary France*. University Park: Pennsylvania State University Press, 2013.
Thomas Hudson, 1701–1779, Portrait Painter and Collector: A Bicentenary Exhibition. London: Greater London Council, 1979.
Thompson, E. P. “Time, Work-Discipline, and Industrial Capitalism.” *Past and Present* 38, no. 1 (1967): 56–97.
Thompson, Krista. “The Sound of Light: Reflections on Art History in the Visual Culture of Hip-Hop.” *Art Bulletin* 91, no. 4 (December 2009): 481–505.

Townsend, Joyce. *Turner's Painting Techniques*. London: Tate Publishing, 2005.
Trengrove, Leonard. "Chemistry at the Royal Society of London in the Eighteenth Century: IV, Dyes." *Annals of Science* 26, no. 4 (1970): 331–53.
Trevor-Roper, Hugh. *Europe's Physician: The Various Life of Sir Theodore De Mayerne*. New Haven, CT: Yale University Press, 2006.
Tuana, Nancy. "The Weaker Seed: The Sexist Bias of Reproductive Theory." *Hypatia* 3, no. 1 (Spring 1988): 35–59.
Turner, A. J. "A Forgotten Naturalist of the Seventeenth Century: William Cole of Bristol and His Collections." *Archives of Natural History* 11, no. 1 (1982): 27–41.
Turner, H. D. "Robert Hooke and Theories of Combustion." *Centaurus* 4, no. 4 (1956): 297–310.
Uglow, Jenny. *The Lunar Men: Five Friends Whose Curiosity Changed the World*. New York: Farrar, Strauss and Giroux, 2002.
Wade, Nicholas J. "Accentuating the Negative: Tom Wedgwood (1771–1805), Photography and Perception." *Perception* 34, no. 5 (2005): 513–20.
Waggoner, Diane, and T. J. Barringer. *The Pre-Raphaelite Lens: British Photography and Painting, 1848–1875*. Washington, DC: National Gallery of Art, 2010.
Walker, Richard. "Henry Bone's Pencil Drawings in the National Portrait Gallery." *Walpole Society* 61 (1999): 305–67.
Wall, Jeff. "Photography and Liquid Intelligence." In *Jeff Wall*, edited by Thierry de Duve et al., 90–93. London: Phaidon, 1996.
Wallert, Arie, and Willem de Ridder. "About Skin: Technical Examination of Paintings by Adriaen van der Werff." In *Studying Old Master Paintings: Technology and Practice*, edited by M. Spring et al., 180–85. London: Archetype, 2011.
Wallis, George. "The Ghost of an Art-Process practiced at Soho, near Birmingham, about 1777 to 1780, Erroneously Supposed to have been Photography." *Art Journal*, August 1, 1866, 251–55.
Walpole, Horace. *Anecdotes of Painting in England*. 4 vols. Strawberry-Hill: Thomas Kirgate, 1771.
———. *Horace Walpole's Correspondence with William Mason*. Edited by W. S. Lewis, G. Cronin Jr., and C. H. Bennett. 2 vols. New Haven, CT: Yale University Press, 1955.
Walton, Kendall. "Transparent Pictures: On the Nature of Photographic Realism." *Critical Inquiry* 11, no. 2 (December 1984): 246–77.
Ware, Mike. *Mechanisms of Image Deterioration in Early Photographs: The Sensitivity to Light of W. H. F. Talbot's Halide-Fixed Images 1834–1844*. London: London Science Museum, 1994.
Waterhouse, Ellis. *Painting in Britain: 1530 to 1790*. Baltimore: Penguin, 1953.
Waters, Colin N., Jan A. Zalasiewicz, Mark Williams, Michael A. Ellis, and Andrea M. Snelling. *A Stratigraphical Basis for the Anthropocene*. London: Geological Society, 2014.
Webster, Mary. *Johann Zoffany 1733–1810*. New Haven, CT: Yale University Press, 2011.
Wedgwood, Josiah. *The Selected Letters of Josiah Wedgwood*, edited by Ann Finer and George Savage. London: Cory, Adams & Mackay, 1965.
Wedgwood, Tom. "Experiments and Observations on the Production of Light from Different Bodies, by Heat and by Attrition." *Philosophical Transactions of the Royal Society of London* 82 (1792).
Wendorf, Richard. *After Sir Joshua: Essays on British Art and Cultural History*. New Haven, CT: Paul Mellon Centre for Studies in British Art/Yale University Press, 2005.
———. *Sir Joshua Reynolds: The Painter in Society*. London: National Gallery, 1996.
West, Benjamin. *A Discourse, Delivered to the Students of the Royal Academy, on the Dis-*

tribution of the Prizes, December 10, 1792, by the President. Humbly Inscribed by Permission to His Majesty: To Which Is Prefixed the Speech of the President to the Royal Academicians on the 24th of March 1792. London: Printed by Thomas Cadell, printer to the Royal Academy, 1793.
"Who Discovered Photography?" *Saturday Review of Politics, Literature, Science and Art* 16, no. 419 (November 7, 1863): 613–14.
Wilding, Nick. "Graphic Technologies." In *Robert Hooke: Tercentennial Studies*, edited by M. A. R. Cooper and Michael Hunter, 123–34. Aldershot, UK: Ashgate, 2006.
Willis, Thomas. "Of Fermentation; or, The Inorganical Motion of Natural Bodies." In *Dr. Willis's Practice of Physick, Being the Whole Works of That Renowned and Famous Physician*. London: T. Dring, C .Harper, and J. Leigh, 1684.
———. *Two Discourses concerning The Souls of Brutes*. Translated by S. Pordage. London: Thomas Dring, Ch. Harper and John Leigh, 1683.
Wilson, Michael I. *Nicholas Lanier: Master of the King's Musick*. Aldershot, UK: Scolar, 1994.
Winckelmann, Johann Joachim. *The History of Ancient Art*. Translated by G. Henry Lodge. 4 vols. Boston: Little, Brown, and Company, 1856.
Wind, Edgar. *Hume and the Heroic Portrait*. Edited by Jaynie Anderson. New York: Oxford University Press, 1986.
Woodcroft, Bennet. *A Sketch of the Origin and Progress of Steam Navigation from Authentic Documents*. London: Taylor, Walton, and Maberly, 1848.
Wright, Michael. "Hooke's Longitude Timekeeper." In *Robert Hooke: New Studies*, edited by Michael C. W. Hunter and Simon Schaffer, 63–118. Woodbridge, Suffolk: Boydell Press, 1989.
Wunsch, Oliver. "Diderot and the Materiality of Posterity." *Early Modern French Studies* 40, no. 1 (2018): 63–78.
Wyman, Colin. "The Early Techniques of Transfer Printing." *Transactions of the English Ceramic Circle* 10, no. 4–5 (1980): 187–99.
Young, Edward. *Conjectures on Original Composition: In a Letter to the Author of Sir Charles Grandison*. London: Printed for A. Millar, 1759.
Young, Hilary, ed. *The Genius of Wedgwood*. London: Victoria & Albert Museum, 1995.
Zalasiewicz, Jan, et al. "The Technofossil Record of Humans." *Anthropocene Review* 1, no. 1 (2014): 34–43.

INDEX

Page numbers in italics refer to illustrations.

Adlard, Henry, 154, *156*
Agrippa, 129
Aikin, John, 153–54
air: chemistry of, 9, 115; as method, 31, 33, 137, 165–78, 183–84
Albert, Prince, 177
alchemy, 22, 70–74, 84, 100, 129; and homunculus, 86
Alchymist, The (Jonson), 74, 77
Alchymist, The . . . (Wright), 27–28, 64–65, 74, 77–78, 154
Alpers, Svetlana, 4
American Revolution, 107
Analysis of Beauty, The (Hogarth), 44, *45*
André, Philipp, 95–96
Anecdotes of the Late Samuel Johnson, L.L.D. (Piozzi), 89
Angel of the Resurrection (West), 95, *95*, 98
Anthropocene, 22, 137, 179–80, 184; technofossils, 182
Antiochus, 141, 230n40
Antiochus and Stratonice (Eginton), 140–41, 158–59
Archer, Frederick Scott, 172
Archimedes, 98
Aristotle, 36, 75, 129
Articles of Porcelain (Talbot), *164*
Aufrere, Sophia, 69

Bachelard, Gaston, 137
Bachelier, Jean-Jacques, 92
Bacon, Francis, 9, 17
Baker, Robert, 65, 74–76
Baker, Thomas, 52–53
Balduin, Christian Adolph, 37, 40, 65, 124; Bononian stone, 33
Bann, Stephen, photographic exceptionalism, myth of, 228–29n21
Barker, George, Jr., 67
Barker, George, Sr., 67
Barrell, John, 47
Barrett, Ross, 182

Barrow, John, 139
Barry, James, 97
Barthes, Roland, 136
Bassano, Jacopo, 101–2
Batchen, Geoffrey, 10, 94, 135–36, 178
Bates, Ray, 215n212
Baxandall, Michael, slow-motion mobiles, 14
Bazin, André, 136
Beale, Charles, 77
Beale, John, 29, 33
Bear, Jordan, 177, 229n28
Beardsley, Amos, 151, *151*
Becker, Carl L., 104
Beddoes, Thomas, 17, 122, 154, 198n15
Beechey, William, 4, 133, 155, 163, 232n80
Bellicard, Jérôme-Charles, 90, *91*
Belting, Hans, 198n13
Benjamin, Walter, 3, 94, 172, 218n32
Bentley, Thomas, 117, 225n157
Berkeley, George, 124
Bermingham, Ann, 9
Berthollet, Claude-Louis, 174
big oil, 182
Birch, William Russell, 3, 94, 106–7, 130, 134, 152
Birmingham (England), 17, 21, 65, 93, 106, 110, 115, 137, 139, 150, 154, 161, 163, 167–68; "Brummagem Gothic," 231n63; and photography, 159; union organizing in, 152–53
Birmingham and Midlands Institute, 153
Black, Joseph, 115, 167
Blake, William, 2, 126, *127*
Bloomsbury, 182
Board of Longtitude, 53
Boerhaave, Herman, 18, 89, 196n75
Bone, Henry, 104–6, 130
Bononian stone, 27, 33, 36, 40, 65. *See also* phosphorus
Booth, Joseph, 3, 18, 101–4, 107, 110, 130, 134, 141, 149, 181; history, version of, 109; polygraphic print, *103*
Boswell, James, 90
Botanic Garden, The (E. Darwin), 169
Boucher, François, 84–86
Boulton, Matthew, 20–21, 93, 110–11, 139, 145, 153, 155–56, 159, 162
Boulton, Matthew Piers Watt, 21, 140, 159–60, 165, *169*, *170*, 172, 174–75, 178, 180, 184, 236n170, 236n171; aeronautic principles of, 168–69; optics, use of, 163; patent applications of, 168–71, 177; Smith chemo-mechanical assemblage, blocking of, 173; sun pictures, opposition to, 22, 137, 161–64, 183
Boulton, Matthew Robinson, 111, 151
Bourgeois, P. F., 101
Boyle, Robert, 8, 18, 28–29, 101, 124, 191n30; air pump experiments, 31–32; corpuscularian chemistry, influence of, 31, 199n34; fermentation theory, 31; saltpeter, use of, 31–33
Brandt, Henning, 65
Brewster, David, 150
Bristol (England), 9, 17
Britain, 5, 7, 9, 15, 17, 19–20, 47, 57, 59, 62, 65, 78, 87, 101, 107, 136–37, 173, 181, 198n13; painting, uncertainty of in, 77; public culture of chemistry, 196n78. *See also* England; Scotland
British Association for the Advancement of Science, 172
British Enlightenment, 14, 17, 19–20, 22–23, 28, 87, 135–36, 164–65, 179, 181
British Institution, 131–32
British school of painting, 9, 97, 101, 132, 164–65
Brockliss, L. W. B., 196n78
Bull Ring riots, 153, 163
Burdett, Peter Perez, 76, 117, 120; aquatint process, 118
Burke, Edmund, 53, 94, 107, 116; sublime, analysis of, 81
Burning of the Houses of Lords and Commons, The (Turner), 131

camera obscura, 4, 5, 6, 10, 39, 172; transfer prints, 232n74
capitalism, 180
Caravaggio, 80
Cardwell, Donald, 165
Carnot, Lazare, 174
Carnot, Sadi, 165, 174
Cascariolo, Vincenzo, 123–24; Bononian stone, 27, 65
Catalogue of the Valuable Collection of Paintings by Ancient and Modern Masters of Sir Thomas Lawrence, 69
Catherine the Great, 232n74
Cavendish, Henry, 115
Cellio, Marc Antonio, 36–37, *37*

Chabelnik, Nina, 239n25
Chambers, Robert, 44
Chapelle, Brianne, 239n25
Charlebois, Laurence, 239n25
Charles I, 26
Charles II, 26, 41
Chartist riots, 134, 152–53, 163
Chemical Revolution, 16, 93, 100, 129
chemistry, 10, 20–21, 27, 49, 57, 63, 65–66, 72, 80, 89–90, 122, 128–30, 137, 183; and art, 84; changing domain of, 60; character of, 17, 178; and chymistry, 16–18; corpuscularian, 31, 34; industrial arts, 18; lithography, 96, 109; and painting, 3, 136; philosophical, 115; and photography, 134–36; pictorial, 92; practical, 62, 114; radical politics, linked to, 94; steam-engine research, 167, 178; of substitution, 22; unstable alloy of, 50; as Ur-science, 60–61
Chisholm, Alexander, 62, 121–22, 154
Cicero, 98
Cicero Discovering the Tomb of Archimedes (West), 98
climate change, 180. *See also* global warming
Cole, William, 7–8, *8*, 9–10
Coleridge, Samuel Taylor, 17, 125
Coloritto (Le Blon), 44
Commercium Philosophico-Technicum (Lewis), 61, *62*
Comte de Caylus, 60, 90–92, 99, 118
Cooper, Anthony Ashley, 78
Coram, Thomas, 58
Cosway, Richard, 98, 219n45
Cotton, William, 57, 59–60, 62–64
Courtenay, Frances, 56–57
Coutts, James, 1–2
Crewe, John, 81
Critique of Pure Reason (Kant), 172
Cromwell, Oliver, 17
Crutzen, Paul J., 179–80
Cullen, William, 60

D'Agoty, Jacques Fabien Gautier, 84–85
Daguerre, Louis Joseph Mandé, 10, 134, 148, 159, 172, 174
d'Alembert, Jean le Rond, 71
Darwin, Charles, 15, 123; Darwinian evolution, 194n57
Darwin, Erasmus, 15–16, 110, 122–23, 129, 167, 169–71, 183
Darwin, Robert Waring, 123
Darwinism, 16
Daston, Lorraine, 19
David, Jacques-Louis, 16, 141
David Garrick with Edmund Burton and John Palmer in "The Alchymist" (Zoffany), *75*
Davy, Humphry, 17, 40, 94, 122, 128, 146, 148–49, 154, 167
Death of General Wolfe, The (West), 79, *79*, 80
de Duve, Thierry, 182
De Marchi, Neil, 77
Demidowicz, George, 234n119
Derham, William, 198n11
Description of a Chart of Biography, A (Priestley), 107–8
Diamond, Hugh, 146–49, *150*
Diderot, Denis, 91–92, 94; and posterity, 104
Die spätrömische Kunstindustrie (Riegl), 183
Discourses on Art (Reynolds), 1–2, 4, 47, 58; "Discourse VI," 3, 49, 70–74, 84, 86
Dollond, Peter, 209n101
Donald, Diana, 233n96
Dossie, Robert, 60–61, 90
"Dream in Two States, A" (T. Wedgwood), *123*
Dr. John Mudge (Reynolds), 54
Duchamp, Marcel, 184; readymades of, 182
Duclos, Samuel Cottereau, 199n34
Duke of Cumberland (Henry Frederick), 68
Duke of Richmond, 59
dung, 71, 84
dyeing and marking, chemistry of, 7–10, 17, 27, 41, 61–62, 181–82

Earth, 130, 137, 163, 183–84
Eastlake, Charles Lock, 4, 67, 132–34
Eastlake, Elizabeth Rigby, 134, 158, 172–73
Eder, Josef Maria, 40
Edgcumbe, Dick, Jr., 57, 60
Edwards, Steve, 134, 163
Eginton, Francis, 110, 139, *139*, 141–42, 145–46, 149, 159, 161–62
Eginton, John, 110
Elcho, Lord (Francis Charteris), 2–3
Elémens de chymie-pratique (Macquer), 60
elemental art history, 182–84
Ellis, William, *152*
Empedocles, 137
emulation, 52–53, 70, 73–74, 83, 86

enamel painting, 3, 14, 21, 77, 93, 104–5, 118, 130; artistic conservation, as suited for, 106
encaustic painting, 21, 60, 90–92, 94, 118–20, 130
Engels, Friedrich, 152
England, 3, 9, 19, 26, 49–50, 96, 136, 152. *See also* Britain
Enlightenment, 7, 90, 104, 154, 158, 165, 178
Enquiry Concerning Political Justice, An . . . (Godwin), 116
Erasistratus, 141
Erasistratus the Physician Discovers the Love of Antiochus for Stratonice (West), 141, 142
Ericsson, John, 168
"Essay on Time" (T. Wedgwood), 124, 126–27
Esther (Tintoretto), 25–26, 28, 41
Etruria, 15, 62, 93, 111, 116–17, 121, 126, 175
Eurasia, 180
Europe, 27–29, 71, 179–80
evolution, 15
Experiment on a Bird in the Air Pump, An (Wright), 65
Experiments and Considerations Touching Colours (Boyle), 8

Fairholt, Frederick William, 158
Farington, Joseph, 49, 73, 98–100, 132
Feingold, Moti, 192n42
Ferdinand VI, 53
fine arts, 41, 93, 178, 184; industrial arts, and problem of emulation, 86; natural philosophy, 86; and time, 87
fire: chemistry of, 30, 32–33; poetics of, 70–73, 124, 130, 183–84
Fisher, Edward, 51
Forty, Adrian, 107
fossils, 36, 38, 40–41
Foucault, Michel, 78, 94
Fouque, Victor, 174
Fowles, John, 22–23
Fox, Charles James, 101
France, 26, 58, 83, 118, 136, 152
Frank, Robert G., Jr., 28, 31, 199n30
Frankenstein (Shelley), 21, 129–30
Franklin, Benjamin, 60
Frantani, Chloë, 239n25
French Revolution, 53, 94, 107
Fried, Michael, 196n81
Friedman, Milton, 177
Fulton, Robert, 176
Fundamenta Chymiae dogmaticae et experimentalis (Stahl), 60
Fuseli, Henry, 97

Gainsborough, Thomas, 58
Galison, Peter, 19
Galt, John, 80
Gamble, Ellis, 44
Gardner, John, 60
Garrick, David, 74
Gaskell, Peter, 152
Gell, Alfred, 136
genius, discourse of, 70–72, 86, 96
George III, 53, 93
Gerard, Alexander, 70–71
Gernsheim, Alison, 164
Gernsheim, Helmut, 164, 218n32
Gibbs-Smith, Charles, 168
Gillray, James, 99–100
Glasgow (Scotland), 167
global warming, 180, 182. *See also* climate change
Godwin, William, 116, 122–23, 129, 180
Golinski, Jan, 30, 196n78
Gombrich, E. H., 47
Graces Awakening Cupid (Ryland), 147
Graces Awakening Cupid and Nymphs Adorning the Statue of Pan (Eginton), 141, 145
Grand Tour, 19
Greatorex, Ralph, 31
Green, Guy, 117
Green, Valentine, 92
Greenland, 179–80
Grew, Nehemiah, 13, 28, 39, 41; camera obscura, 39; fossils, view of, 38
Griffiths, Antony, 102
Guerlac, Henry, 115
guild corporations, collapse of, 77
gun-cotton (nitrocellulose), 171–72, 175
Gunther, R. T., 191n27
Gwatkin, Theophilia "Offy" Palmer, 133

Hales, Stephen, 61–62, 63
Hamilton, Sir William, 118, 206n56
Hamilton, William, 100–101
Handmaid to the Arts, The (Dossie), 60
Hannibal, 81

Hanson, Craig, 9
Harrison, John, 53
Harvey, William, 28
Haydon, Benjamin Robert, 4, 66, 132–34, 155
Hayman, Francis, 59
Hazard, Ebenezer, 110
Head of Minerva (Vien), 90
Heidegger, Martin, 196n81
Henry VIII, 81
history painting, 78–80
HMS *Maidstone* (ship), 203n18
Hogarth, William, 46, 50, 58, 62–63, 71–72, 74, 83, 86–87; skin tones, 44–45; on time, 47
Hogarth's Act, 45
Holmes, Leonard Troughear, 19
Holmes, Robert, 19
Holmes, Samantha, 239n25
Homer, 86
homunculi, 84–87, 94, 129, 181, 183
Hone, Nathaniel, 72–74, 80–81, 86, 177–78
Hooke, Grace, 19
Hooke, Robert, 19–20, 27–28, 31–33, *37*, 40–41, 65, 181, 183, 192n43, 201n59; air pump, design for, *32*; Bononian stone, 36; chemo-mechanical model of time, 34–38, 41, 125; combustion experiments, 36–37, 200n42; Comet, Mark II, *13*; "Cometa" demonstration, *12*; comets, studying of, 10–14, 17, 50, 193n48; and fossils, 36; memory, physiology of, 34–35; time, understanding of, 34–38; watch mechanism, design for, *35*
Hoppner, John, 98
Houghton, Walter, 27
Hudson, Thomas, 55–57, 78, 204n35
Hunt, Robert, 158–59, 172
Hunter, John, 134
Huxham, John, 51–53, 78
Huygens, Christiaan, 35

imitation, 52, 70, 73–74
industrial arts, 86, 93, 184
Industrial Revolution, 7, 17, 22–23, 109, 163, 180; as phrase, 153
Infant Academy, The (Reynolds), 83–86, *85*
Infant Johnson (Zobel after Reynolds), 82, *82*
Infant Jupiter (Reynolds), 81
Interior of the National Gallery of Practical Science, Adelaide Street (Scharf), *138*
International Exhibition of Industrial Arts and Manufactures (1862), 163
Iran, 83
Irons family, 52, 55, 57, 64, 204–5n35
Italy, 47, 58

James Watt & Co., 110–11, 116, 168; and ink, 113–14; portable copying press, *111*, *112*
Jee, Edward, 110
Jefferson, Thomas, 110
Jeffrey, Francis, 235n141
Jeffreys, Thomas, 106
Jervais, Thomas, 233n109
Jia, Tony, 192n42
Jodrell, Richard Paul, 66
Jodrell, Vertue, 66
John Hall and Sons, 171
John Hunter (Reynolds), *135*
Johns, Adrian, 113
Johnson, Elizabeth, 205n53
Johnson, Samuel, 18, 53, 78, 82, 90; chemistry, definition of, 196n75; chemistry, interest in, 89; visual art, ignorance of, 89
Jones, Colin, 196n78
Jones, Rica, 77
Jonson, Ben, 74
Josiah Wedgwood and Sons, *117*

Kant, Immanuel, 86–87, 125, 172
Katz, Jonathan, 192n42
Kauffman, Angelica, 73, 97, 101, 110, 141–42, 157
Keir, James, 110–11, 113
Kelsey, Robin, 3–4
Kepler, Johannes, 39
Keppel, Augustus, 47, 57–58, 70, 203n18
Kircher, Athanasius, 41
Klein, Yves, 23
Koselleck, Reinhart, 221n92
Kowalchuk, Anya, 239n25
Krafft, Johann Daniel, 29
Krauss, Rosalind, 136
Kritik der Urteilskraft (Kant), 86
Kunckel, Johann, 33
Kunstwollen, 16, 135, 183

Lady Frances Courtenay (Hudson), 56, *56*, 57
Lake Price, William Frederick, 134

Lancashire (England), 17–18, 109
Landseer, Edwin, 233n96
Landseer, John, 96
Lanier, Nicholas, 26
Laoköon (Lessing), 78
Laughing Girl, The (Reynolds), 101
Lavoisier, Antoine-Laurent, 16, 93, 100, 129, 167
Lawrence, Thomas, 4, 67
Laxton, William, 132, 133
Le Blon, Jacob Christoph, 43–44, 83, 85
"Lecture of Light" (Hooke), 33, 35–36, 40
Lely, Peter, 77
Leslie, Esther, 22
Leslie, Charles Robert, 100
Lessing, Gotthold Ephraim, 78
Lewis, Samuel, 145
Lewis, William, 61–62, 121; sheep-marking preparation, *63*
liberalism, 78
Life and Opinions of Tristram Shandy, Gentleman, The (Sterne), 20, 84–85
Lincoln, Abraham, 180
Liotard, Jean-Etienne, 104
lithography, 21, 93–95, 123, 174, 218n32; as innovation, 96
Lloyd, Thomas, 159
Locke, John, 124
London (England), 9, 20–21, 26–27, 45, 47, 50, 54, 58, 64, 85, 93, 107, 134, 138, 142, 163
Lorrain, Claude, 101
Lot and His Wife (Nollekens), 59
Loutherbourg, Philippe-Jacques de, 97, 102
Lunar Society, 21, 65, 93, 110, 115–16, 150, 153–57, 159, 167, 169

Macardell, James, 92
Mackintosh, James, 124–25, *125*, 126, 226n171
Macquer, Pierre-Joseph, 60
Macrae-Korobkov, Sarah, 239n25
Magnus, Albertus, 129
Majault, Michel Joseph, 92
Malone, Edmond, 100
Malone, Thomas Augustine, 159, 161, 163
Malpighi, Marcello, 27
Malta, 173
Man and Nature (Marsh), 180
Marcuzzi, Gabby Herie, 239n25
Margaret (ship), 25–26
Maria, Countess Waldegrave, 70
Marsh, George Perkins, 180–81, 183
Mason, William, 77
Materials for a History of Oil Painting (Eastlake), 4, 132
Matthew Boulton (Beechey), *157*
Mayerne, Theodore de, 26–27, 41
Mayow, John, 198n15
McClellan, Andrew, 90
McKendrick, Neil, 126
McNeill, John R., 180
mechanical art, 102–3
Mendeleev, Dmitri, 137
metallurgy, 55, 63–65, 71–72, 74, 90, 96
Meteyard, Eliza, 153–54, 158, 178; "Frog Service," 232n74
Michaels, Walter Benn, 136
Micrographia (Hooke), 32
Miller, David Philip, 167, 178
Miss Irons, with contemporary defacement (Rennell), *51*
modernity, 77, 181–82
Molyneux problem, 124
Morse, Samuel F. B., 177
Mudge, John, 54–55, 63, 72
Mudge, Thomas, 53–54, 74, 205n48
Mudge, Zachariah, 53
Muirhead, James Patrick, 153
Müntz, Johann Heinrich, 60, 66, 91

Napoleon, 235n141
Nasim, Omar, 236n171
National Gallery (London), 48, 132
National Gallery of Practical Science, 138
National Gallery of Scotland, 2–3
natural philosophy, 28, 30–31; and fine arts, 86
Newcomen steam engine, 165
Newman, William R., 16, 84
Newton, Henry Charles, 182
Newton, Isaac, 44, 55, 61, 71, 86
Niépce, Claude, 174
Niépce, Joseph Nicéphore, 10, 150, 178, 183; heliography, 149, 172, 174; Pyréolophore, 174
Nollekens, Joseph, 59
North Carolina, 118
Northcote, James, 2, 10, 14, 18–19, 57, 63–64, 68, 74, 89, 92–93, *94*, 100–101, 129, 132, 182, 212n149; patina, embracing of, 90

Northcote, Samuel, 63–64, 74
nuclear weapons, 180
Nymph Adorning the Statue of Pan (Eginton), 146
Nymphs Adorning the Statue of Pan (Ryland), 148

Objectivity (Daston and Galison), 19
Oldenburg, Henry, 29
Old Masters, 9, 57, 59, 64, 90
On Aërial Locomotion (Boulton), 168–69
Opie, John, 98, 100–101
Oxford (England), 28, 31

Paine, Tom, 107
Painter-Stainers' Company, 77
painting: ancient wax techniques, 91–92; conservation of, 215n205; facture of, 1, 43, 56–57, 60, 64, 66, 81, 90–92, 102, 104–6, 131–32; oil, fragility of, 14–15, 26, 90, 104; photography, differences between, 6, 136; supports and grounds of, 10, 14, 43, 47, 90–92, 99; temporal limits of, 230n40
Paracelsus, 17, 26, 31, 84, 129, 215n208
Pasquin, Anthony, 99, 101–2
patent law, 177
Patent Museum, 21, 137, 177; sun pictures, 136
Patent Office Library, 177
patina, 15, 45–47, 90
Paulson, Ronald, 82–83
Paulze-Lavoisier, Marie-Anne-Pierrette, 16
Pearson, James, 93
Pencil of Nature, The (Talbot), 122, 150, 176
Perkins, Jacob, 138
Petronius, 72
Petty, William, 8
Philadelphia, 107
Philosopher Giving a Lecture on the Orrery (Wright), 65
Philosophical Treatises of the Royal Society of London (Oldenburg), 29
phlogiston, 16, 92
phosphorus, 20, 27–31, 33, 37–38, 40–41, 65, 72, 74, 176, 181
Photographic Society of London, 132, 148, 157–58, 161, 163, 232n74
photography, 3–4, 7, 21, 40, 93–94, 122, 130, 173, 175; and chemistry, 134–36; deterioration of, 5–6; first trials of, 154; fossil fuels, 41; gun-cotton, 172; machinery, as subset of, 163; origins of, 178, 181; and painting, 137; painting, differences between, 6, 136; protophotography, 10, 128; sun pictures, 137, 149–50, 158–59, 161–64; wet-collodion process, 172
"Photography and Liquid Intelligence" (Wall), 22
Pictorial Conjuror, The (Hone), 72–74, 80, 86
Pictures by the Late Sir Joshua Reynolds (exhibition), 49
Piozzi, Hester Lynch, 89–90
plastic arts, 78
Platonism, 53
Plenderleith, Harold, 84
Pliny the Elder, 8, 39, 72, 90, 118, 191n29
Plot, Robert, 7–8
Plymouth (England), 9, 19–20, 50, 52, 54–55, 64, 85
Pneumatic Institution, 17, 122, 154
Point of View Taken from a Window of Le Gras in Saint-Loup-de-Varennes (Niépce), *175*
pollaplasiasmos, 93, 96–97, 101, 104, 109, 123, 127–28; as mechanical art, 102–3
polygraphic painting. *See* pollaplasiasmos
Polygraphic Society, 101, 103, 110, 141
Poor Law, 165
Porter, Roy, 19
Prestonpans (Scotland), 181
Price, Edward, 21, 136–37, 140–42, 145–46, 149–50, 154–57, 160, 163, 178; dishonesty, accusations of, 161–62
Price, Sarah, 81
Priestley, Joseph, 18, 65, 94, *108*, *109*, 115, 123, 152–54, 167; time, linear vision of, 107–9, 124
Principe, Lawrence, 16
print, facture of, 95, 112–19
prints and print-making, 12, 43–44, 50–51, 92–93, 95–96, 101–2, 107–18, 122–24, 154, 158
Provis, Ann Jemima, 98–99
Provis, Thomas, 98–99
Pugin, August Welby Northmore, 231n63

Ramsay, Allan, 59
Raphael, 26, 59
Reflections on the Revolution in France (Burke), 94, 107

Remarks Concerning Certain Photographs Supposed to Be of Early Date (Boulton), 161–62, 236n171
Remarks on Some Evidence Recently Communicated to the Photographic Society (Boulton), 160–61
Rembrandt, 84–86, 184
Rennell, Thomas, 18, 50–53, 64, 74; as experimentalist, 57
reproduction: chemical, 83–84, 86; sexual, 84–86, 129
Restoration, 41
Rev. Thomas Smart (Reynolds), 58
Reynolds, Joshua, 1, 7, 9, 12, 14, 17–20, 22, 41, 65, 91, 94, 96, 129–31, 136–37, 161, 164–65, 176, 181, 184, 203n18, 204n31, 206n56, 207n74, 209n102, 210n119, 216n2; abstraction, drive for, 82; alchemical plagiarism, charges of, 177–78; annotations of, 67–69; artistic study of, 58; background of, 52–53; camera obscura, 4, 5, 6, 10; chemical accident, by fire, 72; chemical instability, association with, 3; chemical materials, making and thinking with, 49–50; chemical replicas after, 3, 93, 101, 104–7, 110, 233n109; chemical replicas of, as photographs, 21; chemical transformations, as center of grand-manner art, 71, 73–74, 93; collecting artifacts of, 4, 57, 66–67, 133–34, 155; Devonian connections of, 52–58, 63–64, 132; encaustic technique, 60; first painting of, 57–58; imitative copying, 52; infant portraits, 21, 80–84, 86–87, 214n191; ledger books, *67*, *68*, *133*; painting, and clock-making, 85; paintings, deterioration of, 5–6, 47–49; paintings, instability of, 2, 77–78; paintings, preserving of, 93; poetics of chemical change, embracing of, 77; and print, 92–93; reception of, 2–3, 4–5, 15, 49, 58, 63, 68–70, 77–78, 80, 89–90, 100, 131–34; secret chemical experiments of, 2; and serendipity, 72; shortcuts of, 49; Society of Arts, break with, 62–63; Society of Arts, involvement with, 59–60; and temporality, 86–87; temporal possibilities of painting, and chemistry, 80; and time, 82, 84; unconventional painting materials, attraction to, 58, 183
Reynolds, Samuel, 53, 132
Rhodes, Barbara, 113–14
Rice, Danielle, 92
Richard, first Baron Edgcumbe, 57
Richardson, Jonathan, 56, 78
Riegl, Alois, 16, 106, 182–84
Rigaud, John Francis, 98–99
Rights of Man, The (Paine), 107
Rippingille, E. V., 131–32
Rising, John, 100–101
Robespierre, Maximilien, 116
Robinson, Henry Peach, 5–7
Romanticism, 86
Rome (Italy), 180–81, 183
Rossi, Paolo, 201n59
Royal Academy of Arts, 1, 3, 9, 20–21, 47, 49–50, 63–67, 69, 70, 72–74, 79–81, 93, 96, 98–99, 104, 141, 156–57, 230n40
Royal Incorporated Society of Artists of Great Britain, 65–66
Royal Institution, 96
Royal Society of London, 9, 13–14, 17, 26–27, 30, 32, 34–35, 38–41, 44, 49, 51, 55, 60–61, 176, 181; new science, 28
Rubens, Peter Paul, 27, 57, 84, 101
Russell, William, 106–7, 152
Ryland, William Wynne, 142

Sackville, John Frederick, 77–78
Sadler, John, 117
Sahlins, Marshall, 181
Salmon, Nicholas, 126
Satyricon (Petronius), 72
Saunders, David, 102
Scandinavia, 179
Scharf, George, 138
Schulze, Heinrich, 40
Schupbach, William, 52, 204n35
Science Museum (London), 142
Scientific Revolution, 34
Scotland, 2, 110. *See also* Britain
Scott, Sir Walter, 173
Scratch, Annabel, *108*
Self-Portrait as a Deaf Man (Reynolds), 48, 215n212
Semper, Gottfried, 183
Seneca, 83
Senefelder, Alois, 96, 107, 109, 172, 174, 181
Serres, Michel, 165
Shadbolt, George, 158
Shapin, Steven, 94
Shee, Martin Archer, 49

Sheehan, Tanya, 228–29n21
sheep, marking ownership of, 61–62
Shelley, Mary, 21, 129–30, 180, 183
Siegert, Bernhard, 197n87
Silk Road, 49
Silverman, Kaja, 136
Simond, Louis, 2
Sir Watkin Williams-Wynn with His Mother (Reynolds), 48
skin, 43–44, 118
slave labor, 17–18
Smart, Thomas, 57
Smiles, Samuel, 153
Smith, Francis Pettit, 21–22; attack on, 161, 173; chemo-mechanical assemblage, 173; as collector, 136–39, 141–42, 145–46, 150–51, 153–54, 156, 160, 163, 165; as engineer, 136–39, 159, 165, 168, 170, 177–78; gun-cotton, 236n171; silver plate controversy, 161–62; sun pictures, 140, 157–59
Smith, John Raphael, 92
Smith, J. T., 2, 93, 106, 132
Smith, Pamela H., 30
Smith, Robert, 55
Snyder, Joel, 164, 178
Society for the Encouragement of Arts, Manufactures and Commerce, 20, 47, 50, 61–64, 65–66, 71, 74, 87, 98, 106, 177, 211n137, 219n45; Committee on Chymistry, establishing of, 60; experimental materials, encouraging of, 59–60; growth of, 59
Society of Artists of Great Britain, 63, 118, 230n40
Society of Arts. *See* Society for the Encouragement of Arts, Manufactures and Commerce
Society of Dilettanti, 58
Soho Manufactory, 21, 93, 110, 113, 136, 139–40, *140*, 141, *149*, 151, *151*, 154, 161–63, 168, 173, 177, 179
Solkin, David, 79
Soubas ab Bassera, Kerim, 83–84
South Kensington Museum, 177
Specimens of Polyautography (André), 95–97
Spencer, Charlotte, 80
Spencer, George, 80
Spencer, Henry, 80
Spicer, Henry, 106
Sprat, Thomas, 200n42
Stahl, Georg Ernst, 60, 92
staining, 44–45, 118
steam engines, 10, 17, 137, 152–53, 165, 168–71, 173, 177, 179–81; chemical principles of, 167, 178
Steffen, Will, 180
Sterne, Laurence, 20, 84–85
St. Helens (England), 181
Stilfragen (Riegl), 182–83
Stockdale, James, *151*
stone, colors and markings in, 36, 38–40, 44, 83, 121–22, 223n137
Stothard, Thomas, 98
Stratonice, 141, 230n40
Streeter, William W., 113–14
Stuart, James "Athenian," 47
Stuart monarchy, restoration of, 26–27
Stubbs, George, 225n157
Studio Experiments in Colours and Media (Reynolds), 66
Study for the Alchymist (Burdett), 76, *76*, 118, 232n74
Study of Sir Joshua Reynolds (Bone), *105*
sun pictures, 140–42, 145–48, 155, *156*, 157, 159–60, 171, 173–74, 176, 184; aerial perspective of, 177–78; attacks on, 161–62, 183; chemical composition of, 236n171; as early photographs, 149–50, 158; origins of, 158; and patents, 177; and photography, 163–64; Watt's copying machine, 158
Sun Pictures in Scotland (Talbot), 150
Sydenham, Thomas, 17
synthetics, 77
Synthetic Worlds (Leslie), 22
Szychowski, Janek, 12, *12*, 13, 192n42

Talbot, William Henry Fox, 10, 15, 21, 40, 93–94, 122, 128, 134, 150, 159, *176*, 178, 183, 232n74, 236n172; calotype production, 175; chemical research of, 175–76; patents, 177
Talley, M. Kirby, 77–78
Tate Gallery, 48
Taylor, Tom, 100
technicshen Reproduzierbarket (Benjamin), 3, 96
technological reproduction, 94
temporality, 35, 86, 172; of fine art, 50, 87; nature of, 13; as phosphorescent, 125
temporal linearity, 124
temporally evolving chemical objects, 10, 14, 22, 77, 87, 94, 178, 181

Teniers, David, II, 154
ten Kate, Lambert, 87; mathematical harmonics of, 43–44
Test Acts, 30
textile industry, 17–18, 61–63, 109–10, 181
thermodynamics, 165
Thrale, Henry, 89
Three Ladies Adorning a Term of Hymen (Reynolds), 48
time, 2–3, 15, 41, 47, 50, 104, 130, 178, 196n81; artists, relationship of, 80; and beholder, 81–82; chemical means, 94; chemical preparations, 107; and chemicals, 184; cup of, 125–29; fine arts, 86–87; human cognition, 125; language of, 127; linear vision of, 107–9, 124–26; meditation on, 76–77; in painting, 78, 80, 83–84; sexualized ontology of, 129; understanding of, 38
Time Smoking a Picture (Hogarth), 46, *46*
Tintoretto, 25–26, 98
Tirrell, David, 192n42
Titian, 2, 26, 28, 99; "Venetian secret," 96–98, 100–101
TITIANUS REDIVIVUS (Gillray), 97, *99*
Tomory, Leslie, 236n170
transfer printing, 117–18; "pounce" layer, 223–24n137
Tresham, Henry, 98
Turner, J. M. W., 131–32, 165
Turner, Matthew, 76, 232n74
Twickenham (England), 17
Two Boys Blowing a Bladder by Candlelight (Burdett), 118, *119*

van der Werff, Adriaen, 43–44
Van Dyck, Anthony, 27, 57, 80, 84
van Eyck, Jan, 49
van Helmont, Johann Baptista, 38
vanitas tradition, 76–77
van Loo, Carle, 84
Van Miegroet, Hans J., 77
Vegetable Staticks (Hales), 61
Venetian School of Painting, 21, 93, 99–100, 127–28; polygraphic replications, connections between, 101
"Venetian Secret," 97, 98–101
Venice (Italy), 98, 101–2
Venus and Adonis (Eginton), 143, *144*
Venus and Adonis (West), 141
Vien, Joseph-Marie, 90
Views of the ruins of the principal houses destroyed during the riots at Birmingham (Ellis/Witton), 152
Virgin and Child with Canon Van der Paele (van Eyck), 204n31

Wall, Jeff, liquid intelligence, 22, 183–84
Walpole, Horace, 4, 68–70, 72–73, 81
Ward, Seth, 52–53
water, 115, 167, 183–84
Waterhouse, Ellis, 57
Watt, James, 17–18, 127–29, 134, 137, 139, 145, 153, 159, 162, 165, *166*, 168–70, 178, 222n113, 230n47; copying press, 21, 110–13, 115–16, 158; latent heat, experiments of, 115; patents, 177; steam engine, 173, 179–80, 235n141; as triumphant engineer, image of, 167; and water, 115
Watt, James, Jr., 94, 111, 116, 230n47
Wedgwood, Josiah, 15, 62, 67–68, 93–94, 111, 115–17, 120, 126, 154, 232n74; aquatint, 21, 118; color change, as chemical process, 121–22; Etruscan vases, 118
Wedgwood, Thomas, 10, 15–17, 21, 40, 93–94, 116–17, *125*, 137, 146, 149, 153, *155*, 158–59, 175, 180–81, 183, 224n146, 225n157, 226n171, 226n174; dream states of, 122–23; *nidus*, 14, 126, 129; on phosphorescence, 123–24; photography, first trials of, 154; protophotography, 128; sun pictures, *156*; time, cup of, 125–29
Werrett, Simon, 30
West, Benjamin, 4, 18, 21, 79–80, 85, 93, 95–96, 99, 101, 107, 110, 141, 157–59, 164–65, 176–77; knowledge, recovery of, 98; on Titian, 100
Westall, Richard, 98
Westley, Joseph, 140
Whatman, James, II, 113
Whitbread, Samuel, 216n2
White, George, 73
Williamson, Joseph, 52–53
Willis, Thomas, 28, 32, 36; fermentation, concept of, 30–31, 33, 199n30
Wilson, Peter Cecil, 204n35
Wilson, Richard, 58
Winckelmann, J. J., 180–81
Wind, Edgar, 79, 81–82
Winsor, William, 182
Winter (Loutherbourg), 102
Wolfe, James, 79

Wollstonecraft, Mary, 129
Woodcroft, Bennet, 177
Woodmason, James, 113–14, 222n113
Worden, Daniel, 182
World War I, 182
Woulfe, Peter, 47, 59–60, 66
Wright, Joseph, of Derby, 21, 27–28, 57, 66, 72, 78–79, 118, 154, 225n157; "candlelight" scenes, 65; reception of, 65, 74–75; time, meditation on, 76–77

Yonge, James (1646–1721), 19–20
Yonge, James (1748–1799), 19–20
Young, Edward, 70
Young Fortune Teller, The (Reynolds), 80
Young Nobleman and His Sister, A (Reynolds), 80

Zervigón, Andrés Mario, 228–29n21
Zoffany, Johann, 74
Zoonomia (E. Darwin), 15, 123, 129